The Emotional Mind

The Emotional Mind

THE AFFECTIVE ROOTS OF CULTURE AND COGNITION

Stephen T. Asma
Rami Gabriel

Harvard University Press

Cambridge, Massachusetts · London, England

2019

Library of Congress Cataloging-in-Publication Data

Names: Asma, Stephen T., author. | Gabriel, Rami, author.
Title: The emotional mind : the affective roots of culture and cognition /
Stephen T. Asma, Rami Gabriel.
Description: Cambridge, Massachusetts : Harvard University Press, 2019. |
Includes bibliographical references and index.
Identifiers: LCCN 2018046433 | ISBN 9780674980556 (hardcover : alk. paper)
Subjects: LCSH: Emotions. | Human evolution. | Emotions and cognition. |
Social evolution. | Evolutionary psychology.
Classification: LCC QP401 .A76 2019 | DDC 612.8/232—dc23 LC record
available at https://lccn.loc.gov/2018046433

For Tom Greif and Jaak Panksepp

Contents

The Emotional Mind

Introduction

THE AFFECTIVE ROOTS OF
CULTURE AND COGNITION

WHEN DARWIN WROTE the *Origin of Species,* he famously closed the book with the provocative promise that "light will be thrown on the origin of man and his history."[1] In his *Descent of Man* and his *Expression of Emotions in Man and Animals,* Darwin began, as promised, to throw some of that light—especially regarding the emotional and cognitive similarities (homologies) of mammals.[2] But shortly after this beacon, all went dark again. The rise of positivism in the early twentieth century, paired with the turn toward genetics and the ascent of behaviorism, effectively lowered the curtain on biological speculations about the evolution of the mind.

When researchers finally turned again to the mind in the mid-twentieth century, it was the computer that both sparked the cognitive sciences revolution and served as its exclusive investigative heuristic. Yet, for all the successes of artificial intelligence (and they are impressive), our understanding of biological minds seems to have gotten lost in the shuffle. While algorithmic digital computation produces problem-solving machines, such problem-solving—confusedly called "intelligence" by the dominant paradigm—lacks the obvious motivational goads and other affective triggers observed in real animals. In fact, artificial intelligence and artificial life research has lost interest,

1

unapologetically, in the biological creature. And, more surprising—with a putative biological orientation, evolutionary psychology gained popularity in the 1990s by actually ignoring the evolved nature of brain and body in favor of describing computational modules to explain human behavior in some largely mythical Pleistocene. Indeed, contemporary moral psychology and its philosophical counterpart often continue this modular approach, assuming the existence of innate normative switches in the human mind and discounting the emotional nature of ethical actions.

With much less fanfare, the late 1990s saw the recognition of an affective science, especially in the pioneering work of Jaak Panksepp, Antonio Damasio, and Richard Davidson. Affective neuroscience isolates emotional brain systems (largely those regions of the brain we share with other mammals) that undergird adaptive behaviors in vertebrates. With the help of neuroscientific and behavioral research, we are beginning to appreciate how the ancestral mammal brain is alive and well inside our higher neocortical systems. Unlike the computational approach to mind, the *affective turn* is deeply rooted in what we know about the brain as a biological reality. In the first decade of the new millennium, affective (or emotional) studies began to trickle into disciplines like ethology (e.g., Frans De Waal), economics (Daniel Kahneman), therapeutics (Jonathan Rottenberg), and even pharmaceutics.[3] But the time has finally come for a full-scale exploration of the evolution of emotions and mind in the biologically rooted human being.

The Affective Roots of Mind

In this book, we will argue that emotional systems are central to understanding the evolution of the human mind (as well as that of our primate cousins). We bring together insights and data from philosophy, biology, and psychology to shape a new research program.

For at least 200 million years (and that is a conservative figure based on the rise of mammals), the emotional brain has been under construction. By comparison, the expansion of the "rational" neocortex (around 1.8 MYA), which is the focus of the cognitive approach, is a latecomer on the scene, and the development of our language-symbol system is younger still. In the suite of adaptive tools, the emotions have been at work eons longer than rational cognition, so it makes little biological sense to think about the mind as an idealized rational cost-benefit computer, projected into deep time.

A sufficient account of the evolution of mind will have to go deeper than our power of propositional thinking—our rarefied ability to manipulate linguistic representations. We will have to understand a much older capacity—the power to feel and respond appropriately. We will have to think about consciousness itself as an archaeologist thinks about layers of sedimentary strata.[4] At the lower layers, we have basic drives that prod the animal out into the environment for the exploitation of resources. Thirst, lust, fear, and so on are triggers in evolutionarily earlier regions of the brain that stimulate vertebrates toward satisfaction and a return to homeostasis. Subsequently, the brain of a mammal creates a feedback loop between these ancient affective systems and the experiential learning and conditioning that the creature undergoes. And, finally, another feedback loop exists between the neocortical "rational" cognitive processes and the aforementioned subcortical triggers and learning systems. As Jaak Panksepp argues, there are bottom-up causes of mind (i.e., those that push the organism to satisfy specific physiochemical requirements) but also top-down causes (i.e., those that regulate limbic experiences through neocortical cognitive and behavioral strategies).[5] Conscious subjectivity does not suddenly arise at the top arc of this feedback circle; rather it exists throughout creatures of the mammalian clade as a foundational motivation process related to biological homeostatic triggers.

· · ·

Affective science can demonstrate the surprising relevance of feelings to perception, thinking, decision-making, and social behavior. The mind is saturated with feelings. Almost every perception and thought is valenced or emotionally weighted with some attraction or repulsion quality. Moreover, those feelings, sculpted in the encounter between neuroplasticity and ecological setting, provide the true semantic contours of the mind. Meaning is foundationally a product of embodiment, our relation to the immediate environment, and the emotional cues of social interaction—not abstract correspondence between sign and referent. The challenge then is to unpack this embodiment. How do emotions like care, rage, lust, and even playfulness create a successful social world for mammals, an information-rich niche for human learning, and a somatic marking system for higher-level ideational salience?

Additionally, these remarkable adaptive emotional systems are suffused with a deep animating power, only dimly understood and alternately called intentionality, conative drive, wanting, seeking, motivation, or the will. From the ancients to the present, we have struggled to understand the goal-directed striving of organisms. Aristotle posited a species-specific and fundamental teleology, Spinoza an essential conatus, Schopenhauer a will to life, Freud an energizing Id, and now we have the motivational functions of mesolimbic dopamine. Our book will acknowledge this biological aspect of embodied mind, track its evolution into behavioral and even cultural pathways, and stand it against the disembodied computational paradigm.

In addition to looking at the fundamental conative structure of the mind, we will track the way that emotional drives were decoupled from specific targets and became available for multi-target and multi-purpose adaptive uses (e.g., oxytocin bonding broadened in the *Homo* genus to include alloparents). This is a crucial aspect of the evolution of the mind. How did vertebrate "reflexes" (which are automatic and require minimal subjectivity) evolve into mammalian "capacities" (which are optional, albeit dispositional)? While the evolution of a symbol system certainly gives Anthropocene humans a way of representing behavioral options, we will argue that pre-symbolic humans (probably pre-sapiens) had nonlinguistic "grammars" (based on association, simulation, memory, etc.) as well as cultural mechanisms that fostered the development of such flexible capacities.

In the last few years, our picture of *Homo sapiens* success has grown more comprehensive. Older simplistic models of selfish genes, sudden brain expansion, and even the central dogma of biology have been complicated by upgraded biology (e.g., evo-devo, epigenesis, cladistics, etc.), but also we have come to appreciate the constitutive role of social and cultural forces on the developing mindbrain, and vice versa. We view the relationship in a way that is best described by the word "dialectical," which has fallen out of vogue (if it was ever *in*). Other candidate phrases for capturing the interpenetration of causality—like "feedback loops," "generative entrenchment," "enactive embodiment," and "emergent holism"—are equally fraught, but we'll avail ourselves of them occasionally because a better terminology has failed to develop.

Recent Insights

Paired with the rise of the affective sciences, three insights, among others, have come to the fore to enable our position in the decades since Stephen Jay Gould and Richard Lewontin pushed for a non-reductionistic biology.[6] One is the empirically strengthened fact of *neuroplasticity,* in which it has become clear that ontogenetic experience has formative and re-formative influence on the mindbrain. Secondly, we have greater appreciation for the *autonomous levels* of scientific subjects and methods. Reductionistic consilience—the collapse of psychology to biology, to chemistry, to physics—we know now, is a parlor game not worth pursuing. Thirdly, we have the recent emergence of *extended mind* theory. Starting with the philosophical work of Andy Clark and David Chalmers, subsequent life scientists, anthropologists, and ecological psychologists have argued that the external environment of objects and social hierarchies function as part of the animal's mind.[7] We will explore and exploit these insights to help us make our case for a mind that evolved through constant engagement with its physical and social environment.[8]

If we think of evolution as a mosaic of developmental systems, then we see that populations (e.g., early humans, but also nonhuman primates) have recurring stable resources, some of which are genetic, some phenotypic, and some environmental. Exciting work in anthropology has emphasized how adaptive behaviors can be drawn from the social and cultural spheres as well.[9] We no longer need to commit to the increasingly outmoded dichotomy of genes vs. environment. An affectively grounded associational system (of emotional learning) in pre-sapiens is precisely the sort of plastic system that can be shaped by selection (with multiple levels of selection) into stable biocultures. These biocultures, in turn, help determine which genetic traits spread through a population.

Some researchers have suggested that hominin evolution exploited a unique "cognitive niche."[10] According to these proponents, the niche includes the coevolution of intelligence, language, and sociality. More recently, Whiten and Erdal have refined this claim by showing how the cognitive niche was fundamentally or primarily *social*.[11] We will be arguing that humans evolved in an *emotional niche,* having affective features homologous with other primates but also having unique affective capacities. We want to provide a key ingredient to the socio-cognitive niche, namely affective or emotional modernity. How did humans become emotionally modern, and what advantages flow from such a transformation?

In alliance with an *extended mind* approach—hitherto focused on information—we argue that emotions are also distributed beyond the organism itself. Some instinctual homologous affects are generated deep inside the animal in response to perceptual / motor stimuli, but even amygdala-governed fear is based on the connections it has with specific environmental experiences. Pavlovian associations only become adaptive when, in the course of development, the "random" external stimuli are paired with painful or appetitive events in such a manner that goal-directed behavior (intentionality) is improved. This environmental and developmental influence alone should give us pause when assuming that emotion is inside the head of the animal, for as soon as we ascend into the social affects of the upper-limbic and cognitive affects of the neocortex, we see that emotions are managed as much outside the animal—in social custom and cultural niche—as inside the animal. Moreover, the animal itself does not perceive the world as filled with neutral objects that then get affective salience assigned to them. The associational work is largely invisible to the animal, and the world itself is perceived as populated with threat objects, appetitive objects, and so on. An animal's affective taxonomy of objects and social con-specifics is relatively stable and also revisable, but the animal's "world," or *umwelt,* is intrinsically emotional.

In humans, the emotional life is more complex—in the sense of cognitive influence, the intermixing of emotions, executive control, and so on—but is also more extended in and through the environment. Our social and cultural world is designed to trigger and manage affect, partly because this is the most expedient means of triggering prosocial behavior, but also because we are connoisseurs of emotion and pursue their intrinsic as well as instrumental values.

The impressive achievements of a human cognitive niche are often heralded, but the emotional niche has gone unsung. Yet, the advances of complex tool industry, for example, could not have happened without parallel advances in *Homo* emotional life. During the Acheulean (to 100 KYA) and Mousterian periods (160 KYA to 40 KYA), families (whether nuclear or common) were required to be domesticated (i.e., emotionally modern) enough to learn and to, eventually, patiently teach the skills involved in flint-knapping. In this interpretation, social skills and language as a communication system itself have certain affective prerequisites; sophisticated language may be the result rather than the cause of emotional modernity.

In short, while impressive research has been emerging in disparate fields such as neuro-ethology, ecological psychology, physical anthropology, the evolution of culture, enactive psychology, and the philosophy of biology, no one has yet characterized an affective paradigm that draws together these data and projects a fruitful way forward. Our book hopes to provide such a conceptual roadmap.

Our approach to the mind is heavily indebted to the revolutionary affective neuroscience paradigm of our late mentor Jaak Panksepp (1943–2017). Following Darwin's conviction that the difference between human and other animal minds is one of "degree" rather than "kind," Panksepp thoroughly investigated and conceptualized the common emotional systems in all mammals. In the course of this book, we will show how the homologous mammalian systems, such as FEAR or CARE, animate human mental life and are, in turn, redirected to and constrained by cognition and culture that are uniquely human. All mammals, according to Panksepp, share seven foundational affective systems: FEAR, LUST, CARE, PLAY, RAGE, SEEKING, and PANIC / GRIEF.[12] Each of these has specific neural electrochemical

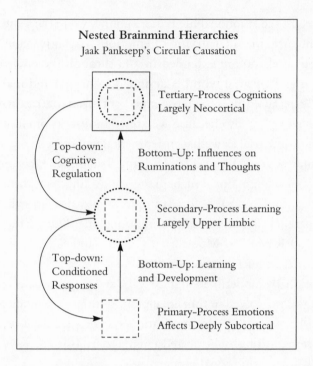

Nested Brainmind Hierarchies
Jaak Panksepp's Circular Causation

Tertiary-Process Cognitions
Largely Neocortical

Top-down:
Cognitive
Regulation

Bottom-Up: Influences on
Ruminations and Thoughts

Secondary-Process Learning
Largely Upper Limbic

Top-down:
Conditioned
Responses

Bottom-Up: Learning
and Development

Primary-Process Emotions
Affects Deeply Subcortical

pathways, with accompanying feeling states and behavior patterns. Although other researchers, like Paul Ekman, draw the map slightly differently regarding what counts as a "basic emotion," continued affective science will iron out the best taxonomy in the decades to come.

As Panksepp argued, however, human beings are not just an assembly of mental modules or even emotional circuits. The affective systems are hierarchically structured in three layers of interpenetrating brain activities: primary, secondary, and tertiary functions. Mindbrain processing that is stacked like a layer-cake—or perhaps like nested Russian dolls, with one inside another—is a better metaphor.[13]

At the very bottom or the "core" are the instinctual drives, like fight-or-flight, or as we will explore in this book, intentional seeking. This primary-process layer is housed largely in subcortical areas of the brain. Panksepp describes primary-process emotions as (1) sensory affects (sensorially triggered pleasant-unpleasant feelings); (2) homeostatic affects (hunger, thirst, etc. tracked via brain-body interoceptors); and (3) emotional affects (emotion-action tendencies).[14] We share these primordial affective systems with all other vertebrates. This layer heavily influences the layer above it, secondary-process emotion, which is more developed in mammals.

Secondary processing includes social emotions, like GRIEF, PLAY, and CARE. It is distinguished from the primary level because it can be sculpted by learning and conditioning. It is the layer of soft-wiring (part native instinct and part learned association), as compared to the hard-wiring of primary-level emotion. Panksepp describes the secondary-process mind in terms of (1) classical conditioning, (2) operant conditioning, and (3) emotional habits. Emotions in primary and secondary layers are largely unconscious, and even when we are regulating them, we do not have clear, introspective, conscious access to their functioning.[15]

Lastly is the top layer of the mindbrain: tertiary-process emotion. This is the layer of mind that most philosophers and psychologists tend to focus on exclusively. Here the emotions are still connected to the primary and secondary processes, but they are intertwined in the cognitive powers of the neocortex. Ruminations and thoughts, underwritten by language, symbols, executive control, and future planning constitute the tertiary level, though they are energized by the lower-level emotion. These ruminations and thoughts also serve as top-down regulators and directors of emotion. At this third level we arrive at uniquely human emotions, like those elaborate and ephemeral feelings so beautifully articulated by introspective literary savants such as Henry James and Fyodor Dostoyevsky. Panksepp says tertiary affects and neocortical awareness function as (1) cognitive executive functions, (2) emotional ruminations and regulations (generally located in the medial frontal neocortex), and (3) free will, or reflective intention to act (frontal cortical executive functions).[16]

The biological and psychological sciences have historically isolated or focused on one layer of mind to the exclusion of others and have thereby presented partial and sometimes conflicting pictures of mind and behavior. Many computationally oriented cognitive scientists tend to focus on tertiary-level processing, while behaviorists focus on secondary-level processing.

Our approach in this book is to show how the lowest layers of mind permeate, infiltrate, and animate the higher layers. The evolution of mind is the developmental story of how these layers emerged and acted as feedback loops on each other. And it's important to foreshadow here that such feedback is not strictly a brain process, but an embodied, enactive, embedded, and socio-cultural process.

While we want to emphasize our shared instincts with nonhuman mammals, we aim to connect this story to a theory of how culture is both a continuation and a transformation of shared mammalian mental capacities. Starting with bifacial tools, between 2.5 and 1.7 MYA, the mind of *Homo* became uniquely extended throughout the lived environment, and these external technologies expanded dramatically—distributing mind well beyond the cranium to the environment and back again. From 50 KYA, the archaeological record reveals a profusion of creative problem-solving technologies, like sewn clothing, heated artificial shelters, watercraft, small game and fishing tools, fired ceramics, portable lamps, and so on.[17] These changes were not just intellectual or computational innovations, but also emotional innovations—involving impulse regulation for example. Moreover, the social environment that allowed for the elaborate apprenticeships of our tech-savvy ancestors (e.g., domestication of anti-social affects, divisions of specialized labor, etc.) created an extended, collective mind. Our claim is that emotions infiltrate all of these complex feedback loops between brain, body, and environment, allowing the crucial space for social and cultural innovation.

Recently, the biological basis of emotions has come into question by thinkers who focus largely on tertiary-level emotions and the cultural aspect of our narratives about emotional life; one example is the well-publicized work of social psychologist Lisa Feldman Barrett.[18] Barrett argues that emotion is inseparable from cognition and that our emotions, like anger or sadness, are very fast mental constructions—almost like real-time, miniature predictive theories about our experiences. Emotions, in this formulation, are intellectual processes, requiring language to carve raw feelings into discrete emotions. Instead of biological or physiological systems, emotions are said to be more like thoughts, and each person learns early in life how to name and organize them into seemingly natural kinds; but in truth, according to Barrett, they are cognitively and culturally constructed conventions.

The argument for this counterintuitive view rests precariously on an analogy with certain top-down aspects of sensory perception. The powerful effects of cognitive suggestion on perception are well known, and psychologists have made careers "priming" subjects to perceive something that isn't there. Cognitive bias in certain kinds of perception is demonstrable.[19] Cognitive expectation can sometimes shape a person's seemingly unmediated perception. Similarly, Barrett claims, our minds suggest ways to categorize feelings into seemingly automatic or instinctual emotions. We simply don't have conscious access to this mental-categorizing activity, and we recognize the emotion only when it is fully packaged by our conceptual activity. This view, according to Barrett, explains the "cultural relativity" and even individual relativity of some emotions.[20]

However, there are several serious problems with this approach, and so our book will argue for the biological roots of emotion and against social constructionism. We will mention only a few objections to this view here and allow the rest of the book to develop arguments more fully.

First, the claim that emotions are shaped by cognitive functions hinges on an analogy of emotional behaviors to perception, but it is a weak analogy. While it is true that some perception is open to top-down structuring (e.g., prejudice influences how I "see" a stranger), the fact is that most perception is not. Recent work from the Yale Cognition and Perception Lab, particularly that by Chaz Firestone and Brian Scholl, shows that most perception is reliably free of top-down influence.[21] Perception is not as penetrable to higher cognitive function as Barrett and others contend. This doesn't mean perception is impenetrable to physiological conditioning and bottom-up influence; if anything, the empirical work seems to suggest that repeatable constraints or influences on perception come not from "beliefs" or language or concepts, as Barrett suggests, but from feelings, emotions, or affects. This means that the permeability of perception cuts against Barrett's argument, since it reveals a physiological causal force (viz. the affective systems) that shapes and constrains perception and cognition.

Secondly, Barrett and other constructionists cannot say how various emotions are differentiated except by cognition, but this explains very little. On this view, we, or our culture, could decide to define fear as happiness, and the redefinition would make it so ipso facto. But fear feels awful and drives different behavior than happiness. Constructionists like Barrett admit to a low-level positive or negative valence of feelings but then ascribe discrete emotions like fear and anger to the cognitive labeling of this low-level feeling. In our view, this theory radically underdetermines the phenomenology, neuroscience, and ethology of emotions. Constructionists like Barrett point to atypical emotional responses in some individual's self-reports and to atypical neural patterning in fMRI (functional magnetic resonance imaging) studies and erroneously conclude that the emotions are not universal and natural but are relative matters of cognitive taste. But exceptions prove the rule, in this case, and are not anomalies that unravel the biological paradigm of emotions. The brain is plastic enough to account for diversity without having to throw out the biology of emotions. The reason we classify a handful of behaviors, expressions, and feelings as "anger" is because an identifiable physiological pattern underlies them, and such patterns evolved in mammal brains to aid their survival.

Third, the argument of constructionists might work for very rarified neocortically enmeshed emotions like "ellipsis," but that is not the correct model for primary-level affects (genetically engraved and largely dedicated to a suite of behaviors). "Ellipsis" is a word coined to describe the sadness a human feels when he contemplates that he will not live to see the future. An unscientific but intriguing online site, *The Dictionary of Obscure Sorrows,* details many of these tertiary-level emotions, including the emotion of "liberosis"—a desire to care less about things, as when an adult longs to be a child again. Barrett's theory of emotions might indeed help explain some of these more cognitive, discursive, and culturally bounded feelings, but mammalian affective systems (primary and secondary levels) are very different (albeit sublated within the tertiary level), not to mention far more numerous and common.

Fourth, neuroscience disagrees with the constructionists. Over the last two decades, neuroscientists like Panksepp, Damasio, Davidson,

LeDoux, Berridge, and many others have mapped clear neural path-ways for basic emotions. Constructionists argue that the facial ex-pressions of emotions like anger are more diverse and variable than previous researchers suggested; but even if this were true (and the jury is out), it doesn't follow that the underlying brain and body processes for anger are diverse. Extensive research on the amygdala, for example, reveals that fear has a clear brain signature. And precise, localized electrical stimulation of the brain (ESB) reveals specific affective and behavioral responses in animals. Mammalian fear, for example, is a natural kind (similar to that of a biological species) with diverse vari-ability (again, similar to a biological species).

Fifth and most important, we must consider the constructionist view in regard to nonhuman animals. Barrett's theory, which makes emo-tion dependent on higher conceptual cognition, the understanding of cultural context, and language, renders emotion in nonhuman animals and even babies unintelligible. In response to this problem, Barrett falls back on the old epistemic quandary that diagnosing an emotion in a nonlinguistic subject is very difficult. We cannot know if they are having emotions, she suggests, unaware of philosophical literature on the problem of other minds. But the clear (albeit unowned) implication of her conceptual act theory of emotions is that animals and babies do not have emotions because they lack language.[22] This seems remark-ably inconsistent with evidence from animal studies and developmental studies, as well as common sense. In fact, it strikes us as an unhealthy return to both behaviorist and Cartesian agnosticism about animal consciousness; and we will argue in this book for the reality of animal emotions.

Our Approach: Chapter Synopses

In Chapter 1 we chart our epistemological orientation. The mind is too frequently conceived as a computer system of circuits and mod-ules, but it is a biological system first and foremost. We begin with a short history of recent metaphors of mind, especially from em-pirical psychology. We describe how an affective perspective rectifies mistakes in behaviorism and cognitive science, while also building on the breakthroughs these views made possible. We describe a non-modular model of affective mechanisms and its implications for an evolutionary approach to cognitive psychology. Most models of value generation are based either on behaviorist conditioning paradigms or cognitive, rational cost / benefit decision-making. But the former mechanical associations are too dumb, and the latter discursive and computational reasoning is too smart. We argue that the neuroplastic brain generates and assigns affective values with "pushmi-pullyu" representations and somatic markers long before the mind engages in propositional manipulation of the external world. Buttressed by interpretation of experimental findings on attitudes and unconscious reactions in a prosopagnosic patient, this new paradigm describes a model of *intentions in action*.

In Chapter 2, we lay out our ontological positions with an emphasis on key issues in the philosophy of biology, including teleology, intentionality, and the causality of developmental feedback processes. Our debt to Aristotle and Spinoza reveals our deeper perspective on the intrinsic properties of biological entities. Spinoza saw nature in fairly mechanical deterministic terms, but he recognized that living things share a simple goal-oriented tendency; namely, they strive to survive. He called this animation principle of living systems "conatus" (striving) and considered it the very essence of all biological creatures.

Biological systems are intentional. Equipped with conatus, proto-representational abilities, and homeostatic processes, creatures with locomotion pursue "maximum grip" on their environments. Conative aboutness in dynamic ecological systems is a kind of intentionality that occurs prior to symbolic representations. This line of thinking has been neglected, however, because earlier theories were tangled in theology, and Cartesian and recent digital notions of mind have failed to incorporate this aspect of embodiment. Biological teleology, and emotional intentionality in particular, need to be worked out *before* the representational theory of mind, which is in turn derivative of those earlier forms of goal-directness.

In this chapter, we reconceive teleology in terms of a post-Darwinian ontology. These ontological considerations will lay the foundation for those telic features of mind, articulated nicely by William James in

1890: "[C]onsciousness seems to itself be a fighter for ends. Its powers of cognition are mainly subservient to those ends, discerning which facts further them and which do not."[23]

After these more explicitly philosophical chapters, in Part 2 we offer a three-chapter section on the evolution and development of social intelligence. Here we outline a model of social intelligence and the foundation it provides for the phylogenetic and ontogenetic cultural accumulation of information. This model includes an emphasis on the features of mental life that are homologous with nonhuman primates. Much of our understanding of primate ethology is gained through the work of primatologists Frans de Waal, Craig Stanford, Robert Sapolsky, and Sarah Hrdy. Drawing on the ecological psychology of researchers like Louise Barrett and the situated cognition approach of Andy Clark, we describe the environmental contours of primate emotional life. We also discuss a set of perceptual, emotional, and social processes as the proximal causes that enable culture and cultural learning, as investigated by philosopher Kim Sterelny and psychologist Darcia Narvaez and colleagues.[24]

The first chapter of this section (Chapter 3) provides a theoretical model for the complexity of social processes using the notion of affordances as a conceptual lever to transform the scope of perception and the role of emotions therein. The second chapter of the section (Chapter 4) delivers a phylogenetic story of social intelligence through a comparative analysis of primate social behavior in its ecological and emotional context. The final chapter of this section (Chapter 5) tracks an ontogenetic narrative—the developmental psychology of social intelligence in humans, with particular attention to the infant-caregiver relationship.

In Part 3 of the book we offer chapters on the affective roots of human culture, tracing affective systems through transformations that enabled representational mental processes, social organization, religion, and art. Our key principle is the notion of *decoupling:* when an affective response, or a perceptual representation, is freed from necessity and attaches onto, and expands into, other functions. Put another way (following philosopher Ruth Millikan), the human mind evolved the ability to separate or disconnect the *indicative* from the *imperative* functions of an image, sound, or memory. This provided enough distance from automatic action-responses that representations could be manipulated (i.e., counterfactuals arose), and a "second universe" slowly emerged inside the head of *Homo*.

Most contemporary work on the evolution of mind fails to address the way in which intellectual representations originated in earlier animal abilities. In this section of the book, we articulate an empirically informed model of how primates transitioned from bodily simulations (the beginnings of decoupling) to symbol systems, and how those eventual symbolic systems still bear the mark of their affective roots.

Chapter 6 is an exploration of representation and imagination. Representational abilities were decoupled from perceptual tasks and allowed an expansion of involuntary and agent-directed simulation possibilities. These representational processes remained nested in the brain processes we described above but, through the pressures of expanding social groups and cognitive needs, acquired new possibilities. The decoupling of representations (in simulations that include valence tags) marked the key shift from here-and-now perception and action demands onto more long-term tasks, which are generally related to social needs and require inhibitory abilities. We postulate an arc from stimulus-response reflexes to affordance-based perceptual systems, associational conditioning in mnemonic systems, cross-modal imagination (enabled by decoupled task grammar and analogical modeling), and, finally, inferential symbolic capabilities. In this trajectory, we will describe how affect can be an organizing principle in the associational and decoupling stages. The transition we sketch here is from direct perception's automatic behavioral affordances to bodily simulations for action and perception in spatial navigation, and from affective reconsolidation of memory in dreams to conceptual and linguistic symbol systems used in the executively directed simulations and novel compositional activities of voluntary imagination. Spatial navigation and the dream state observed in mammals may have served as the crucibles for this evolutionary transformation.

The decoupled simulators and executive processes that constitute cognition and representational mind were necessary for the evolution of the concepts we discuss in Chapter 7. Dialectical cycles between these mental abilities and complex social processes presumably led to our infinitely iterative natural language and culture. The evolution of representations and imagination runs through the valley of the decoupling of affect and leads to the mountaintop of symbolic communication. Analytic philosophy, with its focus on the *lingua mentis,* has erroneously treated the conceptual realm as a syntactic and semantic system that is entirely independent from the body, perception, emotion, and even developmental learning. This nonbiological approach has led to distorted theories of concepts, in our view, and we argue for a more embodied approach.

We discuss how imagination mediates between perception, memory, and judgment unconsciously or preconsciously, and we trace the roots of these ideas from Aristotle, Kant, and, more recently, Mark Johnson. Dreams and mind-wandering are examples of detaching affect from agent-directed goals and here-and-now perceptual-motor tasks, allowing synthetic mental processes for affect organization and representation. Our prototype of unconscious imagination is dream decoupling, wherein objects are assigned new emotional valence tags (or old ones are reinforced), leading toward the creation of more appropriate or adaptive social behavior simulations for use in waking life. We propose the thesis that imagination acts as a middleman between sensory-motor capacities and the cognitive-linguistic mind with its behavioral affordances in body grammar and task grammar.

In Chapter 8 we examine social structures, considering, in turn, the role of affect in cultural evolution, and specifically, collective social organization in the rise of civilization. We draw on archaeological findings that suggest that human social organization has taken many shapes, including hunter-gatherer bands, the tribe, the chiefdom, the state, as well as complex variations and hybrids of each.

As social institutions become a part of our lived environment, culture serves as a secondary niche for the species. This unique ability we have to learn from others and to transmit and reproduce ideas and all forms of knowledge opened a world beyond genes. While sociocultural arrangements are, first of all, responses to ecological pressures, with each solution making survival possible in the given ecological circumstances, the most telling part of the anthropological record demonstrates how the ecological pressure of population growth called forth new technology (such as hunting tools), new forms of social organization (such as the nuclear family), and political regulation (for example, the creation of the chiefdom). What we add to the consideration of the question of society's evolution is how these changes are both bound to, and grow out of, our affective needs and motivations. Indeed, social norms—from reciprocity to ritual restrictions—are ultimately forms of affect management.

The resource-sharing, nomadic movement and the face-to-face social interaction of hunter-gatherer groups reinforce cultural adaptations that are based on complex cooperative strategies. Likewise, changing social and ecological factors—such as an increase in population, the establishment of defensible resources brought on by sedentism, and the firm establishment of kin groups—are foundational to the rise of civilization in the agrarian state. These changes suggest types of societies that tend to be stratified, centralized, and often organized with abstract ideological norms such as laws and ritualized customs.

We argue that affective adaptation to the specific ecological and social topography of human groups is a causal factor in the creation, maintenance, and eventual change of the social norms that define culture and organization. The cultural and ecological factors of band-sized populations, face-to-face interaction, and continuous foraging played a role in orienting our affective motivations, the bandwidth of our empathic broadcast, and the relations between emotion, action, and inhibition. This orientation created the cultural adaptation of some versions of enforced sharing that take hold and serve as a structural force that contributes to the shape of social organization. If social and ecological circumstances change, our motivations, feelings, and group affiliations will recalibrate accordingly. The particular changes that took place in the shift from hunter-gatherer band to agrarian state in the Upper Paleolithic (50 KYA till 10 KYA) were influenced by what has been termed a "release from proximity" (i.e., a loss of immediacy).[25] Under these circumstances, enforced sharing may be abandoned, empathy may take on new forms, and abstract ideology may become decoupled from more immediate social norms to serve as a more prominent source of social organization. In this chapter, our goal is to propose an interpretation of the archaeological evidence of social organization that takes into account the role of affective mediators of social bonding.[26]

In Chapter 9, our most speculative chapter, we explore religion, mythology, and art. Here we argue for an affective bridge that explains not only ontogenetic value generation but also sociocultural adaptations (and exaptations) such as religion and art. The pictorial and narrative faculty of *imagination* (under increasing voluntary control during the Upper Paleolithic) furnishes a bridge between passive sensory memory and associationism on the one hand and active adaptive appraisal (or judgment), as well as mimetic cultural codification of survival strategies, on the other hand.

From rather different perspectives, researchers like Scott Atran, Denis Dutton, Arnold Modell, and Steve Mithen have all recently challenged Steven Pinker's famous suggestion that art is evolutionary "cheesecake," a nonadaptive byproduct of big-brain ingenuity.[27] We stake our position in this debate by using evidence from affective neuroscience and anthropology, revealing the emotionally therapeutic, prosocial aspects of religion, mythology, and art. In those rare cases where researchers acknowledge an adaptive natural history of religion, for example, they tend to offer cognitive interpretations (i.e., primitive religion is crude proto-science that helps early man make predictions about and understand nature), but this is an incomplete picture and rides atop what we believe to be the principal role of religion, which is to shape social solidarity through ritualized affective sculpting.

An affective approach to culture also helps us understand some stubborn contemporary confusions. The New Atheists, like Richard Dawkins and Dan Dennett, are evaluating religion at the neocortical level—their criterion for assessing its claims is the hypothetico-deductive method of verification. However, we argue religion may fail at the bar of rational validity because it is the wrong bar for evaluating religion. The limbic brain, built by natural selection for solving survival challenges, was not built for rationality. Systems that culturally manage our emotions, like religion, were selected for because they helped early mammals flourish. William James understood the tension between passional and rational agendas long before we had a neurological way of framing it. James recognized that faith is not *knowledge,* in the strict sense, but asserted that since it is deeply meaningful (as a felt emotion), it is important to see how and why faith might be warranted. He also understood, as did thinkers like Antonio Damasio, that secular reason is more feeling-laden than we usually admit—there is a sentiment in rationality. The recent debate about religion, like polarizing political rhetoric, lacks James' refined understanding of the real stakes involved, and it lacks the appreciation of the affective roots of religion, myth, and art.[28]

As such, we focus on the role of the spiritual emotions of awe, wonder, and transcendence in art and religion. Taking their energy and drive from basic affective sources like the SEEKING and PLAY systems, these spiritual emotions functioned to temper intense feelings of FEAR and GRIEF in the context of the neocortical imaginative elaboration of culture. The emotionally saturated state engendered in spiritual emotions is immanent in our neuropsychology, but the ways in which we communicate our transformative experiences in art are unique.

Taking place in the association-rich cultural spaces of affective modernity, shamans, artists, and mystical individuals sought to materialize their unique experiences in aesthetic ways that engendered faith, belief, and further transcendent states. While scientists seek the cognitive processes behind wonder and curiosity, we also urge consideration of the emotional landscape that gives this sphere of culture its indelible importance to individuals and groups. As others have said, art and religion coalesce formats of emotional manipulation that are prosocial, but the mystical experience, as William James wrote in 1902, produces an "added dimension of emotion . . . and a new reach of freedom for us . . . performing a function which no other portion of our nature can so successfully fulfill."[29]

1

Why a New Paradigm?

LIKE ANY OTHER FIELD of inquiry, psychology is slow to change but is always changing. The insight of this book is that research undertaken during the last couple of decades in the life sciences, evolution, and neuroscience suggests it is time for psychology and philosophy to alter fundamental assumptions concerning the nature of mind. In the early twentieth century psychologists favored the methods of behaviorism, while in the late twentieth century the methodology of cognitive science came to the fore; each in its own way was startlingly successful, but the time is nigh for us to synthesize and supersede. Behaviorism enabled empirical precision of great subtlety, best suited to nonhuman animal experiments. Cognitive science enabled systematic modeling of internal information states, best suited to computers and some higher human cognitive skills. It is now clear that both approaches reveal different levels of mental functioning; but while the former is not flexible enough to explain the adaptability of the mind, the latter is neither subtle nor tender enough to explain the heat of consciousness.

Behaviorism and cognitive science are comprised of methodological frames and epistemological assumptions that do not adequately take into account the role of the emotions in the mind. The case we put forward tips its hat to the work of psychologists of the twentieth century, but it stakes its roots in the psychology of the nineteenth century, when phenomenology, philosophy, and an appreciation for cultural and anthropological insights played more significant explanatory roles. Maintaining strands from post-Darwinian natural philosophy, our goal is to bring them back in touch with the even older paradigms of Aristotelian-via-Spinozistic neutral monist ontology as well as with the contemporary life sciences, paleoanthropology, animal ethology, and social and affective neurosciences.

With the development of our understanding of affect in the brain, we believe a middle way between behaviorism and cognitive science has become possible. While sufficient empirical evidence has not been previously available to explain how cognition is embodied and how endogenous innate associationist systems are able to point at the world, in our opinion, the great breakthrough of the affective sciences is that it points to how the "middle systems" of affective processes can be trained through ecology and experience—that is, how they enable complex associationist / behaviorist abilities and, at the same time, give cognition its bodily basis in the sentient values that determine behavior and thought.

Although many questions remain, the time is ripe to develop a philosophy of the affective sciences that may enable empirical psychology to synthesize a robust evolutionarily valid framework of behavior and motivation, and of thought and action. Such a frame will allow for more fruitful connections to research on human culture and, subsequently, the humanities. In this chapter, we provide a paradigm for integrating the prevailing methodologies of psychology around the function of emotions.

Paradigms of the Mind

The paradigm that has dominated scientific attempts to explain how the mind works is called "associationism." We know it in its more recent derivation from the work of the British empiricists and subsequently from James Mill and his son John Stuart Mill.[1] It was thereafter used as the basic model to describe how the mental imprint of past events structures future behavior. The essential claim of this approach is: *Ideas* are copies of sensations. All ideas, whether they are complex, simple, or combinations thereof, are derived from sensations (i.e., *impressions*), which are more "vivacious" than ideas. Further considerations relate to the strength of associations; Mill considered association a relation between permanence, strength, and facility, but associations could also be labeled in terms of their similarity, contrast, and contiguity.[2] While J. S. Mill pictured the mind as active and underscored the emergent properties of associations, Alexander Bain emphasized how habits formed the basis of learning.[3] The mind in this latter formulation is essentially passive—it reacts through conditioned reflexes—and this position is what provided the bedrock for associative learning and conditioning theory to be elaborated upon extensively in controlled laboratory experimentation.[4] Through the work of John Watson, Donald Hebb, and B. F. Skinner, this approach developed directly into the empirical paradigm of behaviorism that sought to clarify basic learning processes in nonhuman animals.[5] We do not see the behaviorist adoption of associationism as exhausting the insights of this intuitive theory; in fact, we believe that whether affective and cognitive processes require associationist architecture at some level is an empirical question.[6]

With access to laboratories and lab animals, behaviorism focused on stimulus-outcome relationships. Findings from this approach are now successfully applied to methods used in clinical pathology (e.g., fear extinction, relapse reduction, anxiety treatment, trauma and learned-helplessness therapy, learning in autistics), health and addiction (e.g., treatment for drug abuse and eating disorders), and social learning and motivation.[7] While some behaviorists have been committed to a seeming denial of interior mental states, claiming emotion was not the cause for behavior but rather a second-order probability of behavioral output, others, more recently, have maintained that conditioning is crucial for social cognition and emotional learning.[8] Further, some argue that behaviorism contains in its methodology the cognitive-science seed of modeling repetitive processes while rebuffing Cartesian dualist tendencies—including homuncular tendencies—in psychology.[9] In fact, the pragmatic fundamentals of behaviorism resemble the ecological and embodied perspectives that we will be drawing from throughout the book.

Ultimately, behaviorism has failed to empirically account for serial behaviors, including language and skilled-motor programs, because (1) feedback is provided too quickly to build a reactive model and (2) a hierarchical plan of action seems necessary to explain serial sets of behavior.[10] Additionally, behaviorism's explanatory use of operational definitions is largely tautological and therefore unable to provide causal explanations.[11] According to behaviorist methodology, a given behavior can be defined only in terms of dispositions toward that action, not in terms of intent; further, there is no level of causation and no mechanism other than training intervals. And yet behaviorism remains the standard methodology of the neurosciences.[12] Learning paradigms based on conditioning and its reinforcement programs, not to mention methodologies like maze-running and learned helplessness, seem to effectively reveal and elicit reliable patterns of behavior in non-human mammals—two species of rats and one pigeon species, to be exact.[13]

We do not find this psychological approach to be passé due to its limited purview. Conditioning processes are undoubtedly an important element of midbrain and limbic functioning, for example, as evidenced in the seminal work on fear conditioning and the amygdala by Joseph LeDoux and Antonio Damasio.[14] As the field moves forward, we will need to include relevant insights from this paradigm in characterizing primary and secondary levels of affect and what form they take in a dialectical relation to tertiary representational processes. A notable sore spot in this empiricist vision concerns whether innate mental structures are possible; this discussion opens the door to the postulation of inborn modules in the cognitive sciences, to which we now turn.

The rise of the cognitive sciences coincided with the advent of computers and the rise of a rationalist *geist,* not to mention the putative failure of behaviorism as a complete approach to the study of the mind. The key to this "cognitive revolution" was the excitement of attributing the causal locus for behavior to the modeling of mental "representations" as internal information states. Although seldom invoked, associationism remained important to models of schemata-based learning theory.[15] And the behaviorist methodology of eschewing introspection in laboratory testing remained an unsung legacy in cognitive science, where behaviors—as reaction times or verbal responses—were the only legible variables from which cognitive architecture could be deduced.

The central theoretical tenets of cognitive science devolve upon mental representations as mediators between input and output. As a legacy of logical positivism in early-twentieth-century Continental philosophy, representations were widely considered as propositions. The cognitive sciences modeled the mind upon computational processes, therein de-emphasizing a range of human abilities, such as affect, context, culture, and history. Lastly, the field saw itself as a form of interdisciplinary studies (viz. melding computer science with psychology).[16] Ironically, while likening the mind to a computer made the modeling of mental states in the language of logic and computer code possible—to great fanfare and successful grant applications—this approach also demonstrates the limits of cognitive science.

The "computational paradox," or the trouble with the cognitive sciences, is that this logical, systematic view of human cognition does not adequately describe very much of human thought and behavior. Further, cognitive science is bound between two sciences that do not use the language of representations: neuroscience at the lower limit and anthropology at the upper. Nevertheless, with the neuroscience revolution afoot, cognitive science was in a position to serve as the theoretical arm of brain imaging, and thus cognitive neuroscience came to serve as the default paradigm for the interpretation and perpetuation of the developing technologies used in fMRI, PET, and CAT studies.[17]

Just as behaviorism was the subject of sustained critiques that diminished its scope to predominantly nonhuman experimentation, cognitive psychology experienced its own set of critiques that delimited its own explanatory breadth.[18] Critics pointed out its nonbiological approach, its reductionism, and its disinterest in phenomenology and ecological context, among other things. Nonetheless, cognitive psychology is presently the paradigm of choice for academic psychologists extrapolating empirical data on human behavior into models of neural processes. But the debate continues over whether a level of explanation that demands mental representations is required in addition to neuroscience; that is, whether molar rather than molecular levels of analyses are possible.[19] Furthermore, cognitive science has, in its mature phase, ventured into extrapolating on the ways in which perception and action are embodied, to the extent that some of its own fundamental tenets concerning the need for mental representations may be undermined.[20]

The most important critique of the cognitive sciences has been made by perceptual psychologist J. J. Gibson, who introduced ways of conceptualizing behavior and its causes in terms that are neither fully representational nor fatally behaviorist. The concept of *affordances* (i.e., those potentialities for action inherent in an object or environmental cue) effectively permits analysis of behavior without the need for belief-desire reasoning and representations. Rather than the computational machinery in the head of the perceiver, Gibson essentially sought the invariants, or behavior-causing stimuli, in the environment that could be detected by perceptual systems toward action-responses.[21] If we add the post-Darwinian New Synthesis (and the Extended Evolutionary Synthesis) to elaborate upon adaptational reasons for the existence of such invariants, a robust counternarrative emerges.

We find Gibson's ecological psychology approach compelling for the following reasons: (1) it gives a greater parsimonious role to the perception–action system, which restores the body to an important role; and (2) it suggests that, to a perceiver, there is no such thing as neutral information, which leaves room for affective motivational processes.[22] Also, ecological psychology underscores how a purely computational approach fails to clarify the motivations of intentionality, which we suggest is largely derived from affective and perceptual systems. The burden of evidence for ecological psychology is on how to cash out a scheme of intentionality without inferential representational processes. We take up this issue now.

Psychology and the Emotions

We believe the function of the mindbrain is not easily understood as either behavioral or computational, as it is comprised of a set of bodily abilities and reactions that seem inherently experiential and innate (contra behaviorism) while maintaining behavioral fluidity (contra cognitivism).

Emotions offer a unique conundrum for psychologists: they have resisted cognitive interpretation, have been studied successfully with behaviorist methods, but cannot be understood separately from their phenomenological aspect.[23] They play a role in cognition but can be set through conditioning; they are conscious but can be generated unconsciously.[24] Subsequently, some emotions seem unique to humans, while others seem homologous with nonhuman animals.[25] While emotions have been characterized as appraisals, or as a set of basic emotions, we tend toward considering them both as a neurophysiological state that tracks the relationship between the organism and its environment and as a primary form of sentience—in other words, as the intentional core of the organism.[26] This characterization of emotions can also be described as a background bodily feeling, as it is by Sartre, Damasio, and Colombetti.[27] The debate concerning the nature of emotions is taking place in the context of the burgeoning empirical fields of social and affective neuroscience.[28] The major contribution of our book is to put forward a philosophy of affective neuroscience that clarifies the exact role of emotions in a way that may orient future empirical work. In the second part of this chapter, we describe how empirical results can be interpreted in such a framework.

Some reasons to consider affect at a functional level, separate from cognitive and behavioral processes, are: (1) the effects of motivation on behavior and attention (sometimes referred to as perceptual salience), (2) the nature of internal monitoring of homeostatic states that may be implicated in the intentional directedness of affective experience and subsequent action, and (3) the role of valence on the content of mental processes.[29]

Our conceptualization of affect will include both the primary-level emotions as well as emotion schemas, or secondary-level affects, which involve interaction between sentient feelings and associative and / or cognitive processes.[30] We espouse Panksepp's system of seven primary affective systems in the midbrain and limbic system as primary-level emotions (i.e., fear, lust, care, play, rage, seeking, and panic / grief), as it goes a long way toward reconciling innate adaptations into a paradigm of causal motivations for behavior.[31] And we are in accord that primary emotions engage deeper and older brain areas and may be activated without the intervention of the cortex and conscious processes.[32]

The first several chapters of this book will focus on what we consider the understudied range of abilities made possible by primary- and secondary-level emotions. In the latter part of the book we will turn to tertiary-level, linguistic emotions, which are imbued with cultural significance but remain tethered to the first two levels of affect and, of course, to the conscious mind embedded within a social environment.[33]

We argue that affective neuroscience calls for a new metaphor of the mind. The major metaphors used in the psychological sciences over the last approximately 150 years portray the mind as an animate being with neural or physical elements, spatial processes, and systems. Metaphors in the psychological sciences may enable findings from other disciplines to play a role in the formulation of research programs and thus transform methodological approaches. The new metaphor must accommodate continuous forms of measurement both as a way of including neurotransmitter and hormonal levels, as well as a way of integrating longitudinal life-historical, cultural, and environmental factors. A model that takes into account emotional and cultural factors would be openly discursive, allowing us to consider how affect, ideas, and social encounters sculpt the sentient mind. It must include a modified ontological paradigm that takes into account teleological and developmental biological processes, which we seek to provide in Chapter 2. And finally, the metaphor must aim for a dialectical integration between associationist, behaviorist, introspectionist, and cognitive mental function. This book is a step in the direction of formulating such a metaphor, to serve as an anchor for theorizing in the psychology and the philosophy of mind.[34] We believe a more appropriate metaphor of mind will include notions of organic and biological framing, which disappeared during the ascendancy of mechanical metaphors in the early modern period and subsequent cognitive revolution.[35] Furthermore, as we discuss in the final section of this chapter, computational metaphors at use today have great difficulty portraying non-propositional intentionality and meaning due to their continued use of the machine bias. In Chapter 3 we put forward a model of social intelligence that may more accurately describe the role of affect in social behaviors.

Intentions-in-Action: Internal and External Content

A key aspect of integrating the affective sciences with experimental psychology is demonstrating the intrinsic functional connection between processes, specifically, how affect colors our ideas and impressions. Models of the role of valence in the mind in behaviorism are largely a story of unconscious conditioning paradigms, while in the cognitive sciences they are treated in terms of rational cost/benefit decision-making. And yet, behaviors stemming from the emotional mind are not entirely mechanical in their associations; but nor are they entirely discursive and computational. From the simple homeostatic state of thirst or hunger to the existential feeling of social isolation, we argue that affective values are internal signals that phenomenally manifest bodily needs. The set of processes we hypothesize are responsible for the affective functions of the mind are primary-level affective mechanisms, secondary-level mechanisms like pushmi-pullyu representations, vertical associations, somatic markers, and finally cognitive representations at the tertiary level.[36] Our argument for how affective values provide intentional motivation requires naturalizing intentions by exploring how they function in action.

Recent work, derived from Merleau-Ponty's fundamental philosophical arguments concerning our conceptions of the body as a subject of action rather than as an object, representation, or reflection, suggests that the body is the organ of intention.[37] Using this approach, we can tease apart a pragmatic representation of an action goal from a semantic representation of the object.[38] Seminal empirical work on neural pathways by Ungerleider and Mishkin, and on patients with object agnosia by Milner and Goodale, does in fact suggest a distinction between "what" (i.e., semantic and descriptive) and "where" (i.e., pragmatic and operational) pathways.[39]

The question of how an action is intentional requires elaborate theoretical calisthenics in a cognitive paradigm that has place for neither the wide range of abilities enabled by basic affective mechanisms, nor the embodied nature of cognition. Accordingly, in the cognitive paradigm the concept of intention has been split by philosopher John Searle into prior intention, which triggers an action, and intention-in-action, which plays a continuing role in the control of the execution of an action.[40] This view of intention is a form of internalism, where the cause or intention behind an action is a mental idea.[41] A cognitive, internalist conception of action requires internal representations that mediate the relation between the body and the goal of its action. On the other hand, an externalism of action that we favor—where the cause or intention is in the environment rather than in the head— would take into account the goal-directed nature of action and how external, contextual constraints play a role.[42] Drawing from the work of Marc Jeannerod, Elisabeth Pacherie brings forward a helpful conceptualization of action. She describes how, if object attributes are treated as affordances that activate predetermined actions (i.e., as causally indexical), the motor representations of a goal state can be considered as something between a sensory and a motor function.[43] We believe this in-between category is part of the secondary-level affective mechanisms qua action processes, and we elaborate on this model in Chapters 3 and 5.

The enactive approach considers all living systems based on their autonomous and adaptive natures, wherein cognition entails affect by virtue of its capacity to bring forth a world of sense—an "umwelt"— that has special significance for the enacting organism. Researchers are increasingly adopting O'Regan, Varela, and Noë's enactive view of perception in order to refine an embodied model of cognition that includes content externalism.[44]

Are such goal-oriented action processes manifested homologously in human and nonhuman animals? It might be the case that similar associational and perceptual processes are at work when the pertinent information that sets the affordances is available in the environment, but that over and above this system, the action sequences in humans can be decoupled from their affordance triggers such that the motor plan can become an internal mental simulation that is modifiable in line with broader long-term plans.[45] At some point in the evolution of mind the externalist processes described in enactive and ecological psychology may become integrated with cognitive representation when a shift occurs between immediate pictorial affordances and the greater reflective voluntary control made possible by representational processes. Of course, the range of psychological abilities made possible by simple perceptual abilities devolves upon one's notion of memory in its declarative and, more importantly, nondeclarative function.[46]

In this regard, Dreyfus has been a trenchant critic of the computational element in cognitive science. He claims skill acquisition is a better model for framing how we learn and execute skillful action.[47] Consider how skill acquisition depends largely on unconscious learning processes in procedural memory.[48] Merleau-Ponty includes similar claims in his notion of "intentional arc," where body skills are not stored as representations but rather as dispositions within a "motor intentionality" that describes the body's implicit relatedness to the world in a concrete and practical "reaching out."[49] He uses the term "maximal grip" to refer to the body's tendency to refine its responses so as to bring a motor situation to some sort of optimal gestalt that enables the appropriate "fine-grained" disposition. In such a scheme, goal states serve more local affordance-activated functions rather than structuring behaviors via an internal template of action.[50]

The embodiment of intentionality in action is possible when we consider its intrinsic function in body-world loops of goal-directed action. In this regard, what primarily characterizes an action is its goal, of which the action is a function.[51] Some goal-directed actions require further scrutiny; for example, actions mediated by mirror neurons that seem to have evolved not only as part of imitative learning systems but rather as part of social intelligence systems. Mirror neurons thus enable subtle dispositional and nonverbal information in social companions to be sussed out via *intentional attunement*.[52] As Gallese et al.

have empirically demonstrated, mirror neurons (so-called F5 neurons) detect actions oriented toward a goal, rather than elementary movements at the level of neuromuscular programs. Using the goal as the context of any given action clearly delineates the intrinsic, implicit, meaningful, and focused nature of intentions-in-action.[53] Or, as Gallese puts it,

> What functionally makes such a state a goal state is the fact that its internal representation is structured along an axis of valence: it possesses a value for the system. A value is anything that is conducive to preserving an organism's integrity (e.g., homeostasis), to maintaining integration on higher levels of complexity (e.g., cognitive development and social interaction), and to achieving procreative success. Therefore, the reward system is a second important element of the way in which a goal representation can be implemented in a causally effective way. Goal states imply values on the level of the individual organism, and values are made causally effective through the reward system.[54]

This consideration of intentions-in-actions, or biological aboutness, requires a further commitment to a biological teleology we describe in Chapter 2 as manifested in a conative SEEKING system. Such an ontological shift makes possible a paradigm that emphasizes the role of affective mechanisms as motivational and directive within the embodied nature of cognition.[55]

While actions are intentional vis-à-vis their goal-directedness, moods are intentional in that they are open to the world.[56] Primary- and secondary-level affective processes create response biases in action outcome selection; for example, in inferential processes. In an oft-cited vignette in *Descartes' Error* (1994), Damasio suggests that his patient could not make adequate decisions without an affective valence that would act as a phenomenal weight in the decision process. That is to say, affective mechanisms are the core of value generation, of the valence that directs, slows down, speeds up, and gives meaning within decision-making and action release. They also play a role in guiding an organism's response toward associative or true reasoning processes.[57] Valence thus may be manifest in the following manner: (1) as background bodily feeling, (2) as response / reward bias, (3) as core affect, and (4) as unconscious emotion.[58]

. . .

Although it does seem to have a role in mediating conscious decision-making, primary-level affect is decidedly non-representational. By contrast, the secondary-level affect is where primary-level processes interact with associations or mnemonic schemas. The secondary level serves as a process node where cognitive functionality coincides in a feedback relation with evolutionarily older processes. This territory has been assumed representational despite the lack of definitive empirical evidence.[59] Gaining a more clear understanding of the nature of this level is an area ripe with research opportunities.[60]

To keep intentions–in–action at the level of secondary affective mechanisms, what we need is a process whereby an organism can self-correct its real-time activity toward a goal, without the counterfactual goal being mentally represented in the animal. The cerebellum is the most likely candidate for such sequencing processes, as this area is known to make inner models that continuously update and error-correct responses within reciprocal cortico-cerebellar connections. Modeling, prediction, and organization of sequences of events and behaviors are involved in many important processes, including tool use, language comprehension and production, learning of procedural sequences, recognition of correct spatial and temporal relations among actions, temporal organization of verbal utterances and planning of speech, mental rehearsal, and story comprehension in terms of order of events. Lastly, such cerebellar real-time modeling is crucial in primates for extractive foraging and tool use, suggesting an ancient homologous system.[61] The advantage of conceptualizing intentions–in–action via multimodal cerebellar models as internal impressions is that we get dispositional goal-state processes intrinsically bound up with secondary-level affective content.[62] Taking the cerebellum into account allows us to emphasize the external constraints on action.

To clarify with an example, let's say homeostatic imbalance makes an animal thirsty. Could a gestalt-like multimodal cerebellar model of drinking and an affective model of homeostatic satiation arise from external affordances and thereby give direction to action? Further, how could such a process order movements in real time to replicate a process of remembered satiation? Such a multimodal mnemonic dispositional mechanism would mediate between inner and outer states until a given homeostatic goal was satisfied and then removed from the workspace of consciousness.[63] We feel a model emphasizing the role of embodiment and cerebellar sequencing, along with affective motivations, goes a long way toward solving these problems.

Other avenues of research include nonverbal body communication and work on change and repetition in automatic motor commands and control.[64] Even with this range of affective processes and their intrinsic intentionality which we will expand upon in the next few chapters, tertiary-level affect and its aboutness, most observable in linguistic behavior, seems to require further internal representational structures for generative and compositional capabilities.[65]

Our aim is not to snuff out cognitive explanations but to enable more clear delineations between levels of mental function within a theoretical framework that includes the functional role of affective sentience. In support of that goal, the next section suggests how an affective paradigm relying on secondary-level affective mechanisms can explain empirical data more sufficiently than either a behavioral or cognitive framework.

A Case Study of Emotional Intelligence

We illustrate secondary-level affective processes with this empirical work on how a prosopagnosic (i.e., face-blind, or PA) patient can have accurate affective responses to faces he does not consciously recognize via an intentional affective system tuned through evaluative conditioning.[66]

The general paradigm employed consists of two parts: in phase one, a PA patient is shown a series of faces of familiar people and asked to rate each on a series of likeability scales; in phase two, a PA patient is presented names of the same people whose faces were shown in phase one and asked to rate each on the same scales. Ratings in phase one were then correlated with ratings in phase two; the degree of correlation reflects the degree of accuracy of covert recognition via affective processes, since they reflect the PA patient's explicit opinions of the target.

The first experiment suggested that MJH, a patient diagnosed with face-blindness, was able to make accurate ratings of faces of people that he knew personally in phase one.[67] Moreover, his likeability ratings for faces of people he knew and liked were highly correlated with his explicit ratings for the same targets, even though he could not explicitly recognize a single face. Another experiment showed that MJH was able to accurately rate faces of people he knew personally and whom he disliked, that is, the prosopagnosic subject could accurately rate faces of familiar people in positive and negative directions, even though he could not explicitly recognize the faces.

In the experiments, patient MJH is not phenomenologically aware of facial identity but nevertheless demonstrates affective responses that correspond to his explicit ratings of the person represented. Ratings in the face condition occur in the absence of overt recognition of facial identity, while ratings in the name condition are explicit. The experimental results provide evidence that, notwithstanding MJH's overt non-recognition (i.e., lack of awareness of identity), information about the face's identity is available to an affective reaction system.[68] MJH's affective reactions are not tertiary-level instances of conscious recognition, since he fails to experience familiarity with faces and entirely lacks the ability to recognize familiar faces. Thus, the affective abilities demonstrated in the experiments are distinct from recognition.[69] We suggest these findings reveal accurate appraisals caused by secondary-level affective processes.

Affective responses and feelings may be instantiated through three different mechanisms: primary-level affect, which refers to the triggering of innate, sensorimotor programs essential to bioregulation; secondary-level processes triggered by the mapping of stimulus features onto acquired schematic structures that have been previously associated, through conditioning, with particular emotional responses; and tertiary-level processes of controlled appraisal of the stimulus through subjective assessment. An affect-as-information interpretation suggests that people generally form overall evaluations based on their momentary feelings toward the target and appear to do so in an informed manner.[70] But in the case of a prosopagnosic patient, the emotional appraisal of each face is determined by secondary-level affective object-centered evaluation (i.e., attitudes) and possibly by a primary-level core affect sensation as a reward/response bias, which is potentially elicited by covert recognition of facial identity.

The fact that some fundamental form of recognition occurred at a nonconscious level and was not affected by brain damage to overt recognition processes deficient in PA patients suggests that affective appraisal functions at more than one level. The purpose of residual affective signals may be as a somatic marker, or associative mnemonic schema, in the form of a basic preference response. Affective reactions to faces, and thus facial identity, may serve the purpose of preparing one for possible contingencies that may unfold in the social situation at hand. Such a somatic marker provides a phenomenal heuristic summary of the situation, event, or person one has come in contact with. This summary can then be used as a context upon which subsequent behaviors and responses take place. Affect as a somatic marker may be a functional process shared across species. Its role qua sentient tone may be to convey a heuristic summary of situations, events, and the social environment presently experienced by a creature. Such an imperative, as opposed to indicative, conscious categorizing of experience, must be responsive to the idiosyncrasies of social and ecological experience and be appropriately connected to subsequent action. In sum, this example of covert recognition in a prosopagnosic patient suggests that perception and affect are bound and actionable before tertiary-level conscious appraisal.

Affective Mechanisms and Modularity

To investigate further how an affective paradigm clarifies and sharpens our model of the mind, we will contrast it with the computer metaphor of the cognitive sciences. In doing so, we will demonstrate how the latter overemphasizes tertiary-level processes, neglecting the constraints of evolutionary considerations.

The theoretical foundations of the contrasting approaches express their epistemic range and teleological assumptions. Embedded deeply in the core of cognitive science, the Computational Theory of Mind (CTM) conceptualizes the mind as a collection of modular processors of information.[71] The foundational interpretation of modularity of mind is philosopher Jerry Fodor's original description of basic perceptual processing via modules and transducers.[72] Cognitive psychologists, from David Marr onward, agree that the mind is a computational device consisting of three units: transducers, modules, and central processors.[73] Transducers transform perceptual input, modules process this transduced input with algorithmic equations, and central processors integrate the various module-processed signals.

On the other hand, Affective Neuroscience (AN) is a description of the mind that focuses on basic affective mechanisms in the midbrain and limbic regions of the mammalian brain; that is to say, its epistemic sources rely more on biology, anatomy, and the electrochemical system of the brain. While the triune brain model of Paul Maclean (i.e., reptilian complex, limbic system, and neocortex) works best as a metaphor and roadmap, rather than as a literal anatomical description, its emphasis on the importance of the first two layers of the brain for primary and secondary processes as opposed to the largely neocortical tertiary processes is a useful heuristic.[74] For one thing, in contrast with the formal theoretical basis of cognitive psychology in computer science, the evolutionary and biological bases of AN make it translatable to other species.[75] Whereas the cognitive sciences postulate the module to be the main unit of mental processing, the main functional unit of AN is the basic affective circuit. Unlike the agnosticism of cognitive sciences and artificial intelligence concerning the material implementation of cognitive models of thought, AN builds its theories first and foremost on the instantiation of affective circuits in neural regions across the mammalian clade.[76]

Comparing Fodor's definitive description of modules as impenetrable information processors to Panksepp's basic affective circuits is helpful in gauging the functional units of mindbrain activity.[77] A number of the characteristics of Fodor's modules and Panksepp's basic affective circuits are similar—namely, both have narrow content, are innate, are composed of primitive, evolved, or unassembled processors, and are hardwired, localized brain circuits. Yet, there are two crucial differences: (1) basic affective circuits share anatomical and chemical resources, whereas modules do not share computational resources; and (2) modules are impenetrable and encapsulated, whereas basic affective circuits are modulated by various feedback loops, including neocortical controls, bodily inputs, and internal as well as external contextual factors. It is due to these major differences that it is not appropriate to describe basic affective circuits as modules. These disjunctions lead us to consider whether basic affective circuits fit more easily into the category of transducers.[78]

If basic affective circuits are categorically unique from cognitive modules, a space opens for further discussion of intentionality and teleology in biology. Consider, first, the major function of basic affective circuits: maintaining life-supportive internal homeostasis through direct connections to appropriate responsive actions.[79] This process consists of a physical transduction of internal and external stimuli into bodily homeostasis states that directly promote life by means of chemical processes and simultaneous influence upon behavior. It is likely this occurs prior to the cognitive computational level of modules, although affective states do become represented in secondary and tertiary-level processes. The central function of basic affective circuits that integrate chemical resources and feeling states is the maintenance of organismic coherence in response to endogenous and exogenous stimuli. In terms of information states, this is a continuous variable, which may be considered as a nonlinear, dynamic, psychomotor attractor landscape. Some theorists relate dynamic systems theory to affective science as a method to understand behavioral coupling as entrainment, the self-organization of complex systems, and the emergence of behaviors in dynamical systems.[80] In biological causation, for example, we find more "phase transitions" than the binary on / off algorithms assumed in CTM. Changes in energy flow in a biological system like a brain or body can result in "jumps," "slides," and gradients between discrete states.

More important, basic affective circuits are intentional: they are about the body and about the world, and they are a manifestation of evolved motivational drives as instantiated in a set of internal chemical values. As an affective feeling, this tracking of internal chemical values—for example, of homeostatic levels for thirst or blood–sodium levels—is a functional, primal form of sentience and possibly the foundational kernel of consciousness.[81] Indeed, Mark Solms describes regulation of the organism as an "existential crisis" of survival, a "homeostatic epic." Primary-process homeostatic values can be considered naturalized intentions manifested in the organism's actions.[82] According to Panksepp, "The core function of emotional systems is to coordinate many types of behavioral and physiological processes in the brain and body. In addition, arousals of these brain systems are accompanied by subjectively experienced feeling states that may provide efficient ways to guide and sustain behavior patterns, as well as to mediate certain types of learning."[83] The emotional systems allow us to make adaptive predictions about our physical and social environment.

Getting back to our contrast, what is notable is that basic affective circuits simultaneously do too much and too little to be categorized as cognitive modules. Basic affective circuits at the midbrain level do not easily fit into the CTM because they (1) are like transducers in their relation to external factors but have the extra global function of maintaining homeostasis; (2) are intentional via intrinsic action tendencies, somewhat like evolved reflexes, without being propositional; (3) are penetrable through a set of feedback relations with limbic and cortical regions; (4) are not strictly computational, since they share neurotransmitters and brain circuits with other primary processes and are thus better characterized via nonlinear volumetric dynamic systems theory, and finally, (5) have modulatory or contextual scenic effects on neocortically generated processes and behaviors. While in their immediate role of facilitating survival, the extent of their functioning is narrow, subsequent associations may be broadened by learning across secondary and tertiary levels.

It is for these reasons that the CTM may be the wrong kind of explanation for midbrain and limbic functions of the mind, in which case we need to buttress our understanding of the mind with the biological epistemology of the affective sciences. An accurate framework for interpreting the function of basic affective circuits may necessitate both biological aboutness and a teleological understanding of embodied motivations.

The CTM and its modular framework do not have the explanatory power to portray the functioning of basic affective circuits. This may be because the CTM is essentially a psychological theory about a fundamentally different level of the mind; namely, tertiary-level processes. Additionally, studying tertiary-level neocortical processes with current neuroscience methodologies will not work for basic affective circuits, since the latter take place in midbrain structures and require analysis of neurotransmitters in real time.[84] Not to mention the complexity of dialectical relations between levels of mental functioning and connectivity.[85] This confusion of levels is one reason we seek to forge a model of the mind that can include distinct and mixed explanatory levels, allowing room for the methods of behaviorism and cognitive science as mediated by affective neuroscience.

Basic affective circuits do not fit into the Computational Theory of the Mind because they seem to function with the properties of penetrability and intentionality. This is an awful lot of power and scope for a set of unconscious chemical processes in the evolutionarily oldest part of the mammalian brain.[86] It would mean that, in tracing objects of interest, basic mammalian affective systems interpenetrate with sensorimotor and conceptual systems to perceive conspecifics and predators in real time using mnemonic and predictive abilities.

Insights gained through the affective sciences actually invert the traditional causal arrow; basic affective circuits energize, orient, and direct the higher-level processes and behavior of the animal toward necessary and desirable elements in the world that will help achieve homeostasis and satisfy basic bodily needs and psychological drives. According to this characterization, modules and central processors are, in fact, at the behest of basic affective circuits. Although the feedback loop between them certainly serves to contextualize, inform, and modulate in both directions, the evolved needs of more primitive brain areas and functions are essential to the second-to-second survival of the organism in a way that neocortical processes are not—for example, see the apparent behavioral complexity of decorticated rats.[87]

AN is a study of objects that are alive, whereas the cognitive sciences describe models that may or not have minds but have functions subtended by algorithms.[88] AN bases its evolutionary analysis of the human mind on homology (as well as adaptation), characterizing elements of the mind as evolved functions and strategies. This emphasis differs from the Panglossian adaptationism favored by the cognitive sciences, which seeks to understand the mind through reverse engineering and logical derivation of computational circuitry.[89]

A deeper philosophical difference between the two is that biology is an historical science, whereas the cognitive sciences are nomothetic (i.e., lawlike). The goal of the cognitive sciences is to construct models of the mind (toward weak artificial intelligence), whereas the goal of AN is to reconstruct genealogical relations and isolate actual brain networks that concurrently generate adaptive behaviors and affective phenomenal experiences. A recent cogent description of the mind, offered by Proust, illustrates a biological approach to the mind in line with AN: "mental function (i)s a progressively differentiated, but initially global, capacity to store previous dynamics in existing brain matter in order to predict the environment and to adjust to it . . . one could say that the distal function of the brain is to orient its growth so as to 'resonate to' the environment—or to be 'dynamically coupled' with it in a flexible way. . . . The function of the brain is to ensure a cognitive dynamic coupling with its environment, driven by inputs and biased by innate motivations."[90]

In sum, the basic functional units in cognitive and affective sciences differ in terms of their respective epistemological bases, relations to evolutionary theory, ultimate explanatory goals, and levels of analyses. We contend CTM and AN describe different layers of metaphysical complexity; the cognitive psychology of tertiary-level processes may be a special science, but at the same time, it has failed to provide clear linkages to how the human mind supervenes on biology in its evolutionary origins.[91] In contrast, evolutionary biological processes are what are being described in AN. A new dialectical metaphor of mind must be developed to serve as the meeting ground of two epistemological approaches: the psychological and the biological study of the mind.

Emotional networks in the midbrain and limbic areas are among the biological founts of affective consciousness; their function is to embody homeostasis and survival via internal affective values. With sufficient feedback linkages to secondary- and tertiary-level processes and engagement with the external environment, intrinsic values are then shaped through learning. That is, the human mind (and possibly the mind of other mammals) is intrinsically biological as well as intrinsically psychological.[92]

In discovering the inadequacy of the dominant paradigms, we are at a stage where psychology can grow past our previous metaphysical mistakes. Since basic affective circuits are not adequately described in the language of cognitive science or behaviorism and vice versa, the paradigm we forge in this book takes the stand of relying on a distinctly biological epistemology to contextualize the affective roots of culture and cognition. A biological, evolutionary epistemology can span the gap of the mindbrain problem by supplementing computational and behavioral models with affective mechanisms as intentional body-world loops. In sum, a new paradigm allows us a consilient pathway to understanding motivation, feeling, thought, and behavior.

2

Biological Aboutness

REASSESSING TELEOLOGY

IN THE HISTORY of philosophy and psychology, teleology has often been considered a unique feature of the conscious mind. The human being can call up a mental representation of a goal, like a migration destination or a task to complete and then organize his behaviors toward achieving it. Such paradigm cases of teleological causation are often contrasted with the push-pull of mechanical causality or the stochastic causality of physics.

However, an accurate rendering of the evolution of mind cannot start with this representational paradigm of modern cognition, but must instead build up from humbler predecessors. In fact, reading nature through the lens of representational teleology is precisely the anthropomorphic mistake that produces centuries of natural theology (wherein God sets the goals of natural processes). Instead, what is needed is a non-representational teleology, a biological aboutness or intentionality.

Our view of biological aboutness takes its start from the unique Aristotelian naturalism, which views organisms as holistic functional systems. Any purpose in nature, for Aristotle, was species-specific and not cosmic. Moreover, his talk of "final causes" (*teloi*) refers to the organism's relevant functions. To this Aristotelian foundation, we will add the scaffolding of Spinoza's *conatus,* a word that means "striving." Conatus characterizes the idea that "each thing, as far as it can by its own power, strives to persevere in its own being," and this foreshadows our argument that all mammals strive for this survival via affective or emotional strategies.[1] Finally, our view of biological aboutness is completed by the affective system of "seeking" (Panksepp's term) or "wanting" (Berridge's term). This is a master emotion and behaviorally drives the animal toward a variety of adaptive goals, including resource procurement, mating and other social relations, and even informational acquisition and closure.

This biological aboutness is a much stronger foundation and provides a better anchor for subsequent debates about human conscious intentionality. While there is plenty of good work on intentionality, we fear that too many theorists have started with the late-arising, human, mental intentionality and tried to build a phylogenetic foundation under it. We are moving, instead, from the bottom up, and this chapter will give us a more realistic groundwork for the discussion of human mental life that follows.

Mind is an expression of a very old biological intentionality (entirely natural; not supernatural) that can be found throughout the animal kingdom. This biological intentionality will be affectively (emotionally) structured long before it becomes cognitively structured in the *Homo* lineage. Adaptive affective systems take specific targets (e.g., the endogenous arrow of LUST drives the organism toward completion of a procreative goal), and the animal's life cannot be abstracted from such teleological projects (drives); neither should animal strategies be solely modeled on conscious representational intentionality (e.g., a human-like pursuit of a mental image of a counterfactual state of affairs). A brief conceptual history of teleology will be helpful to make sense of these claims.

In this chapter, we will tour three teleology traditions. This ontological foundation is crucial to our overall project of placing intentionality inside the body and building a vision of the affective roots of culture and cognition. These three traditions are logically distinct, but the history of biology reveals profound confusion and conflation among them. The traditions are as follows: (1) natural theology, (2) goals in natural processes ("autopoiesis"), and (3) goals inside agents. This last tradition explores goals that guide animal behavior and can be of two major types: (a) representational and (b) non-representational. Most traditional philosophy has failed to even acknowledge the non-representational forms of intentionality. Our claim is that there are at least two forms of non-representational intentionality: (i) perceptual affordances and (ii) affective or emotional intentionality. The first category, perceptual affordances, will be explored in more detail in Chapter 3, and the second category, emotional intentionality, will receive a zoological treatment in Chapter 4. In the current chapter, however, we will endeavor to rescue a scientifically respectable form of natural teleology from both theology and vitalistic obscurantism. To that end, we will eventually arrive at an ontologically robust and contemporary notion of conatus, or seeking.

Teleology and Theology

In his 1790 *Critique of Judgment,* Kant famously predicted that there would never be a "Newton for a blade of grass."[2] Biology, he thought, would never be unified and reduced to a handful of mechanical laws, as in the case of physics. A Newtonian law (e.g., the law of inertia) applies universally to material objects, and we need only to chart the variable initial conditions to produce a certain or very probable prediction. But a living thing, according to Kant, always has its own agenda—its own invisible intentions; like grass, it seems that its organic structure has been intentioned or purposefully organized. When a plant grows a thorn or when a dog barks, we ask, "What is it for?" but when a ball rolls down an inclined plane—or my pen falls to the floor—we do not say, "What is it for?" The question "What is it for?" applies to living structures in a way that has no corollary in physics. We cannot seem to expunge teleology (goal-directedness) from our understanding of living systems.

Some scholars have misunderstood this teleological argument, and the confusion contributes to perennial debates about purpose in nature—including the recent debate surrounding Thomas Nagel's *Mind and Cosmos.*[3] We often rehearse an old debate that we may be doomed to repeat ad nauseam unless we gain some fresh perspective on teleology and intentionality.

The natural-theology tradition has dominated in the West along with claims that adaptation in nature must be the result of a supreme designer, because chance alone cannot account for gills in water, lungs on land, complex eyes, and cell flagella. A designer-god, according to this view, gives nature its purposive teleological structure, and that's why a mechanical science will be incomplete. This, in a nutshell, is the natural-theology tradition of teleology, and though it goes back to Plato's *Timaeus,* its heyday was in the eighteenth and early nineteenth centuries—and it lives on in thinkers like William Lane Craig.[4] Even Darwin, before he went on *The Beagle,* read and admired the natural theology of William Paley (1802), who likened nature to an elegant and adapted watch: like any good watch, a system that contains parts that fit other parts and has a function (i.e., telling time) presupposes a designing intelligence—a watchmaker.[5]

Darwin killed this tradition. In truth, David Hume's *Dialogues Concerning Natural Religion* had already fatally wounded the tradition with a scathing book-length *reductio ad absurdum* of the design argument.[6] But Darwin's theory of chance variation and natural selection finished the job, by adding alternative measurable mechanisms of "design" to Hume's merely skeptical critique. The accumulation and spread of heritable traits, by the mechanical operations of genes, proteins, geology, climate, and so on, slowly shape organisms to fit their environments— making them appear designed. The reality of limited resources serves as the editing gauntlet, preserving a tight gap between organic structure and function. Philosophically speaking, Darwin changed a priori design (God's plan) into a posteriori adaptation.

A small but vocal minority of Darwinian opponents in the academy, however, failed to appreciate the de-purposing of nature (while the majority in the general population have also failed to appreciate this aspect of the Darwinian revolution). Some, notably the recent Intelligent Design proponents like Michael Behe, William Dembski, and Phillip E. Johnson, continue to attempt to revitalize natural theology for pseudoscience journals and high school textbooks (e.g., *Of Pandas and People*).[7] Others, like Alvin Plantinga, have tried to weave teleology into Darwinism by arguing for divinely "guided mutations."[8] In contemporary schoolboard and courtroom debates, biologists and philosophers are still called upon to root out the natural-theology fallacy before it gets into classrooms (although, of course, it's always free to flourish in temples, churches, and mosques).

This rather melodramatic debate in the political and cultural sphere has obscured the deeper philosophical issues of intentionality in biological systems. Some concepts of teleology have nothing to do with natural theology and religion, but they get caught up, dismissed, and eradicated in the confusion.[9]

Other teleology traditions include Aristotelian teleology, holism, unity-of-nature teleology (cosmic scale), conatus / SEEKING systems, autopoiesis (self-making), and various nuances within and mixtures among the categories. In addition, the popular anthropomorphic tradition of natural theology—which projects human-like mental intentionality into nature—gets confused together with these other traditions. Discussion of the unique purposiveness of mind has been forced by these old paradigms into a false dichotomy that treats said purposiveness either as illusory (mere epiphenomena) or as mysteriously superadded late in *Homo* development; therefore, sifting out these traditions and rescuing the scientifically useful forms of teleology that are compatible with our Darwinian naturalism is crucial for uncovering a new paradigm of the evolution of mind.

Aristotelian Teleology

Aristotle was a step closer to the paradigm we will be presenting below; he saw teleology in nature because natural processes always unfold toward some goal—for example, acorns develop into oak trees.[10] But simultaneously, parts of animals are for the sake of their compositional wholes—osseous tissue is for the sake of bone, blood is for the sake of circulation, and the tooth is for the sake of mastication. The organism is thus a "Russian doll" of nested teleological relations of structures and functions. Aristotle refers to these ends / goals / wholes as final causes, defining a final cause broadly as the "end, for the sake of which a thing is done."[11]

Aristotle's teleology is uniquely difficult to appreciate because hundreds of years of medieval theology misinterpreted it as mental and theological. For example, Aquinas's famous view was that God's mind put the goals into nature—"*Ergo est aliquid intelligens, a quo omnes res naturales ordinantur ad finem, et hoc dicimus Deum.*" But that was not Aristotle's view, despite generations of schoolmen who tried to "baptize" him. Then, after the scientific revolution, both religious and secular scholars came to think of nature as a giant machine; so as with all machines, its goals would need to be installed by some kind of designing mind.[12] Again, this was not Aristotle's view.

Instead, Aristotle thought of teleology as a feature of nature, in the same way that we think of gravity as an impersonal, undesigned, aspect of matter. Unlike all Western monotheisms, Aristotle's prime mover god was not a *creator* god, so he did not install teleology into nature (like Plato's *demiurgos*). Moreover, Aristotle did not simply theorize teleology, but empirically observed regular, goal-directed behavior and structure while studying and dissecting animals in Atarneus and Lesbos. The principle of teleology was descriptive in his biology but also normative in his recommendations for good science.[13]

Aristotle was critical (*Phys.* II) of the simple versions of evolution that he saw in work by Empedocles (and Democritus), because he thought material bits could not clump together into sustainable organisms unless matter had specific recipes built into nature itself (the formal / final causes). Unlike Plato, who thought the goals of natural processes were incarnated *ideas,* Aristotle refused to postulate cosmic archetypes and saw teleology instead as a way of describing the regularity of biological procreation, behavior, and anatomy. If he had known about DNA, he may have embraced it as the means by which information informs matter.

Notice, however, that we still have Aristotle's final cause question: How does common stuff (e.g., DNA or stem cells) get differentiated into diverse organs and organisms? The DNA alone is not enough, and after we cracked the genome, we realized that we needed to study development more carefully, so we are finally discovering Hox genes, evo-devo, and epigenetic processes that regulate all that DNA potential into actual organs, structures, and behaviors. Those regulatory causes were the aspects of life that Aristotle called teleological.[14] He may not have accepted the slow evolutionary change of those top-down regulatory structures, but this scarcely matters to the local and practical scientific domains of animal ethology, anatomy, or physiology.

Unlike natural theology, Aristotle's methodological teleology is not incompatible with Darwinism. Aristotle thought that one cannot do biology by talking about whirling atoms (like Democritus) and that one needs to contextualize *why* this organ or behavior fits with the animal's structure / function and environment.[15] That question or issue only reverts to divine psychology if you are a natural theologian, but for Aristotle and Darwin, it reverted to the unique living conditions of the organism. Aristotle explained human teeth by looking at their function of mastication, but he recognized that other vertebrate teeth were also used as weapons and explained their unique structures accordingly.[16] Most phenotypes don't make sense unless we relate them to their functional milieu.[17]

Aristotle objected to the reductionist atomism of his predecessors on the grounds that it was the wrong kind of explanation or account. It is too low-level to shed light on the organism's embryogenesis or behavior. His critique is not a call for supernatural amendments to science, but a call for autonomous levels of scientific explanations. That Aristotelian ideas can reconcile with modern biological science is evidenced by the fact that some scholars see Nikolaas Tinbergen (1907–1988), the father of ethology, as an Aristotelian, at least in part. Tinbergen famously formulated the "Four Whys" as a methodological system for investigating animal behavior / psychology: (1) mechanism, (2) function, (3) phylogeny, and (4) ontogeny. These whys are not unlike Aristotle's four causes: material, final, formal, and efficient.[18]

The fact that, in general, embodied animal minds regularly and predictably orient themselves toward specific foods, social interactions, pleasures, and goals is a metaphysical point—not merely an epistemic point. This more full-blooded aspect of Aristotelian biological metaphysics does not stand or fall with the other outmoded aspects of his ancient philosophy. Yet contemporary science is wary of any such metaphysical commitment, preferring an epistemic (instrumental) teleology that is agnostic about its own foundations. Daniel Dennett's *intentional stance,* for example, is a way of acknowledging the goal-directed, belief-desire motivations of organisms while avoiding any metaphysical underpinnings.[19]

The problem with Dennett's intentional stance is twofold. First, though it is a helpful methodological and epistemic position that allows investigators to treat animals "as if" they had intentions, it remains at the heuristic level and cannot help us think about the metaphysical or causal aspects of intentionality in nonhuman animals (viz. the development of intentionality in biological matter per se). Second, and more problematically, it assumes a species-centric anthropomorphic cognitive version of intentionality, but thereby puts "the cart before the horse" regarding the older biological intentionality of all animals. Unlike the intentionality of modern philosophy, in which the subject rips away from the real-time world and holds a mental representation of a future state before his mind's eye (guiding his actions toward that ideal future state), the *biological aboutness* that we are proposing is submerged in the real-time environment of the creature (how behavior stems from reactions to external ecology and internal affective states). Dennett's approach to intentionality, in fact it's conceptualization in most post-Cartesian philosophy, denies a completism in physicalism; it must be superadded by reason alone, but this is ultimately a provincial move considering the advances of neo-Darwinian and dialectical biology.[20] Biological aboutness, by contrast, is compatible with the associationist and conditioning processes common in animal life, as larger life strategies of the creature emerge out of the associationist processes that imbue salience to specific experiences and behaviors.[21]

We can, according to Dennett, treat animals as if they were designed (even if we *really know* that natural selection blindly sculpted them) and animal behavior as if it consciously intends a future outcome (even if we *really know* that unconscious mechanical antecedents produce the behavior). This strategy has great advantages over previous tendencies to anthropomorphize creatures, but it fails to acknowledge that goals may have real causal power in organisms that have affordances or other ways of reading their environment.[22] Moreover, it commits its own anthropomorphism by treating animals as if they had humanlike minds, even when the intentional stance is only pro-visional and metaphysically neutral.

Aristotle's ontological intentionality clears a middle way between anthropomorphizing animals (as human-like minds) and reducing them to mere stimulus-response machines (that merely seem inten-tional). Our project, in this book, is to travel this middle way to find the ignored intentional foundations of embodied mind.

Holism and Final Causes in Biology

Setting aside his temporal teleology (e.g., acorns become oak trees), it behooves us to examine the holism tradition that Aristotle created. Holist teleology claims that biology must recognize the causal rela-tionships of cells inside tissues, organs, physio-systems, organisms, and environments. Aristotle didn't know about cells and would have started with tissues, but we can now take the Russian dolls down to infinitesimal levels. The medieval metaphysicians pursued this avenue, called *mereology,* but derailed the inquiry by trying to find the "principle of individuation" that would determine which of these nested levels was the true "substance." Eventually, analytic Anglo-American phi-losophy became reinterested in holism in the twentieth century, but only as a logical and linguistic problem.[23] Continental philosophy, on the other hand, has had a longstanding obsession with the metaphysics and epistemology of biological holism.[24] Goethe, Kant, and Hegel were deeply interested in the way that biological form seemed to govern simpler physio-chemical processes, and they tried various ways of organizing nature without any appeal to natural theology.

Why can't biology succeed by dissecting everything down to chem-istry? Because we cannot understand DNA methylation, for example, without understanding *what it is for.* This is not the same as finding out what *effects* it has months or years later, although that is important too. We also need to know what *beneficial effects* it has for the organism. One of the effects of methylation, after all, is that it's involved in most forms of cancer—an obviously deleterious suite of diseases.[25] But evolution is a cost/benefit process, and methylation is also a crucial regulator of gene transcription. It aids in our individual survival (its regulation seems to respond to environmental challenges in embryological development) and in our species survival (carrying epigenetic information across generations). We're just learning about the mechanisms of DNA methylation, but we fully expect to find survival benefits for most methyl group switches. There aren't just *mechanical effects* of methylation, but *advantageous effects—adaptive effects*. And even when we don't know what they are, we are vigorously working to find them.

Scientifically decomposing cells and genes into their parts wins us many prediction victories, but it does not tell why a trait *persists,* why it continues in a population. For that, we need to ask what the organic structure or behavior is for. The oxidation of carbon to produce carbon dioxide, for example, is just simple antecedent / consequent causation, and no one in chemistry would claim that the carbon loses electrons "for the sake of" becoming carbon dioxide. But in biology we have to acknowledge that most specific traits or behaviors are for the survival of the organism or population (unless it is vestigial or a spandrel).[26] Sex, for example, has the advantageous effects of offspring fitness (through variation and hybrid vigor). This adaptive effect explains why the mutation was selected for and why it persists. This contrasts strongly with how the mutation itself arose—since such a chemical transcription glitch needs no "why question" answered. The glitch has no purpose and no regular directional pattern that cries out for special explanation.

For holists, the attempt to find the "end, for the sake of which a thing is done" applies to the structures as well as the processes of biology. A leaf is unintelligible without understanding something about trees, a heart is incomprehensible without the circulation system, a brain makes little sense except in the body of a creature that can move, and on and on. Pursuant to such systems-based analyses, these teleological wholes refer to the ultimate teleology—which Aristotle, sounding very Darwinian, described as "the most natural of all the functions of living creatures, namely to make another thing like themselves."[27]

The holism school wants us to remember, amidst all the real successes of reductionist science, the validity of higher levels of causation and reality.[28] Holism is a kind of causal pluralism, gently reminding the atomic and genetic determinists that organisms and ecologies are not just epiphenomena. In an analogous fashion, our claims in this book connect plural levels of causation in the mind, from conatus to ecological psychology, cognition, and the feedback loops of culture.

Teleological statements are explanatorily robust in biology, but is it real or sham teleology? Kant argued that our reason cannot help but project purpose into biology, and we should accept modest teleological claims as "regulative principles" (which we might call "instrumental principles" today). By this logic, it's scientifically respectable to claim that hollow bird bones are for the sake of flight. But the mind does not stop there, according to Kant, and naturally goes on to project a whole system of purposes into the biosphere—and here is where the natural theologian (then and now) starts to salivate. But the extrapolation quickly grows silly: grass is for the sake of cows, cows are for the sake of human food, and so on. Like Voltaire, in *Candide,* Kant in the *Critique of Judgment* (1790) mentions hyperbolic teleologists who claim that mosquitos help humans wake up and stay active, and tapeworms aid digestion for their victims. These absurd examples remind us to restrict ourselves to *local* teleology, dispensing entirely with global or cosmic purpose.

When we are doing biology, Kant argued, we need to subordinate simple physics / chemistry explanations to functional teleological explanations. We need both levels of causation and explanation, and one level does not reduce to the other. Many biologists and philosophers, following Kant, have argued that we can pretend that things are for the sake of goals but that this is just methodologically helpful, not real. This approach—the "as-if" teleology tradition—helped the Germans employ means / end explanations, but avoid the temptations of natural theology.[29] As mentioned above, Dennett's intentional stance is a more recent species of this view. Can we, however, go beyond the purely instrumental justification to a kind of teleology that is, well, real?

The answer is complicated. Consider teleology as analogous to "free will" for a moment. Free will is heuristic or instrumental in common life—we consider ourselves free when we are posed with decisions every day and we assume free will in our fellow citizens when they behave in myriad ways. Free will has explanatory power in formal domains as well. In law, ethics, psychology, and sociology, free will makes sense out of human behavior in nontrivial ways—despite the fact that there's no metaphysical or scientific evidence for free will per se. Every act of free will could be reduced, in principle, to a series of synapse firings, but these are, from the vantage of social science, only the conditions of human behavior, not the relevant causal levels. Which level is the most real? That's a bad, albeit tempting, question.

If a *neuroscientist* asked me why I do philosophy, I might report that "certain neural pathways were sculpted in my developing brain, such that cingulate, prefrontal, and parietal activity easily trigger my hedonic dopamine system." When my *friend* asks me why I do philosophy, I'm likely to say something like, "Solving conceptual puzzles and reflecting on weird stuff is deeply satisfying for me." When the dean of my college asks me the same question, I'm likely to trot out something pious like, "Philosophy improves critical thinking and shapes students into better citizens of our democracy, and I want to be a part of that mission."

These explanations are not in competition with each other. One of these accounts is not the "correct" one, usurping the others or reducing them to mere figments. They are all compatible, and they are all true. Likewise, if geneticists give a molecular account of human skin color differences and evolutionary biologists give an adaptive account of skin color, they are not competing to be the real explanation. Here are three compatible accounts: (1) A purely mechanical account of small changes in the melanocortin 1 receptor gene (MC1R) can tell us how melanin concentrations can produce darker or lighter skin coloration. (2) People living in an intense solar region will survive better if their skin is darker because carcinogenic UV-B radiation is blocked by increased melanin pigmentation. (3) Around 1.2 MYA, about 300,000 years after *Homo* lost its body hair, group migrations started new environmental selective pressures—lighter skin evolved in less insolated regions (allowing necessary vitamin D production), and darker skin evolved in the populations of high-insolation regions. So we have three levels of explanation; a chemical story of DNA to melanin, an ontogenetic story of the dangers of skin cancer, and a phylogenetic story connecting human migration and ecology. Notice that the first biochemistry explanation may work fine without teleology but the two adaptive explanations are strongly teleological—not in the sense that skin cells "foresaw" the goals, but in the sense that the *distribution* and *persistence* of these phenotypes (and their genes) only make sense if they are "for the sake of" survival (excepting the usual caveats about spandrels or founder effects).

A theory of the evolution of mind need not resolve the ontology of wholes versus parts, but it does need to toggle between the levels. In trying to understand fear in humans, for example, it would not be enough to point to the amygdala's role in fear feelings and behaviors. The amygdala is necessary but not sufficient for a strong explanation, as we need to understand many aspects of the whole limbic and endocrine systems (e.g., feedback communication between other brain structures like the thalamus, hippocampus, and cingulate gyrus, etc.)—and this is not to mention that the body contains the brain as a part and the ecological environment contains the individual body as a part. Michael Ghiselin has even argued persuasively that the phylogenetic species itself is an *individual* rather than a collective or a kind, which means that we stand in relation to *Homo sapiens* as blood cells stand in relation to the circulation system. We are parts.[30] These considerations make holist teleology scientifically reasonable, but, of course, natural theology has no intrinsic relation to these concerns whatever. The contemporary legacy of holism includes the resistance to greedy reductionism and greedy consilience, wherein higher levels of biological causation are reduced to lower levels of chemistry and physics. In the area of mind, holism similarly resists the facile reduction of cognition to informational computation. Orienting a holistic analysis of mind toward the affective / emotional systems helps us appreciate the aboutness and teleology of pre-representational levels of biology.

Autopoiesis

Well before the Darwinian revolution, the inexplicable strangeness of matter's self-organization (autopoiesis) did not go unnoticed. Yes, environmental conditions dispose of, or "edit" out, organisms and populations with deleterious traits, but do we need a better science of the "proposal" step? From body plans to brains, matter crystalizes and canalizes into repeatable structures. Do we need a better science of form or self-organization itself?

Many thinkers, like Darwin's friend Richard Owen or the American naturalist Louis Agassiz, thought that the development and anatomy of the animal form represented the incarnation of divine ideas in physical matter. The common vertebrate structure that we share with dogs and fish reveals, according to these Platonic thinkers, a coherent archetype or leitmotif that God installs in nature. Then mutation and natural selection go to work to spin out biological variations.

This unverifiable speculation is no longer a scientifically respectable position, but it remains a popular, if inarticulate, assumption in mainstream theistic culture. Still, the question of organization has not folded neatly into neo-Darwinism, and some clever twentieth-century thinkers, like Darcy Thompson, Stephen Jay Gould, Stuart Kauffman, and William Wimsatt, have suggested (and modeled) ways that material systems tend toward specific workable structures.[31] Wimsatt calls these self-ordering aspects of complex material processes "generative entrenchment."[32] There's nothing occult about this; it might be best thought of as an attempt to articulate the logic of a middle level between genetics below and selection above. Kauffman, for example, has shown that systems of dynamic material variables will coalesce around predictable states, according to rules of Boolean logic.[33] He and others have suggested that some science of self-organization will need to join natural selection to give us a more accurate understanding of biological form.

Like vitalism before it, some of this research seeks to address the development of complexity in animal embryology and origin-of-life studies. It has become scientifically respectable by assuming a materialistic naturalism, but it is still a recent descendent of an older teleological tradition. Autopoietic science tries to understand the way that micro processes are regulated by relative macro states over time, so it treads in the territory of means / end relations.

Biologist Andreas Wagner has recently applied self-organizing logic to the mutation process in evolution. Natural selection edits the variations that mutation proposes, but how does mutation hit on any useful variations when the number of deleterious options approach infinity? Consider the Hox gene circuit that controls for anatomical body plan. For humans and snakes, there are around forty genes in the Hox circuit, and each of these genes has regulatory powers on the others (through activation or suppression), leading to a staggering number of possible combinations.[34] Wagner suggests that the circuits themselves are robust entities that govern the sublated gene levels below them.[35]

There are robust and versatile networks at higher levels than genes—these are multigene circuits, like Hox, or they are metabolic circuits that remain functional and tolerate changes and failures at the levels below them. These higher-level processes are sustained, in part, because there is extensive redundancy of the successful processes beneath them. So, failures and negative mutations do not collapse the higher functions like a house of cards. The relative stability of these higher networks helps explain how phenotypes can continue to express the same trait when the genes below them are changed or deleted. This work suggests that our genes are not as important as we once thought. The function of a gene is a product of the higher-level network that contains the gene. If these networks are being selected for or against, then evolution can make moves that are bolder than the ineffectual and almost infinite incremental moves at the level of A, T, G, and C.

Some theorists have suggested that autopoiesis folds neatly into cybernetics, the science of regulating feedback systems. From the early work of Norbert Wiener and game theory economics, cybernetics explores how systems with multiple parts and processes stabilize (achieve homeostasis states like thermoregulation) and yet continue ongoing information exchange with their environments.[36] To what degree are organic systems similar to cybernetic feedback mechanisms like temperature thermostats?

There are many mindbrain processes, and organic processes generally, that fit cybernetic descriptions. We take the view, however, that while system regulation is attainable by several means, the one that matters most for understanding mammals is affectively based, subjective, homeostatic self-regulation. Inanimate cybernetic examples, like thermostats or automatic pilot, help us understand the way feedback works generally, but animal self-regulation is more often motivated by sub-jective, felt experience and the behavior and chemical changes that can feed back upon the self that generated the actions. In this sense, animals are not like thermostats because the triggers of action (toward and away from homeostasis) are affective experiences (feelings). And the system alterations can feed back upon the well-being and survival of the organism (the ultimate teleological umbrella for the creature). It feels like something for an animal to be activated by lust, for example, and it is that feeling that spurs the animal to track its mate and make many error-corrections to bring about the consummatory feelings produced by successful mating. And more generally speaking, self-preservation has a feeling in mammals, whereas self-preservation in a cybernetic system is without sentience.[37] In fact, there is no experien-tial subject, or self, in an inanimate cybernetic system.

A rich and confusing minority tradition has emerged in biophilos-ophy since Francisco Varela (and Humberto Maturana) first reminded neo-Darwinians to think about development. Varela defined an auto-poietic unity (an organism) as "a network of processes of production (transformation and destruction) of components which: (i) through their interactions and transformations continuously regenerate and re-alize the network of processes (relations) that produced them; and (ii) constitute it . . . as a concrete unity in space in which they (the com-ponents) exist by specifying the topological domain of its realization as such a network."[38]

We count ourselves in the extended family of Varela's intellectual descendants who have all worked to articulate a non-mechanistic or-ganicism for purposive nature, without falling back on any form of vitalism.[39] We agree that mechanical approaches to organic form and function must be supplemented with systems and feedback causality of various levels of ontology. This is an important aspect of our in-terest in biological aboutness, but our main focus in the present book is specifically regarding affective / emotional aboutness. Many of our Varela-inspired "kin" have leapt from a non-mechanistic view of bio-logical causality straight to human conscious mind (complete with representations, concepts, and semantic complexity). For us, this jump is an unhelpful saltation, and we want to describe the layer of affective mind that preceded our linguistic minds in the order of evolution and that still exists just below the level of propositional consciousness. SEEKING is a good example of such an affective feedback system that demonstrates autopoietic unity within a holistic analysis of mindbrain and behavior.

Conatus and SEEKING

Spinoza saw nature in fairly mechanical, deterministic terms, but he recognized that living things share a simple goal-oriented tendency; they strive to survive. To reiterate, he called this animation principle of living systems "conatus" (striving) and considered it the very essence of all biological creatures. It is not the design teleology of the natural theologians, but it is recognition that natural creatures have an essential goal-directed imperative within them that cannot be captured by purely billiard-ball causality.[40] Like every other animal, humans have survival drives that are directed out at the world, but unlike many other animals, humans can decouple from the unreflective immersion.

As Spinoza puts it, "Between appetite and desire there is no difference, except that desire is generally related to men insofar as they are conscious of the appetite. So desire can be defined as appetite together with consciousness of the appetite."[41] This becomes part of a larger psychology of ethics that we're not interested in here, but we are sympathetic to Spinoza's dual-aspect monism. Just as Spinoza argued that conscious mind is one aspect of reality, with matter forming the correlate aspect, we tend to think that animal sentience is identical with specific neurophysiological processes (the affective systems). When the animal is feeling fear (sentience), for example, a specific arousal is happening in the lateral and central amygdala in communication with the periaqueductal gray region.

In the hands of some later theorists, the conatus life force became an occult metaphysical force.[42] Following J. F. Blumenbach, Kant seemed to think that a "formative force" (bildungstrieb) worked inside matter and caused the seeming miracles of animal reproduction and embryology.[43] Embryologist Hans Driesch (1867–1941) even proffered an *empirical* vitalism on the grounds that no matter how much he mutilated a developing vertebrate zygote, it still stayed on course—as if an invisible outside force guided the process.

Modern genetics and stem-cell science have clarified many of these mysteries for us, and occult embryology has rightly gone the route of phlogiston. In the twentieth century, life science tilted heavily toward the issue of genetic code and largely forgot about issues of development. Once biology had the ingredients, it was thought, we would quickly resolve the mysteries of species differentiation and deviation. Only in the twenty-first century, after the genome proved underwhelming, have biologists tried to make up for lost time on the complex issues of *development*. Still mysteries remain, and may do so because the mechanical model itself has run its course.

Philosopher of biology John Dupré has recently considered the mitotic spindle, in which our chromosomes are aligned along a cellular equator and then teased apart to create two separate sets for cell division. The mechanisms that cause such a process are myriad (e.g., centrosome-nucleated microtubules, kinetochore-driven microtubule generation, and so on). But, as Dupré points out, cell biologists can eliminate any of the mechanisms—even the crucial centrosomes organelles—and yet the spindles will still form, albeit more slowly. Dupré makes an argument against reductionism similar to Driesch a century before, but of course he has no interest in an occult vitalism. Rather, he wants to annex the substance-based ontology of microbiology with systems-thinking that refers dynamic processes to their contextual functions.[44] Biologists Duncan and Wakefield suggest that the mitotic spindle cell process is not an irreversible set of mechanical instructions, but a "self-regulating dynamic structure where multiple pathways of microtubule generation are spatially and temporally controlled and integrated, constantly 'talking' to one another and modifying the behavior of their microtubules in order to maintain a flexible yet robust steady-state spindle."[45]

We may feel uncomfortable calling these processes "conatus" or "vitalistic" because such terms are tinged with some historical supernaturalism, but the biological phenomenon they isolate still lacks sufficient explanation. The intuitive question of the conatus tradition (viz. What is organic striving?) has gone unanswered in modern biology, but some legitimate empirical work has emerged to better isolate biological striving in mammals at least. This is the affective neuroscience research on SEEKING.

Instead of philosophizing about conatus as a property of all living systems, neuroscientists today have discovered something like a brain-based conatus system that drives mammalian behavior. In the same way all vertebrates possess a fear system, they also engage in SEEKING behavior—and recently, neuroscience has isolated a foundational motivational drive that underlies diverse searching behaviors (hunting, foraging, procreation, and so on).

Jaak Panksepp calls this modern conatus SEEKING.[46] Alternatively, Kent Berridge has discovered a similar brain-based emotional system that he calls "wanting." Berridge et al. explain, "Usually a brain 'likes' the rewards that it 'wants.' But sometimes it may just 'want' them. Research has established that 'liking' and 'wanting' rewards are dissociable both psychologically and neurobiologically. By 'wanting' we mean *incentive salience,* a type of incentive motivation that promotes approach toward and consumption of rewards, and which has distinct psychological and neurobiological features."[47] This incentive-salience form of wanting, or Panksepp's SEEKING, is below the neocortical circuits of conscious desire. Incentive salience and SEEKING are subcortical systems that seek to reset homeostasis, and while they often align with our neocortical conscious desires, they sometimes lead to irrational wanting—a want for what is not cognitively wanted.

SEEKING is often classed with the emotions, but it is really a master emotion or a drive, a motivational system that organisms enlist in order to find and exploit resources in their environment. It energizes mammals to pursue pleasures or satisfactions, but it is not the same as pleasure. It is that growing, intense sensation of heightened attention and the increasing feeling of anticipation—as if you are just about to scratch a powerful itch. Panksepp calls it a "goad without a goal," but the goad eventually does fasten onto specific goals.[48] Its intrinsic aspect is promiscuous and flexible—motivating different pursuits at different times.

This desire or SEEKING system sparks the ventral tegmental area (VTA) of the midbrain, rising through the nucleus accumbens and extending neurons up to the prefrontal cortex and down to the brain stem. It enlists a dopamine pathway in the brain, and while it strongly correlates with pleasure rewards, it actually spikes highest just *before* you receive the reward—when desire or anticipation is at fever pitch. If you turn on this system (with electrical stimulation), mammals will go into foraging behaviors, exploration of environment, selective attention, pursuits of specific appetite rewards (food, water, warmth, sex, social interaction). This system is not stimulus / response machinery, but an endogenous conatus—energizing the organism toward the goal of survival (via homeostasis). There's nothing metaphysical about this conatus, but it answers scientifically to many of our older intuitions about life's unique teleology; namely, the organism's motivation to move itself, to persistently track an external resource, to error-correct or adapt when suboptimal conditions obtain, and so on.

Crucial questions arise here and may one day unlock some evolutionary puzzles. For example, since this striving SEEKING system is housed primarily in the limbic system and periaqueductal gray, one wonders about the striving we observe in insects. Will we find homologous brain structures and functions, or did feeling-based intentionality in mammals emerge from or parallel (analogously) to a more robotic (non-feeling) conatus in simpler creatures?

Human Intentionality and the Present Debate

Theories about the evolution of the mind, especially those propounded by evolutionary psychology in the 1990s, raised the dander of biologists like Stephen Jay Gould and Richard Lewontin. Loud scuffles ensued throughout the late nineties and early aughts. Gould characterized his enemies, psychologists like John Tooby, Leda Cosmides, and Steven Pinker, as severe reductionists who treated the mind like a machine, and they, in turn, wondered why the great biologist Gould should act as if evolution stopped at the neck and spared the mind its deterministic causation.[49]

Each side had a point, but each side gave in to the melodramatic caricatures of their opponents. In retrospect, Gould was right when he warned, at the onset of evolutionary psychology, that we were entering a new phase of conjectural Panglossian adaptationism— wherein every contemporary cognitive tendency, idea, and prejudice would be read back into prehistory and justified on the grounds of natural selection. His prediction came true in the rise of popular modular theories of the mind that assumed hardwired computational processors existed for everything from language, to morality circuits, to what kind of landscape paintings we prefer. We share Gould's skepticism and think the computationally-minded evolutionary psychologists over-interpreted the data. But Gould went too far himself and cordoned off the mind as a mysterious byproduct of evolution (i.e., exaptations and spandrels), forever above and beyond the mundane causality of biology.

Gould's major objection to facile mental modules was that the *history* of the mindbrain was too complex and contingent to track through phylogenetic timescales with any reliable results. That elaborate contingent history had created, according Gould and Lewontin, an astonishing, general, problem-solving organ. According to Gould, we contemporary scientists are neither able to reverse-engineer the kluge nor to make good predictions about the mindbrain's decision-making powers.[50] Contingency, chance, and idiosyncratic history (phylogenetic and ontogenetic) were major stumbling blocks to an evolutionary psychology. No doubt these same warnings will have to echo down for generations to come, but we maintain that specific theories about evolutionary trajectories can be brought from the *infra dignitatem* of the "just so story" to empirical status by interweaving comparative primate study, anthropology, neuroscience, and psychology (with philosophy doing the conceptual engineering).

We reject Gould's claim that the mind is too complex for scientific deconstruction, and we reject the traditional evo-psych view (e.g., that of Tooby, Cosmides, and Pinker) that the mind is a toolbox (or Swiss Army knife) of simple modules for specific environmental problems. Yes, the mind does have some "mechanical" properties (e.g., visual processing, at least from retina to occipital cortex is relatively mechanical), but the machine and computational model fails to capture the truly remarkable feature of the mind, namely its *teleological* orientation. The mind is historically constituted, as Gould (and Hegel, very differently) would have it, but it is also largely about future states of affairs. This is complicated territory after Darwin, of course, because teleology was then so confused with theological purpose that eliminating the latter from biology also condemned the former.

Traditional evolutionary psychologists proceeded as good Darwinians by creating a research program wherein mind could be treated without purpose, design, and intentionality. And as for the origins of mind and the natural selection of mental faculties, such an approach was not ill-conceived. But while the origins and the selective pressures that build the mindbrain are more easily susceptible to non-purposive causation (e.g., chemistry), the mind itself is an intentional teleological system. Denying this fact only postpones the problem.

We need a theory of mind, then, that does not deny intentionality to mind by stipulating purely mechanical or computational modules sculpted by external forces. We also need a theory that does not idolize the mind as a mystical layer of Cartesian consciousness.

Emergence theory has been a way of steering between the Scylla and Charybdis.[51] Blind physiochemical processes may coalesce into increasingly complex systems, such that the resulting whole contains new properties that are absent in the constituent parts. The oft-quoted example is how properties of "liquidity" emerge in water from the parts oxygen and hydrogen, which, in isolation, lack liquidity. In this way, it is suggested, an intentional conscious mind may be an emergent property of a sufficiently complex system of synaptic processes. In *weak emergence,* the new properties are supervenient on the parts, have no metaphysical autonomy, no causal power on the lower-level parts, and can, in principle, be captured or simulated by a computational model. In *strong emergence,* the new properties are supervenient, may or may not have metaphysical autonomy, definitely have downward causal power on their parts, and may not be amenable to computational modeling (at least insofar as those models are analytical and reductionist).[52]

In general, we are not opposed to emergence theories, and an affective mind paradigm does not stand or fall with how this debate gets resolved. But such an approach has an historical (if not logical) tendency to designate a "moment" in the evolutionary process when consciousness "comes online" so to speak. Such designations have had a propensity to follow old cultural prejudices of speciesism, such that human minds attain unique layers of emergent mind, whereas non-humans are said to lack it.[53] Or, there are some contemporary designations that carve an awkward taxon of primates, dolphins, dogs, and some birds. We see at once the almost arbitrary taxonomy of emergent consciousness versus non-emergent mechanical brains. Moreover, we take issue with the "read out" approach to consciousness that such emergence seems to promote—as if dark zombie machinations fill the heads of beasts, but the light of consciousness emerges in humans and allows us to illuminate internal processing.

Our approach is to think about the mind as intentional activity rather than as a light of consciousness or a spectator of processing. The model of mind in philosophy, from Descartes to David Chalmers, stresses the idea of an inner space of consciousness, where abstracted, attenuated, "objects" float about—mirror copies for memory recall or manipulation. In this view, contemporary dualism remains as unpersuasive as the "ghost in the machine" absurdities that Gilbert Ryle lampooned decades ago.[54]

Our approach to mind, in contrast, emphasizes its embodied active involvement with unique ecological contexts. Conscious mind, in our view, is not an on/off light switch that gets switched on during the expansion of the prefrontal cortex or during the evolution of language. Such brain-based and cultural innovations certainly intensify and expand consciousness, but we think it's preferable to start thinking of consciousness as a family of biopsychological processes rather than as a single thing. John Searle typifies many philosophers when he says, "Consciousness is an on/off switch: a system is either conscious or not."[55] We respectfully disagree and prefer to consider degrees of mind.

Ultimately, emergence theory itself may need a refinement that steers it away from the current "on switch" culmination model to a graduated series of incipient steps—wherein each affective system (e.g., lust, fear, care, etc.) has its own levels of emergent complexity. Our approach sees intentionality as foundational and sees subjective conscious awareness as yet another way for an organism to attain "maximum grip" on its environment.[56]

So, we accept that the mindbrain does indeed have unique properties of intentionality, as compared with the entities of chemistry and physics. But aboutness is not just a human layer of mind, nor is it primarily a feature of cognitive representational activity. We suggest that intentionality is *affective* firstly—grounded in the adaptive emotions—and only derivatively *ideational*. Affordances, for example, are forms of prelinguistic, pre-representational intentionality. As such, aboutness is a homologous property across the mammalian clade, and probably all the way down the chain of biological phylogeny.

Some philosophers, notably Dretske, Asma, and Millikan, have forged the promising, albeit still-little-traversed, pathway of biological teleology.[57] Our mental states are representational or *about* external objects and properties because our perceptual and cognitive equipment was sculpted (or "designed") by natural selection to respond to those objects and properties. Natural selection sculpted a mindbrain that extracts information about its environment (both external and internal) by an archaic limning of resource potentials. But we want to add the novel claim that much of this foundational aboutness is a pre-ideational relationship between (a) endogenous mammalian affective systems (e.g., lust, rage, care, etc.); (b) associational mechanisms of conditioned learning; and (c) ecological constants (including, eventually, cultural constants). Sexually dimorphic animals, for example, have evolved perceptual sensitivities that glean information about mating-readiness from estrus-based body changes. Experiential learning strengthens the accuracy and effectiveness of those gleanings and behavioral responses. And in humans, culture lends another ingredient in shaping adaptive sexual aboutness.

One way to capture the matrix of causal threads (a), (b), and (c) above is to characterize them as the constituent factors in producing affordances.[58] According to this precognitive notion of social intelligence, which we turn to in Chapter 3, an affordance is a meaningful interaction between the properties of the environment (including the social environment) and the capacities of the animal. The distance between steps on a staircase, for example, afford the appropriate position of one's foot.[59] And, in the social realm, for example, when a low-ranking ape sees the dominant male, he simultaneously has an indicative (or descriptive) experience of the dominant male and an imperative (behavioral) experience. He perceives the dominant male and simultaneously comports himself accordingly (e.g., heightened attention, subservient gestures, grooming behaviors, and so on). The imperative aspect co-arises with any indicative folk-taxonomic cognition. If conatus or the goal orientation of a seeking system is foundational, then urgent or insistent action (toward food, or copulation, or escape) will probably even precede reflectively accessible representations. We will also explore the way that animal perception motivated by homeostasis and seeking reads imperatives out of local environments.

Consciousness and Intentionality

If nothing else in nature seems obviously teleological, at least my own human mind looks like a compelling token. I have goals, make plans, organize my life toward ends that I can actually reflect upon because I can represent them to myself. The reason why autopoiesis, conatus, and other teleological phenomena have been so easily confused with divine mind is because our best metaphors for understanding directed biological processes are human *craft-analogies* or mental intentions. Humans can think about a spear and then bring the means together to build that end goal. But focusing on this kind of behavior, as we've already pointed out, has the cart before the horse when it comes to biology. Mental intentions should be seen as derivative, rather than constitutive, of biology. We should not follow the medieval temptation to put mind as the instigating cause behind non-random, regular natural processes.

Thomas Nagel is careful not to raise the specter of natural theology in his controversial book, but he has suggested that nature's complexity needs a special explanation (possibly new teleological laws).[60] The problem is that no one has any idea what these might look like, once you've eliminated the historical option of divine mind and the contemporary option of emergent materialism. Rushing into the gap that Nagel is trying to open is a lot of teleological history, theological yearning, and materialist ire (e.g., see Jerry Fodor and Massimo Piattelli-Palmarini's strange and unconvincing critique of natural selection).[61]

However, Nagel's recent debate with critics is about both the challenges of autopoiesis teleology (the least compelling part of his book) and the difficulty in giving a purely materialistic account of subjective consciousness. The conscious mind is teleological. Either it is (a) a subset of ancient biological teleology throughout nature (and this makes consciousness a biology problem), (b) a latecomer special trait that needs unusual explanation, or (c) a primordial irreducible sidecar reality (via dualism or panpsychism). Both (a) and (b) draw on notions of emergence. And we set aside the distracting "idealism" option of natural theology (design on the installment plan), because that "research program" is a blank cartridge.

Nagel, like many philosophers, thinks that there is something special about conscious mind that makes it irreducible to materialism. The irreducibility, they claim, stems from the strange aboutness or intentionality of consciousness and the subjective privacy of felt first-person awareness.[62] Philosopher Dan Arnold, for example, nicely exemplifies this view when he says, "The 'intentionality' of the mental names the fact that mental events can mean or represent or be about other things: it has indeed been proposed as a hallmark of mental states, of states like believing or having an idea, that they thus have 'content.'" Unlike a rock or a plant, "Only mental (and significantly, linguistic) things can thus 'take' parts of their environment as their content, as what they are about."[63]

This view of intentionality has led many philosophers to reject the (a) and (b) forms of naturalism above, in favor of (c). We think this is a mistake, and it flows in part from the difficulties we have had historically in conceptualizing biological aboutness. Our short history of alternative teleology traditions should help us recognize now that biological aboutness is not dependent on mind (i.e., divine design or occult prescient forces). We have shown here that one can be anti-reductionist about biology without nesting such holism in human-like or divine minds.

The defense of biological teleology is not just negative. We do not just have a teleology "of the gaps" that emerges only when mechanical explanations break down. We have increasing positive *evidence* (from post-behaviorist ethology and affective neuroscience) as well as *argument* (from philosophers like Ruth Millikan, Marc Jeannerod, and Fred Dretske) that nature has intentionality well below the level of representational human cognition. Non-representational animal perception, for example, is already loaded with the aboutness of conative drives. Animals are not just stimulus-response machines, but neither are they cognitively sophisticated like we are (with symbolic representations of goals). Rather, dogs perceive leashes as "walk-makers" and peahens see peacock plumage as alluring. Animal perception is already loaded with aboutness—imperatives mixed with information. Latent action possibilities are already contained within the animal's perception of and interaction with the environment. The endogenous drives (e.g., FEAR, LUST) provide the motivational directionality / intentionality that makes a mere *perception* into *opportunity*. This subjective psychological territory is somewhere above secondary level simple conditioning associations but below tertiary level representational reasoning. It seems willfully ignorant or at least provincial to deny that these biological interactions have aboutness or intentionality.

Aboutness in nature does not need to be superadded. It is already everywhere, but our mechanical paradigm of nature and our Cartesian biases oblige us to ignore it. Goal-directed behavior is not just in neocortical representational consciousness, but in subcortical SEEKING systems and direct perceptual affordances. Philosophers like Thomas Nagel and Dan Arnold claim we do not have meaning without reasons, but we do. We have intentionality in high degrees, even before we have language. Not only is the body intentionally oriented to other bodies, but many of our human mental events are also prelinguistic projects. Psychology has begun to develop models of embodied cognition that recognize a "grammar" of visual thinking, or thinking with images.[64]

One reason for trying to explicate a theory of biological aboutness, especially one that is premised on the feelings (rather than on the representational thoughts) of the organism, is that it may help to reveal why AI cognitive science and artificial life (AL) computation only simulate and never actually instantiate life and intelligence. A computational model of organic behavior will not manifest the internally motivated intentionality of the animal because real intentionality is a product and result of subjective feelings, not algorithmic programming. Basic affective circuits are both sentient and intentional. Biological aboutness is a way of describing embodied feeling-states that guide animals toward resources and away from threats, via varying levels of centralized nervous system complexity. The animal is not just a syntax system with arbitrary digital inputs, guided toward some goal that only "means something" to the programmer. The creature is already immersed in a meaningful world of possibilities and constraints, and those meanings are affective experiences that are organized and managed into increasingly complex projects depending on the animal's brain-based powers of planning and executive control.

So conscious teleology is but a subspecies of biological teleology, not a sui generis. But now, what of the other worry—can natural materialism ever explain inner subjective feeling? Without working out a full solution to this perennial problem here, we suggest a general direction for further work.

Once again, philosophers have been too stuck in their reflective theater minds. The problem of subjectivity should be recast around the question of animal pain / pleasure and not around conceptual thought. We are finally far enough along after B. F. Skinner to remember that feelings of pain and pleasure are very widely distributed throughout the animal kingdom. Feelings are adaptations, but they don't motivate the animal in the right direction unless those pains and pleasures are "owned," so to speak. It's easy to see how centralized managing (executive control) of valenced reactions to environments could be selected for in creatures that move. Thinking about subjectivity and even agency biologically will bring this vexed question under the proper aegis—namely, one that explores the evolution and comparative anatomy of animal nervous systems.

Philosophers like Thomas Nagel want to drape the traditional mind / body problem over the whole cosmos. But this old saw is a cultural problem, not a metaphysical one. It was born in the early modern period when everything experienced (thought or pain) was drawn behind a veil of representation into a theater of consciousness and when everything in nature was reconceived as a machine. Its status as a cultural artifact is evidenced by the fact that the mind / body problem never even occurred to Aristotle. Instead of Descartes, we should be bringing Kant, Spinoza, and Aristotle back into our conversation about nature—all of them are far more compatible with Darwin than previously imagined.

As we have seen, there are perfectly legitimate forms of biological teleology, having nothing to do with natural theology. Ultimately Kant was right: there will never be a Newton of a blade of grass. But that's not because biology retains some supernatural vitalism. Rather, it's because (1) biology always requires specific means / end analyses (teleological accounts) rather than just antecedent conditions and (2) the contingent survival history of phylogenies renders all lawlike generalizations relatively trivial.

For their part, philosophers like Thomas Nagel might consider that there are legitimate forms of biological teleology that do not have conscious mind lurking behind them—neither are they working to render the universe susceptible to the birth of consciousness (whatever that means). The order of epistemic and temporal priority is incorrectly reversed in such philosophy. The conscious mind emerges out of earlier forms of biological conatus, not the other way around. The biological goal-driven aspect of life is not a form of vitalism but an accidental marriage of rudimentary nervous system, sensorimotor system, endogenous homeostasis systems, and ecologies of limited resources. Contrary to the belief of Nagel (and other philosophers like Alvin Plantinga), consciousness does not require its own non-materialistic science, which then needs to be pasted onto current evolution science. Against the dualist philosophers who think mind precedes biology, we submit that biological teleology actually precedes the sophisticated purposiveness of representational human consciousness. Mind is a subset of biology, but biology is more multidimensional than we previously imagined.

Conclusion

In conclusion, this chapter has endeavored to sift an ignored idea of teleology from the jumble of historical and conceptual options. With some qualifications, there are essentially three types of teleology that get confused: (1) natural theology, (2) goals in natural processes (including [a] Aristotelian entelechy and [b] autopoiesis), and (3) goals inside agents—goals that guide behavior (including [a] representational intentionality and [b] non-representational intentionality, the latter of which comprises [i] perceptual affordances, and [ii] affective / emotional intentionality). Our claim is that 3b is a domain of inquiry that will unlock many mysteries in the evolution of mind. We will offer specific keys for these mysteries in Chapters 3 and 4.

This chapter has not toured the history of biology just to eliminate dead ends and non-starters. Rather, we have incorporated many features from different teleology traditions, each of which grasped a fruitful aspect of organic causality. This conceptual history helps us to establish how affects can be intentional. They are intentional in at least four ways.

First, affects are adaptations. They are adaptations in two ways: phylogenetically (as evolved dispositions) and behaviorally (as real-time responses that may be a product of genes, learning, or cultural shaping). Affects are adaptations to regular environmental (ecological and social) challenges. As such, they are endogenous feeling / behavior complexes that make sense in reference to the problems they solve. They are *about* these problems. They limn these problems because natural selection and social learning continue to reiterate them.

Second, affects are mediating and motivating causes. They are more than automatic responses to stimuli and more than computational modules and transducers. They shape behavior, have levels of flexibility, and underlie different animal purposes: for example, LUST drives the primate toward his sexual object but is also responsive and revises pursuit in light of new dominance-competition threats, and finally attains coitus and returns to homeostasis. Instead of being epiphenomenal, affects target goals unconsciously when homeostatic imbalances encounter specific environmental conditions. The affective system of SEEKING, for example, successfully takes a variety of objectives. Intrinsically, SEEKING may be a "goad without a goal," to use Jaak Panksepp's phrase, but not for long. Once the system is activated, it fastens onto a goal quickly. Inside the agent, affects are proto-representational, not ideational representations.

Third, affects have the unique intentionality structure that places their raison d'etre outside themselves. From Aristotle, through Aquinas, to Brentano, philosophy has noticed the paradoxical way that future states (completion states) seem to be causally efficacious on earlier states. Terrence Deacon has called this the "absential phenomena"—the not-yet-present final cause organizing antecedent behavior.[65] Obviously, conscious referential beliefs can generate an absent goal in mental space and guide action accordingly, but affect, as we will discuss, can do a version of this too. There is nothing spooky about this feature of affective systems. It just means that animals with locomotion have endogenous systems that target resources in the external environment. In this sense, absential phenomena can be demystified.

Finally, some affects have a classic conscious structure, as emotions that agents are aware of. In mammals (and maybe other vertebrates), there is something that it feels like to be afraid, or enraged, or anxious. Some of the primary level unconscious subcortical affects (unowned by the agent) become conscious in the secondary level of a core self, where conditioned learning happens only because the animal has subjective feelings (e.g., fear via the basolateral and central amygdala). Here the animal has conscious awareness of emotions, or at least feelings that are meaningful and *about* the external world. Affects at this level of mind are referential. And at the highest level of mind, the tertiary, these emotions are fully intermixed with the executive functions of thought and planning. As emotion is shaped by frontal cortex functions, they take on more typical cognitive referential aboutness—more like propositional belief states. Affects become intrinsic features of judgments and deliberation.

Jerry Fodor points out that there are "notable properties of mind" that evade the naturalistic consensus; namely, "mental states and processes are often teleological (they have goals and functions); they are often (perhaps always) conscious . . ."; and finally, they have about-ness.[66] Our own view is that these notable properties do not evade a properly sophisticated naturalism. We agree that the human mind does indeed have these amazing properties, but so do the minds of many nonlinguistic creatures.

3

Social Intelligence
from the Ground Up

HUMAN UNIQUENESS, and our subsequent ecological success, are commonly attributed to cognitive processes.[1] An analogue of this anthropocentric approach is dramatized in Stanley Kubrick's opus *2001: A Space Odyssey* with the *monolith*. The sudden and mysterious appearance of representational mind—as demonstrated through tool use in the film—which does not provide any indication as to how such *de novo* mental abilities are continuous with the native skills of early hominids. A popular story in the psychological sciences is that a set of information-processing modules—for example, the Theory of Mind Mechanism or Bayesian inference models—is responsible for our unique human mental abilities.[2] But what abilities did primates and early hominids possess that may have donated to our unique success? In this chapter we discuss primate capabilities to develop and manage social bonds with conspecifics, and we determine how this alternative story fits into our understanding of the foundation for human uniqueness and ecological success.

We argue that, while cognition is an important factor in the success of our species, we need a deeper understanding of the role of older evolutionary processes. This chapter aligns itself with enactive and embodied cognition approaches by broadening the role and range of perception, while simultaneously diminishing the role of propositional and computational processes.[3] Our goal is to suggest a model of social intelligence that relies on perception and affect.[4] This is not to take away from cognitive modeling, but rather to suggest that analogous evolutionary trajectories in the realm of adaptive social interaction may have been built with a set of mechanisms that rely less on cognition and more on affective and perceptual processes.[5] We claim that perceptual and affective processes enable interaction and communication between primates, subsequently providing aptitudes for cultural learning; and further, that this cultural learning has been essential for the accumulation of knowledge, playing a preeminent role in our ecological success.

Social intelligence is a set of abilities that create, maintain, nurture—and even enable us to revel in our interaction with other creatures. Some examples of social intelligence are shared attention, pretend play, ability to socially organize and be a part of a group, negotiation and mediation, short-term and long-term personal connection, empathy, social analysis, and insight into the feelings and goals of other creatures.

Emotional intelligence is part of social intelligence as it relies upon the secondary emotions: LUST, CARE, PANIC, and PLAY.[6] Some examples of subtle social intelligence in the human realm include knowing your friend is sad, keeping distance from odd-acting—or smelling—strangers on the bus, not making too much or too little eye contact, and nodding when someone is talking and needs reassurance to continue. These social interactions are embodied—the perceptual system being the mode in which they occur—and require motivation from the affective systems. Whether they are conscious or not, social behaviors constitute a type of intelligence insofar as they demonstrate integration of knowledge about the past, various dimensions of motivation in the present situation, and an appropriate understanding of the consequences of action for the future.

As Andy Clark points out, "mind and intelligence themselves [are] mechanically realized by complex, shifting mixtures of energetic and dynamic coupling, internal and external forms of representation and computation, epistemically potent forms of bodily action, and the canny exploitation of a variety of extrabodily props, aids, and scaf-folding."[7] Accordingly, we take the mind to be a developmental process between brain, body, social, physical, and cultural environment. This approach enables the formulation of a bottom-up cognitive model—one that relies upon evolutionarily earlier manifestations of mind and social intelligence—as distinct from a top-down model of the mind.

While cognitive mental processes have been studied and modeled for the last fifty years, behaviors emerging from perceptual and affective processes have not garnered as much attention as top-down cognitive abilities and thus occupy a relatively lower position in our explanatory framework of the mental and behavioral palette of animals. In our argument, we place less emphasis on the role of cognitive modules than on ecological psychological processes like affordances and infor-mation scaffolding.

These arguments are in line with recent work that delimits the role of cognition as the sole or main generator of intelligent behavior. For example, Carruthers and Ritchie claim "uncertainty-monitoring"—a paradigmatic example of propositional metacognition—can be ex-plained sans cognition, through processes that are simply perceptual and affective.[8] In this case, a comparator system controls action through reafferent sensory feedback comparing intended outcomes with on-line sensory information. Joelle Proust develops on this research by delineating the causes of animal behaviors into two types of systems: particular-based representational systems (PBS), which refers to ob-jects and to truth values (i.e., cognitive, propositional processes) and feature-based representational systems (FBS), which rely on protocon-cepts. Feature-based representations are not strictly determined, have no exact boundaries, and use similarity-based conditions of application in analog representation and embodiment.[9] FBS, while still ostensibly representational is not propositional and relies upon a broader, associa-tional framework for understanding animal behavior.

Further, in a formulation similar to our own below, Proust provides a meticulous description of animal behavior where a feature-based representational system identifies affordances, categorizes them for in-tensity, and triggers associated motor programs.[10] In a related vein, taking into account thirty years of empirical work on chimpanzees, Call and Tomasello eschew full-fledged metarepresentational belief-desire psychology in ape behavior, focusing on perception-goal psychology.[11] Similarly, Metcalfe casts doubt upon the evidence for metacognition in monkeys and rats, instead seeking more parsimonious perceptual explanations.[12] On the neuroscience front, rather than focusing on frontal executive processes, Robert Barton's exciting work emphasizes the role of the perceptual genius of the cerebellum in the evolution of the capacity for planning, executing, and understanding complex behavioral sequences.[13] In the field of ecological psychology, we tip our hats to the work of Rob Withagen, Harry Heft, Erik Rietveld, and Julian Kiverstein, among others developing the range of direct perception.

Whereas some goal states can be directed or realized in multiple ways, the goal states we focus on are goal-oriented, motivated by largely innate processes and refined through conditioning. Taken together, this recent work demonstrates the limits of cognitive explanatory frameworks and provides the opportunity for a novel epistemological analysis of the relationship between mind, body, and behavior.

The Toolbox

In using a bottom-up approach, let us assemble a basic toolbox of the mammalian mind with three very elementary mammalian systems: survival drives, sensorimotor systems, and emotions. The specific *survival* subsystem we use is homeostasis: the set of regulatory processes in the brain that maintain equilibrium in an organism through maintenance of the body's internal environment; within this system, there are levers for hunger, temperature, thirst, and sleep.[14] *Perception* is the set of information collected by an organism's *sensory* equipment and subsequently wedded to action in *motor* systems.[15] *Emotions* may be distinctly characterized by function, where they are located in the brain, and when they evolved in evolutionary time. The main distinctions in such a characterization are between primary emotions (sometimes referred to as affective mechanisms: SEEKING, RAGE, and FEAR), secondary social emotions (LUST, CARE, PANIC, and PLAY), and tertiary cognitive emotions (such as angst and aesthetic titillation).[16]

Emotion and homeostasis together create motivational states that engender perceptual and motoric events, for example, SEEKING is a generalized goad toward ferreting out food or sexual partners, while RAGE motivates violent and ultimately self-preserving behaviors. Emotions impel action while implying the past; that is, they are associatively situated by mnemonic associational traces and innate targets that pitch appropriate action sequences.

In this characterization of the basic mammalian toolbox, the body plays the role of a bridge between (1) our homeostatic, affective, sensorimotor processes; and (2) transacting information (through communication and perception) to conspecifics and the environment.

Physical embodiment organizes an organism's abilities by spreading the load of learning, emoting, problem-solving, and eventually cognition across the body, brain, and environment.[17] Essentially, embodiment is a type of holistic way-finding through environmental feature-domains.[18] According to the ecological psychologists, animals make their way in a world where the value of any given object consists of what it affords.[19]

The three processes that characterize the mammalian mind do not require representational mind to function. The traditional definition of "representation" is an integrated set of symbols that may be internally manipulated.[20] More recently, though, cognitive scientists have opened up the definition of representations. The Continental phenomenological tradition from Merleau-Ponty to contemporary French and German philosophers places an emphasis on theorizing the importance of the bodily self. This has led to a contemporary resurgence of interest in ecological psychology, a focus on how representations are action-oriented, and the subsequent elaboration of this idea into enactive psychology.[21]

A representation is as a mental construct to characterize the constant causal connection between the object in the environment and the tracking relation that stands in for it in the mind.[22] In this traditional definition, representations are taken to be internal states that mediate between a system's inputs and outputs in virtue of that state's semantic content.[23] In this case, representational systems connect behavior with environmental features not present in the system. The representation is thus a "stand-in" for the missing feature that guides behavior; it thereby occurs regularly and can be manipulated systematically.[24]

Our own approach to the notion of representation is indebted to Ruth Millikan, who introduced the crucial refinement of the pushmi-pullyu representations or PPRs: "Think of perceptual representations (qua PPRs) simply as states of the organism that vary directly according to certain variations in the distal environment. The perceived layout of one's distal environment is first a descriptive representation. It is at the same time a directive representation of possible ways of moving within that environment: ways of passing through, ways of climbing up, paths to walk on, graspable things, angles from which to grasp them, and so forth."[25]

PPRs are a form of directive and descriptive perceptual signs that fulfill the intermediate role of integrating knowledge from the past into future behaviors. The integration between descriptive and directive aspects is manifested in the mind as valence-weighting within associative networks of option-outcomes (i.e., as somatic markers).[26] A somatic marker structures a creature's response landscape in generally implicit action biases that integrate distal or absent events into current action sequences. Though helpful and pertinent in some circumstances, notably in decision-making, knowing why a sensory percept leads to a particular behavior is not necessary to a creature. Thus, an affordance-based PPR does not necessitate the ability to represent the world via a system of symbols; rather the function of PPRs is to "mediate the production of a certain kind of behavior such that it varies as a direct function of a certain variation in the environment, thus directly translating the shape of the environment into the shape of a certain kind of conforming action."[27] For example, when throwing something at a moving target, we change our posture, positioning, and the velocity with which we project the object. Or, social behaviors with a particular conspecific develop with changes in the dominance hierarchies, as illustrated by Shirley Strum's work on how social order determines communication and grooming behaviors in baboon groups.[28] Or imagine we are visually tracking a stranger: how do our movements and facial expressions differ if the stranger wields a banana rather than a knife?

Drawing upon Millikan's PPRs, we favor a model of nonconceptual content at the subpersonal level with external content.[29] It is at this level that we believe one can make ascriptions of "thoughts" to nonlinguistic and prelinguistic creatures. Functioning largely at subpersonal levels, there are conative connections between homeostatic processes, the SEEKING system, and contextualized perception-action affordance-generated sequences. Distinguishing between *imperative* and *indicative* percepts is another way of recognizing how affordances are a sort-of proto-belief state that is both perceptual and instrumental.[30] Note the relation between this concept and the *feature-based representational system* that functions on proto-concepts discussed above.

Affordances are also analogous to the property of involving objects under modes of presentation, namely the possibilities for action and reaction that the perceived object affords.[31] For example, the same percept, say of a conspecific in estrus, can present different affordances, depending on the occasion of the presentation (i.e., a lower-ranking male attempting to mate a nearby female will respond differently to receptivity-signaling if the alpha male is nearby versus when the couple is alone). The presence of the alpha male colors the perception (i.e., the mode of presentation) of the female in estrus with the affect of anxiety, itself created by past experiences of fear engendered by previous interactions with the alpha male.

A strategy to develop the middle ground between direct perception and representations is being pursued by several researchers.[32] It wisely breaks up representations into linguistic and action-oriented representations. While the former requires full-blown representations of information and the apparatus of modular information processing, the latter are minimally representational.[33] Action-oriented representations (AOR) are temporary egocentric motor maps of the environment determined by the situation-specific action required of the agent.[34] AORs are action-specific, egocentric relative to the agent, and context-dependent; they serve as a functional mediator between perception and action.[35] Taken together, even rudimentary manipulation of AORs (perhaps as a "task grammar") would be an adaptive advance over rigidly automatic responses, especially if the local environment was volatile.[36]

The field of Sensorimotor Enactivism elaborates upon the direct-perception tradition of ecological psychology, arguing that perception is an active, adaptive control point for behavior. Perception is something we do, not something that happens to us. Perception is thus practical knowledge realized in the active life of a skilled animal.[37] In Milner and Goodale's well-known work, this move toward defining perception as an active rather than passive process, as well as the theoretical achievements of ecological psychology, redeem the first explanatory function of perception: to act on the world through distal control of motor output.[38] We are excited by contemporary researchers pairing an active-perception approach with dynamic systems theory to model perception-action systems more accurately.[39]

In accordance with this interpretation of direct and active perception, we argue that the mammalian toolbox may not necessitate representational processes for a wide set of social behaviors that will be elucidated below. Rather, a complex mix of homeostasis, direct perception, and affect may serve as a basis to argue that social intelligence precedes many forms of cognitive control, even though, to a large extent, they subsequently become integrated with higher cognition in some animal minds.[40] The current vogue in cognitive science for thinking about the mind as a "prediction processor" is often focused upon higher levels of representational modeling, but the evolutionarily older affective / affordance toolbox needs to be incorporated if we are to understand real minds.

Affordances as Communication

Animals communicate through chemical signals, sight, touch, and sound.[41] From pheromone emission to social grooming, communicative processes rely upon the basic mammalian toolbox. Communication systems are the crucial element in social interaction; they provide organisms the opportunity to send and receive cues to future behavior and to manipulate conspecific affective states as well as one's own affective states. Consider how social animals have the need to communicate their homeostatic states (viz. in their search for food or drink, or—for social animals—in their sexual and violent encounters with conspecifics). These internal needs are externalized and communicated via perceptual and motor equipment such as body movements, gestures, sounds, facial expressions, eye contact, etc. Furthermore, animals have affective systems to mediate the reception and subsequent production of communicative events. In this way, homeostasis, bodily display, and affect form a unit that undergirds social intelligence.

Across the animal kingdom, we observe nonlinguistic creatures engaging in complex communication for social purpose; to take a couple of obvious examples, the dominance displays of rage that precede possible violent confrontations or the bodily displays of strength and vanity that take place for the purposes of wooing a mate. If we define intelligence as a complex of abilities that requires the integration of adaptive or task-relevant knowledge about the past, along with various dimensions of the present situation and an appropriate understanding of the consequences that a response now will have on the future, these social behaviors do in fact constitute a type of intelligence. We argue that such abilities, and many others, are foundational for complex social relations in animals, including humans, because motivational and emotional state information is actively being shared through the body.[42] Based primarily on perception and elicited and motivated by affective and homeostatic processes, the ability to display and respond to bodily communication as a form of descriptive information (i.e., that this is the state of the animal) and directive information (i.e., that these are the options for a response) is the dialectic of social intelligence.

If social order is an affordance, it has to be understood as an agglomeration of perceptually-mediated dominance and submission cues.[43] But, for a given perceiver, there is no one piece of information that defines a conspecific's position in the social order. It is thus the accumulation of perceptions of behaviors in shifting social contexts, as well as the reception of neuroendocrinological indices, like cortisol and testosterone, that "add up" via mnemonic processes.[44] Here we suggest possibilities for how a set of percepts of a particular conspecific collected across time and place can be mentally associated (i.e., add up) in such a way as to modulate future behavior toward the individual. The reigning methodology in cognitive psychology has been to postulate an internal representational system that individuates and manipulates mnemonic engrams toward creating higher-order data that may subsequently act as top-down filters for further perception of a given conspecific. But our bottom-up approach explains how descriptive and directive percepts are stored and manipulated without the mediation of cognitive processes. We believe that pushmi-pullyu representations (PPRs), affordances, and somatic-marker encodings are the proximate mechanisms for the integrated and intelligent communicative behaviors that make up social intelligence.[45]

Social Intelligence and Perceptual Affordances

Ecological psychology largely focuses on how direct perception enables motor abilities, but we are aiming to characterize the way that direct percepts of social actors may be integrated either without representations or with action-oriented representations and feature-based representational systems. Affect is the key locus in our proposal, as the valence format of somatic markers biases option-outcome scenarios, thereby determining behavior.[46]

Here is a play-by-play: a sensory percept PPR affords a directive set of possible actions. The action that is subsequently released depends upon past experiences in associated circumstances, as encoded in somatic markers. These somatic markers are subsequently manifested as either implicit biases or explicit affective tugs on the field of possible responses. The animal perceives its options vis-à-vis the direct percept and its internal affective reaction. To be clear, this notion of affordance space is not the same as the propositional representations of possible actions that humans seem to use in explicit decision-making. In a model that does not rely upon representations, the appropriateness of a behavior *looks like* an inferential process. In fact, the "intelligence" of the behavior can be a matter of simple operant conditioning of behavior sequences.[47] The sensorimotor relation between PPR and directive would be relayed through the body via affective weights in a somatic marker to determine implicit option-outcome decisions.

Note how notions of agency are transformed in this model where the cause and the effect are simply relations between body, innate drives, and the environment. As Marsh et al. put it, "[T]his approach is dynamic and interpersonal rather than static or focused on individual-level attributes emitted from one individual and detected by another. What is not required is knowledge of an individual's inferences. Rather, possible mechanisms are grounded on lawful principles that generate a flow of information between animal-and-environment and animal-and-animal and that lawfully constrain social behavior."[48]

Let's say, that, upon meeting a friend, I sense that she is sad. In this example, my body uses emotional complexes associated with particular perceptual signals from the past (like the sight of emotional strain in holding back tears, or even more developmentally constructed social conventions of demonstrating sadness) to detect the emotion. Detecting the emotion then opens a set of possible actions in the present that must be decided among, based on respective potential results and on past experiences. This affective-sensorimotor complex is then paired with the particular eliciting stimuli via conditioning processes.[49]

According to J. J. Gibson, perception is both direct and meaningful; it is direct in that it does not require inferences based on intermediaries such as sense data, retinal images, or representations, and it is meaningful in that we perceive a world that is relevant to the activities in which we are engaging.[50] That is, percepts can provide the organism with a set of possible actions that are adequate in relation to our past, present, and future. Social processes can be loaded with intelligence by PPRs in their descriptive (this person is in a particular affective state) and directive (such an affective state affords the following behaviors) aspects.

The perception of conspecifics in acts of communication are directive; they enable possible movements and behaviors based upon changes in the environment to which the behavior is a response; they offer a perceptual description of the environment in terms of the possibilities of action that may ensue—as Millikan states, "[The] contents of the directive and descriptive aspects of the representation are not different but coincide."[51] In the earlier examples of throwing an object at a moving target and perceiving a stranger with a knife versus a banana, the PPR includes coinciding descriptive information concerning the moving target (i.e., He is now moving to my left) as well as directive changes in the way the body responds to produce a successful conforming action (i.e., I have room to maneuver my way out of arm's distance from the stranger). Note the relations between this characterization qua AORs, or *feature-based representational systems*.

PPRs are one of the processes by which complex information can be stored, manipulated, and communicated; they are cognitively simple (i.e., are minimally representational) but perceptually hard (i.e., require direct perception and integration of multiple perceptual sources).[52] Using PPRs and a bottom-up approach that includes affective biasing cues in the form of somatic markers, we can provide causal stories for complex behaviors using only the basic mammalian toolbox. In language we introduced in Chapter 1, PPRs and somatic markers are a form of intentions-in-action that rely upon secondary-process affective mechanisms.

Proxemics

Proxemics is a field that investigates social communication that is bodily. For example, the differences in appropriate social distance across cultures seem unconscious; gauging the appropriate physical distance does not require a step in between the perception and the triggering of a socially appropriate action.[53] The body provides a display upon which social interaction depends; for example, manifesting a negative affect with a scowl is a way of externalizing the subjective, bodily state of "I do not want to be approached," thus suggesting the "stay away" directive. A directly perceived scowl—itself a component of the affective and possibly homeostatic state of the creature—directs, or affords, the behavior of keeping one's distance from the scowler. Or, consider, in many species, the scent of the estrus signals of "approach" or "avoid." In this case, the directly perceived scent of a female in heat affords the behaviors of pursuit, courting, and mounting.[54] A dog's territorial urination demarcates space for other dogs; the scent indicates that this territory does not afford settlement, that the marking dog will return to the territory, and that approaching food sources and sexual mates in this territory is a fraught enterprise. In mammals, an infant's high-pitched squeak signals "pay attention" both descriptively and directively; it describes the state of the infant and directs, or affords, behavior toward the screaming infant. Lastly, grooming, in the form of picking lice from a conspecific's hair, for example, signals social care and deference, and thus affords reciprocal behaviors.

Many complex, socially intelligent behaviors rely primarily on body language.[55] In the field of proxemics, there are some well-studied examples of PPR, affect, and affordance complexes stored in nonverbal communication. In one such example, head angle correlates with perceived dominance: a head tilted upward correlates with social dominance while a head tilted downward demonstrates and is correlated with submissiveness.[56] There is also evidence that humans are considered more dominant when displaying lowered brows and that nonsmiling faces indicate and correlate with social dominance.[57] As affordances, this empirical work suggests that an upward head angle, a lowered brow, and an unsmiling face are directly perceived as information about how dominant that individual is in the observed dyad or group. What is afforded by this direct perception is the set of admissible behaviors toward the dominant individual; for example, "I want to be close to this individual, but how close can I get considering my own position in the hierarchy?" or "Who should I follow when this group breaks up?" The affordance of head angle seems to directly communicate information pertinent to these behaviors.

In humans, gaze acts as a display of dominance: high-status individuals have nearly equivalent ratios of looking while listening and looking while speaking, while low-status individuals look more while listening than while speaking.[58] This cue works in real time; gaze modulates the nature of the interaction, leading one to hold forth and one to observe. This affordance does vary, however, according to the meaning of eye gaze in a given species; for example, silverback gorillas take any eye contact to be a sign of challenge, while with kin in other species, eye contact can be a form of intimacy, though it nevertheless carries information about social position and acts as a social affordance.

In baboons, a longitudinal study demonstrated the effects of social grooming on the establishment of position within group hierarchy, especially when individuals joined new troops and used body language and body contact to insinuate themselves into coalitions.[59] Further studies suggest baboon affect and behavior is informed by maps of rank in a two-tiered hierarchical system that includes both inter- and intra-familial information.[60] Pygmy marmosets also change their vocalizations depending on their position in a social hierarchy.[61]

Of course, cognitive evaluative processes can mediate one's reactions to nonverbal dominance cues.[62] Nonetheless, studying nonverbal behavior allows us to understand embodied intelligence as social adaptations for the purposes of communicating behavioral cues to conspecifics. The communicative external scaffolding of the body extends social intelligence and behavioral complexity through PPRs, affect, and perceptual affordances.

Social Affordance Model

To collect these ideas and furnish a description of the proximal causes of some social mammalian behaviors, we put forward the following model for how social intelligence may function: First, *a message is displayed through the body. Then that message is perceived as a set of possible responses (i.e., as a directive PPR). Each of these possible responses is associated with mnemonically coded affective values (qua somatic markers), which serve as a weighted catalogue of the perceiving creature's history that tips the balance within the set of action tendencies toward a particular action response.*

In this model, affect is a source of bias and functions as a somatic marker in the service of homeostatic dynamics, thus serving as a basic ordering principle of motivating behavior and action selection. The homeostatic system is intimately connected with the affective system in a number of ways; in this model, the salient relation is between homeostasis and the energizing goad of SEEKING, a heightened affective state that directs our perceptual systems to find the ingredient in the environment that may restore our bodies to equilibrium. Affects are thus wedded to motivational processes, providing intrinsic direction for what the perceptual and homeostatic system seeks. Through perceptual attention and behavior (using AORs or FBS), affect makes certain objects, creatures, and colors in the world salient; for example, a potential food source when one is hungry, or a sexual partner when one is in estrus.

Adding perceptual affordances and PPRs to affect that is directed by homeostatic tendencies produces actions that are intentionally motivated to the given local situation. Taking a step further, we can highlight perceivable elements of the environment specifically relevant to social intelligence and call these social affordances.[63] Similar to how a chair as a perceptual affordance is perceived as a thing to sit or stand on, another creature's behavior provides a set of social affordances: a smile affords approach, eye contact affords engagement, a puffed up chest in a possibly violent confrontation affords another to puff up one's chest as well. These interactions are studied abundantly in the field of nonverbal communication and social psychology.[64]

An affordance-enriched conception of perception applies not only to action but to affect as well, since affective information gleaned through perception feeds into possibilities for actions in the social realm. In a similar vein, the capacity to understand social norms is another case where the descriptive and directive functions of a percept coincide.[65] A clear description of the interaction between perceptual plans and affective processes, albeit loaded with the language of representation, is put forward by Carruthers and Ritchie:

> [U]ncertainty-based decision-making may be best understood as of-a-piece with affectively-based decision-making generally. . . . On this kind of account one runs the instructions for a motor action offline, using the efference copy to generate a forward model of its outcome. . . . When attended to, this is globally broadcast as an imagistic representation of the action, which one's evaluative and emotional systems receive and respond to. The result is some degree of positive or negative affect, which provides the motivation to execute the action or to seek an alternative means to the goal (or to pursue an alternative goal). On this kind of account feelings of uncertainty would consist of negatively valenced affect that is caused by the thought of an otherwise-attractive action, and that is directed toward the situation represented in the content of that thought.[66]

While Carruthers and Ritchie rely upon forward models generally thought to be representational, our model requires little to no representationalism. We envision an emulator system that "inputs" the starting or current state of a system and control commands, and "outputs" a prediction of the next state of the system as a set of values for the future feedback that the new state should yield, modeling the target system in real time without being explicitly representational.[67]

Let us consider how much information we can attribute to social intelligence processes by analyzing complex bodily signals. For example, the puffed up chest means something different depending on where the perceiving creature falls in the social hierarchy; it can be taken more or less seriously, as a bluff from a weakling, or as the beginning of another rampage from a belligerent individual.[68] Subordinate animals, for example, demonstrate deferential nonlinguistic signals in the presence of their dominants. Each creature's signals are couched within its social position in a group, and a given creature's social position is determined by its interpersonal history, current social connections and aspirations, and kinship lineage, which may all be subtly coded within affective complexes that weight social-affordance action space. Accordingly, recent work on culture in nonhuman animals demonstrates that adult male vervet monkeys migrate to new groups and conform quickly to the social norms of the new dominance hierarchies. In our view, the vervet monkeys are reading and learning social affordances.[69] In Chapters 4 and 5 we focus on the developmental aspects of social affordances. Mnemonic processes are involved insofar as they track individuals, groups, events, places, etc. in the form of affective values that weight the catalogue of conforming action option-outcomes as somatic markers. Memory in this sense need not be categorized according to a semantic / episodic framework, as it may more appropriately be considered as coded in non-declarative conditioned complexes. The manner in which such social affects are encoded remains controversial, with some claiming that encoding requires cognitive infrastructure and some claiming that conditioning processes are sufficient. Regardless of the cognitive science or behaviorism interpretation, we claim that the role of the emotions is crucial in helping us characterize goals and drives in a more parsimonious model.

Conclusion

Natural selection builds both ecological and social affordances into vertebrate perception as a way of solving the common spatial and social challenges of a given creature's environmental niche.[70] In our case, humans are social animals, and so, interpersonal communication, just as gravity and orientation, is a natural characteristic that we have developed sensory equipment and perceptual affordances to navigate. In this way, the social affordance model above provides an evolutionary story for what enabled the mammalian mind prior to the cognitive *monolith*. Although the sensorimotor capacities for social affordances as nonverbal cues seem to be inborn, the particular skills necessary to navigate the social environment within a given cultural context need to be learned. Just as we seem to have a predisposition to learn language, we have a set of skills that needs to be calibrated according to social mores through ontogenetic enculturation (i.e., cultural learning).[71] As well as providing an evolutionary story for the abilities of a wide range of social creatures, this model leaves room for built-in and learned complexity as well as connections between such abilities and the subsequent development of representational cognitive processes.

In conclusion, we have put forward a model of social intelligence from the ground up that does not rely on cognition or full-blown representations to explain many types of complex social intelligences, such as grooming, dominance hierarchies, and some approach / avoid behaviors. The concatenation of survival processes, motor and perceptual affordance equipment, minimal representations (i.e., AORs and FBS), and affective states provides a substantial complex of abilities to social mammals. It is these social and emotional intelligences that may enable successful social interactions and that, through cultural learning, bestow upon the human species the wherewithal to collectively manipulate their environment. The next Chapters 4 and 5 elaborate upon the relationships between this model of social intelligence and cultural learning from both phylogenetic and ontogenetic angles.

4

Emotional Flexibility and the Evolution of Bioculture

DO NONHUMAN PRIMATES have culture? It may be tempting to believe that cultural adaptation is the exclusive province of humans and to see the evolution of other primates almost exclusively in terms of biological adaptation. In part, this temptation betrays the erroneous "mind first" assumption that culture could not arise until after sophisticated cognitive prerequisites were in place (e.g., deep language grammar, the Theory of Mind Mechanism, propositional representations, etc.). We make no such assumption about culture and cognition, and we suspect that early human culture is less cognitive than we think but more sophisticated with regard to associative conditioning and affective processes.

Culture, in the form of behaviors and practices that get passed down in a pseudo-Lamarckian transmission, can piggyback on genetic evolution—never penetrating down into the thicket of DNA heredity. Cultural evolution is often considered Lamarckian in the sense that useful information, acquired during an organism's lifespan, can be passed down (via language, etc.) to the next generation—without having to wait for a heritable genetic mutation. Of course, calling culture Lamarckian is only metaphorical here. The metaphor helps us understand that adaptive cultural innovations are not bound by the central-dogma causal arrow of strict neo-Darwinism.[1] Given the recent advances in the field of epigenetics, the idea that some acquired traits (not skills, but metabolic physiological effects) could penetrate into the biological mechanism of heredity (epigenome) is being taken seriously again.[2]

Essential adaptations for survival can ride on top of biology for generations without any need for molecular reliability or innate status. Humans acquired fire, for example—which gave them incredible advantages—but then lost it again for generations and finally regained it. Control of fire appears as early as 1.5 MYA in Swartkrans South Africa, then crops up again in Israel and China around 700 KYA, but doesn't appear in European populations until 400 KYA (controllable by all hominins after 130 KYA).[3] Our ancestors lived a long time without fire, and however unpleasant that sounds, presumably we could do it again. Natural selection and genetic mutation do not need to build fire-starting knowledge into human brains like an innate prewired module.[4]

Cultural learning alone is able to canalize deep resources, even modifying brain structures in reliable ways.[5] If we think of evolution as a mosaic of developmental systems, then we see that populations (e.g., early humans, but also nonhuman primates) have recurring stable resources, some of which are genetic, some phenotypic, and some environmental. Adaptive behaviors can be drawn from the cultural well too. We do not need to commit to the increasingly outmoded dichotomy of genes versus environment.[6] Even much simpler classes of organisms (e.g., beetles) have been shown to have substantive nongenetic inheritance (parental effects) when it comes to their relatively "plastic" or malleable traits.[7] An affectively grounded associational system (i.e., emotional learning) in pre-sapiens is precisely the sort of plastic system that can be shaped by multiple levels of selection into stable biocultures. These biocultures, in turn, help determine which genetic traits spread throughout a population.

In this chapter, we sketch a bottom-up emotions-based research program for understanding hominid cultural evolution. In particular, we show how mammalian affective systems (SEEKING, LUST, and CARE) are channeled by ecological demands into sophisticated social traditions. Dedicated emotions can be decoupled from their original target functions and broadened into more plastic, open-ended suites of general responses. We examine the transition of homologous affective foundations into diverse primate cultures, looking at chimpanzees, bonobos, and humans, in particular, with an eye toward constructing an affective evolutionary model.

In what follows, we will be using "culture" in this broader sense of the learned forces that shape animal communities—forces that are transmissible (via emotional contagion, conditioning, imitation or simulation) and develop before the evolution of language. These traditions or folkways are "inherited" in the sense that an individual is born into them, but they're not innate or genetic.[8] In this broader definition, nonhuman primates do indeed have culture.

Chimpanzees and bonobos learn their unique sexual cultures and dominance hierarchies from their peers and elders. Some of their behaviors are species-specific and therefore seem like phylogenetic effects, but some of these primate populations are contiguous enough in their geographic distribution to suggest that some of the widespread behaviors are sustained by culture rather than genetics. Some chimpanzee communities transmit special nut-cracking or termite-fishing techniques to each other, but nearby communities may lack this same technique. These cultural innovations are definitely not species-specific, but tribal specific. Microcultural traditions may persist independent of other conspecific neighbors; so cultural similarity and diversity must be assessed at ontogenetic and phylogenetic levels of primate life.[9]

As an illustrative example, the Sonso chimpanzee community of Budongo Forest, Uganda, was recently observed engaging in fairly sophisticated social learning. Chimpanzee cultural transmission was observed in the wild when members of the group modified their water-collection tools.[10] The members often set water traps to collect sips of drinking water using leaves plugged into tree holes. This method of leaf-sponging is universally common among chimpanzees. But in the Sonso community, researchers observed a dominant male invent a "moss sponge" technique, which other members then slowly adopted over the course of a few days.

Most theories of human evolution focus on the Lamarckian (and / or memetic) qualities of culture—in other words, its ability to pool, retain, replicate, and transmit information.[11] We want to focus, however, on a neglected aspect of cultural evolution: emotional plasticity. Precognitive social learning must be driven in part by emotional (affective) dynamics, since these processes exist across the mammalian clade.[12] Emotional changes and social innovations correlate.[13] *Homo* pair-bonding structures between males and females, for example, change during the Pleistocene, as do family or kinship structures, dominance structures, and even the length of childhood. These changes don't happen because rational agents make utilitarian calculations based on the cost-benefit outcomes. Instead, these changes happen at the meeting place between local environmental resources (e.g., food, shelter, defense, etc.) and mammalian emotional systems (e.g., fear, care, lust, aggression, etc.). This complex story has not yet been told, but in this book, we lay the groundwork for a future research program.[14]

Some recent definitions of culture have sought a diversity of criteria. In order to avoid a species-centric approach for example, cognitive psychologists have suggested several views of culture.[15] Culture exists when repeatable *patterns* emerge, like dialect song types in some passerine birds. Or it exists when certain *technology products* are repeated horizontally (across the same generational conspecifics) or vertically (across multigenerations), as when Western chimpanzees crack nuts, leaving chipped anvils. Or it exists when unique behaviors increase survival skills through imitation, as when rats in Jerusalem pine forests learn to get seeds from cones. And, in a way that foreshadows group identity symbolism, it exists when behaviors / things are *coded with meanings* that did not derive directly from physiology—as when certain chimpanzee groups use different styles of hand-clasping.

This pluralistic approach to demarcating culture is promising, but it still fails to consider what role, if any, is played by affect or emotion. For example, Richard W. Byrne et al. interpret all of the above as the result of "information transmission," where such information is characterized as either cognitive or blindly behaviorist.[16] "Information" is a misleading characterization because it suggests a neutral body of data that is provisionally embedded in creatures, but can be abstracted out and replicated or transmitted like digital content. But affective orientation is not neutral. It is biased, situational, difficult to abstract, and often imperative rather than indicative. Adding the emotional element (i.e., bio-intentionality, contagion, affective memory, flexibility, etc.) can strengthen our understanding of cultural transmission among conspecifics.

Our claim is not that emotional flexibility alone made us *Homo sapiens,* or that it trumps technological innovation. Moreover, it is obvious that neocortical brain expansion and language were paramount. But we are suggesting that emotional modernity is a more crucial piece of the puzzle and needs a greater place in the explanatory matrix. The meaning of "emotional modernity" will be made clearer in the course of this chapter, but a provisional definition is *the suite of emotional tendencies shared by members of the genus* Homo, *rooted in the basic affective systems of mammals but facilitating high levels of domestic activity, including pair-bonding, alloparenting, and apprentice cultural transmission.*

To better articulate the adaptive emotional-cultural nexus, we introduce three distinct emotional systems that all mammals share; namely SEEKING, LUST, and CARE. Then we show how chimpanzee, bonobo, and human adaptive landscapes channel these raw materials (the three emotional systems) into very different sociocultural folkways. Lastly, we consider the brain-based changes (e.g., associative mechanisms) that may underpin the emergence of emotional flexibility (see Table 4.1).

SEEKING

In the same way all vertebrates possess a fear system, they also engage in SEEKING behavior—and recently, neuroscience has isolated a foundational motivational drive that underlies diverse searching behaviors (e.g., hunting, foraging, procreation, etc.). As we saw in Chapter 2, philosophers like Spinoza recognized this drive in all creatures and called it conatus (striving), but affective neuroscience calls it the SEEKING system, or the wanting system.[17]

This system enlists a dopamine pathway in the brain, and while it is strongly correlated with pleasure rewards, it actually spikes

Table 4.1 Neural Correlates for the Seven Prototype Emotions

Affective Prototype	Distributed Neural Networks and Major Structures	Neuromodulators
Generalized Motivational Arousal—SEEKING	Ventral Tegmental Area (VTA) to lateral hypothalamic to periaqueductal gray (PAG), with diffuse mesolimbic and mesocortical "extensions." Nucleus accumbens as crucial basal ganglia processor for emotional "habit" systems and affective learning.	DA (+), glutamate (+), many neuropeptides including opioids, neurotensin, CCK, and many other facilitators
RAGE (Affective Attack)	Medial amygdala to bed nucleus of stria terminalis (BNST) to anterior and ventromedial and perifornical hypothalamic to more dorsal PAG.	Substance P (+) (Ach, glutamate (+) as nonspecific modulators)
FEAR	Central and lateral amygdala to medial hypothalamic to dorsal PAG to nucleus reticularis pontine caudalis.	Glutamate (+) and neuropeptides (DBI, CRF, CCK, alpha MSH, NPY)
LUST (Sexuality)	BNST and corticomedial amygdala to preoptic and ventromedial hypothalamus to lateral ventral PAG.	Sex steroids (+) (T / E), vasopressin, oxytocin
Nurturance / Maternal CARE	Anterior cingulate to BNST to preoptic hypothalamic to VTA to more ventral PAG.	Oxytocin (+), prolactin (+), dopamine, opioids
Separation Distress / PANIC (Social Bonding)	Anterior cingulate / anterior thalamus to BNST / ventral septum to midline and dorsomedial thalamus to dorsal preoptic hypothalamic to more dorsal PAG (close to circuits for physical pain).	Opioids (−/+), oxytocin (−/+), prolactin (−/+), CRF (+) for separation distress, ACh (−)
PLAY / (Social Joy and Affection)	Parafascicular / centromedian thalamus, dorsomedial thalamus, posterior thalamus, projecting to ventral PAG (septum inhibitory re: play).	Opioids (+ in mod. amounts, − in large amounts), ACh (+), cannabinoids (+)

Key: Dopamine (DA), Acetycholine (ACh), Cholecystokinin (CCK), Corticotropin-releasing factor (CRF), Neuropeptide Y (NPY), Diazepam–binding inhibitor (DBI), Melanocyte-stimulating hormone (alpha MSH)

Source: Douglas F. Watt. In memoriam: Jaak Paksepp. *Emotion Researcher* (ISRE's sourcebook for research on emotion and affect). Carolyn Price and Eric Walle, eds. http://emotionresearcher.com/in -memoriam-jaak-panksepp/.

highest just *before* you receive the reward—when desire or anticipation is at fever pitch.[18] If you activate this system with electrical stimulation, mammals will engage in certain types of behavior, such as foraging, exploring their environments, paying selective attention, and pursuing specific appetite rewards (e.g., food, water, warmth, sex, social interaction).

Now consider how this mammalian desire system expresses through the different folkways of chimpanzees, bonobos, and humans. Obviously, SEEKING is a subjective feeling. An entire introspective phenomenology of seeking could be written—indeed, some of it surely has been by our renowned authors of literary desire (from the search for knowledge of Sophocles' *Oedipus Rex* to the destructive search of Mizoguchi in Yukio Mishima's *Temple of the Golden Pavilion*). But animal SEEKING is presumably less phenomenologically urbane, albeit every bit as motivational. An organismic itch that cannot be scratched has a high degree of incentive salience. And this subjective feeling matches those homeostatic imbalances that drive the organism toward resource exploitation and satisfaction.[19]

In chimpanzees, bonobos, and humans, the SEEKING system is dedicated to specific resource challenges. The dopamine-based excitatory system pushes primates to forage, scavenge, and, of course, hunt. Chimpanzee hunting is more common than previously thought. Chimpanzees eat over thirty different species of vertebrates, but the red colobus monkey is a favorite target. A lone chimpanzee, hunting a monkey, has only a 30 percent chance of success, whereas a coalition of ten or more chimpanzees produces a 100 percent success rate.[20] This is one of many indicators of selective pressure for chimpanzee social coalitions. However, these coalitions are uneasy, since males also compete for females and for rank. Frans de Waal points out that human men share this uneasy tension with our chimpanzee cousins.[21] We band together in shared SEEKING projects and also against common enemies, but men also undercut each other in the competition for females. Then, like chimpanzees, we "groom" each other (with language or beer), to smooth over the competition and get back to cooperation.

Such social grooming is enabled not by cognitive processing but by the kind of *affordance* complex that was originally articulated by James J.

Gibson and further developed by Ruth Millikan (see Chapter 3).[22] According to this precognitive notion of social intelligence, an affordance is a meaningful interaction between the properties of the environment (including the social environment) and the capacities of the animal. When, for example, a male chimpanzee sees the swollen perineal skin or buttocks of the female, he simultaneously has an indicative (or descriptive) experience of the female's anatomy change (i.e., a fertility signal), but also an imperative (i.e., lust/behavioral) experience. Presumably, for most nonhuman animals, there is no inferential step from indicative to imperative. For example, a hungry predator needs only to see animal x and animal y as "food" tokens (with animal z as not-food), and further taxonomic refinement is not necessary.

Apes have different hunting and grooming cultures, but affective affordances (sculpted by ontogenetic experience), together with unique ecologies, provide the idiosyncratic behavioral scripts for each community. The details of the ontogenetic development of prosocial affordances (and their brain systems) in humans, is now starting to emerge (see Chapter 5).[23]

Bonobos do not appear to have the same hunger for animal protein as chimpanzees or humans. They do a little collective hunting, but like the rest of us they dedicate significant SEEKING energies to foraging fruits and vegetation. Bonobos are socially matriarchal, compared with chimpanzees. This cultural difference manifests itself in many ways, most significantly in hunting.[24] Chimpanzees, bonobos, and humans all benefit when their respective males hunt. Anthropological data on foraging human societies reveals that 88 percent of a society's protein intake is acquired by men.[25] But nutrition is not enough to explain primate hunting. When chimpanzee males catch prey, they become female magnets—and the males use meat to attract mating partners. Some primatologists are convinced that this is actually the principal reason why chimpanzees and humans evolved hunting—to impress the females.[26] Corroboration of this thesis may come from the bonobos' disinterest in hunting. When bonobo males catch and kill monkeys, the dominant females take the meat. Bonobo males cannot use meat politically when the females simply confiscate the bait. This, in turn, appears to radically reduce male interest in hunting, and it illustrates how carnivore culture is not strictly a matter of diet and nutrition. Females value meat for the amino acids and the boost such nutrition can provide in the gestation of offspring, but male chimpanzees seem to value meat more as sexual currency. The cultural differences among apes are significant, and recent work on bonobos suggests that they don't share food just for reciprocity or sexual access (as chimps seem to), but to assess relationship status. Bonobos will beg for food, and if the possessor shares it, this signals a positive affiliation.[27]

It's important to point out that this hunting looks cognitively sophisticated, but it may not be. Comparative ethology of different primate species and other mammals shows that cooperative hunting does not require cognitive sophistication.[28] Hunting is perceptually difficult but may be intellectually simple. Hunting is more like jigsaw-puzzle building—solving perceptual affordance problems, one piece at a time—than chess, which requires symbolic thinking and rule-based predictive strategizing.[29] Moreover, as we mentioned in Chapter 1, growing evidence from affective neuroscience suggests that some sophisticated social interactions (certainly play, but perhaps cooperative behavior like hunting, as well) remain intact after radical neo-decortication early in life.[30] In other words, removing the "thinking brain" of higher cognition shows just how adept the limbic "emotional brain" can be in organizing social life.

Many behaviors that look cognitively coordinated, like chimpanzee hunting parties, can be explained sufficiently by affective / emotional systems (like SEEKING), which are channeled by ecological and cultural constraints into dedicated action patterns. Early human SEEKING is not a different kind of process, but it received its own cultural channeling and evolved into a feedback loop of social learning. In other words, humans are pre-adapted for learning or research, broadly conceived, because they already possess a powerful source of motivated directionality. Our motivational system (which can be dialed up or down, as well as redirected) is adaptive, selected for, and highly conserved across mammalian clades. But some unique aspects may have emerged in our cultural expressions of SEEKING. When SEEKING produced cultural effects (at first as mere byproducts), some of those cultural effects, like curiosity, became highly influential and acted as a new selective force that fed back on fitness patterns. But whereas this is the case with *Homo sapiens,* many nonhuman primates experience their cultural effects more as byproducts—derived from preexisting adaptations (with little to no feedback strength). In humans, SEEKING can be trained on other human behavior (probably conjoined with separation anxiety and the need for bonding; see Chapter 5). Human infant attention to the caregiver's face, for example, is intense, and a harbinger of imitative things to come. The result is our uniquely strong social learning, which acquires many behaviors (e.g., sharing, aggressive response, cooperation, delayed gratification, etc.) by imitation rather than by instruction.

Exploration of the environment, aroused sensory attunement, selective attention, and stimulus-bound appetitive behavior (i.e., the pursuit of food, water, warmth, sex, social bonds, etc.) form the behavioral suite of the mammalian SEEKING system. But hominid cultural innovations transformed us from omnivores to informavores.[31] After hunting, foraging, and scavenging, our human ancestors acquired a new kind of taste—equally motivated by the SEEKING system, but hungry for information and expertise rather than just nutrients. Our ancient SEEKING system, originally adapted to motivate specific resource exploitation, was more recently exapted into an engine for Lamarckian-style cultural evolution. Chimpanzee and bonobo cultures are comparatively poor in technology and confined to real-time events, but our culture is more cumulative. Our informational pool (as a species) is hundreds of generations deep, whereas our individual information pool is about two generations deep (as is that of primates)— albeit now fed and nourished by the species-level reservoir (and picking up speed through language, writing, and material culture).

Information about tools and techniques can be stored and transmitted cheaply once language and symbolic mind is up and running, but affective learning (originally acquired in trial-and-error experiences) can scale up to social learning without specialized genetic brain changes.[32] This means that our ancestors may have accumulated rich stores of embodied information by unconsciously associating sensory-perception events (witnessing conspecific activity) with internal motor-system mimicry.[33] In this view, other mammals may be good at "learning by doing" (and thereby extracting important ecological information), but pre-humans may have been better at learning by watching others doing.

Learning how to crack open a nut with a rock is impressive for a chimpanzee, but it pales compared to flint-knapping stone tools, making clothes, carving spears, or processing toxic tubers. Human hunting, for example, is extremely skill intensive, taking around two decades of daily hunting for males to become masterful. It is doubtful that our human ancestors could have harvested complex knowledge and skills without the emotional push-pull of the SEEKING system. This is because the immediate environment is not a transparent wealth of resources, so knowledge of such resources needs motivated curiosity, scrutiny, and dogged repetition to either dig out or master. But if chimpanzees and bonobos have this affective prod as well, then why didn't they develop a hunger for information like we did?

Compelling answers might be found in the sociocultural changes that developed during the Pleistocene era. A flexible SEEKING system is necessary, but not sufficient, for explaining human cultural evolution. Philosopher Kim Sterelny, unimpressed with the big-brain theory of human success, isolates some cultural changes that propelled us forward. Sterelny argues that feedback loops emerged between individual adaptations, skill expertise, enriched learning environments, and social learning.[34] These loops (implicated by Oldowan or Acheulean industries) created high fidelity and high-volume information flow across the generations. Accidental tool creation can be mimicked or imitated, but crucial cultural prerequisites have to emerge for informavore evolution. The growth of expertise and transmission requires much safer and longer childhoods. Juveniles need a less hostile environment in order to acquire complex skills, and they need enriched, stable environments (filled with tools and experts) to practice on and interact with. If you put the SEEKING system into an enriched environment, with the safety to explore in open-ended ways, it can begin to absorb large amounts of the local information pool. In other words, juvenile hominins could observe, experiment, and practice skills in secure, low-stakes environs.

The culture of safer childhoods is the result of several factors, including the invention of cooking (around 1.5 million years ago), food storage, and collective hearth innovations. But one of the most important changes that helped create SEEKING informavores was better reproductive cooperation. One ingredient in this increased domesticity may have been a reduction in aggression associated with male competition. In short, males probably reduced infanticidal reproductive strategies as longer-term parenting partnerships formed.[35] How did pre-sapiens human fathers go from baby-killers to bodyguards? This transition is unclear, but tuber exploitation may have played an important role. As pre-sapiens moved from forests to the woodlands, a new sexual division of labor emerged around food sources of the savanna.[36] Men could hunt for protein, and women could exploit the new resource of underground storage organs, creating better resource partnerships (possibly in some early form of long-term pair-bonding).[37]

If our hominid ancestors had "fission-fusion" polygyny cultures like contemporary chimpanzees, then their kin groups did not resemble our contemporary nuclear families. Mothers and infants had strongly bonded relationships, but adults drifted together into subgroups (for hunting or sex) and then separated again almost daily. The size of these groups (thirty to fifty members) varied depending on the available resources.[38] When unattached males entered a new group or found an unprotected mother, they might kill the infant (as chimpanzees and other mammals often do). This terminates nursing, puts the mother back into estrus, and allows the interloper to impregnate the female—hijacking the procreation system for his own gene line. Our hominid ancestors may have lived in such a nasty, brutish, and short world, but eventually males and females stumbled upon a new strategy of semi-stable partnerships. And this solved another important problem for pre-sapiens: paternity. It is much more likely that males will provision juveniles if their paternity is established. Female primates are promiscuous, and males ensure paternity by four possible methods: alpha-male harems (e.g., gorillas), pair-bond mating (e.g., tamarins and humans), band-of-brothers kin groups (e.g., chimpanzees), and indiscriminate mating (e.g., bonobos).

The evolution of the family probably played a significant role in creating the stable, secure environment that was necessary for amplification of social learning and the information culture. As an economic unit, it provided a diverse set of safe spaces for development. Additionally, these longer, safer childhoods must have contributed to the growth of inner-subjective head-space—no doubt leading to greater representational sophistication (and eventually language). And the striking feature of this new social learning is that it evolved into a flexible, multi-modal, and open-ended system. SEEKING plus an information-rich environment produces curiosity for all manner of problem-solving, and both curiosity and the products of skill can co-evolve via natural and artificial selection.

Curiosity is often ignored in philosophical epistemology and the psychology of learning, largely because it is the very condition of knowledge itself. But a handful of theories about the nature of curiosity are worth mentioning, because, ever since Darwin's *Descent of Man,* we have recognized the animal roots of inquisitiveness. Daniel Berlyne famously suggested that curiosity could be divided into perceptual and epistemic types.[39] The perceptual form, shared by most mammals, is, according to Berlyne, an innate "drive" that is aroused by novel stimuli (producing exploratory behavior) and is reduced by continued exposure to said stimuli. Epistemic curiosity, he argued, was a predominantly human form, involving cognitive information; a paradigm case would be solving a puzzle. Moreover, curiosity can, according to Berlyne, be specific (focused on a defined problem) or diverse (unfocused and wandering). When we add the recent neuroscientific findings about the anticipatory energy of the dopamine-based SEEKING system, we arrive at the intense affective feelings underneath this taxonomy of curiosities.

Another feature of curiosity is that it emerges from some perceived incongruity between a thing / event and our default worldview or expectation.[40] Such category violations—category jamming—can be seen in our curious fascination with chimera and hybrids of all kinds. While this category mismatch has long been recognized as a feature of curiosity, it is worth noting that the affective aspect of such mismatches can be positive or negative and all points in between. Indeed, SEEKING itself is a mixture of anticipatory attraction and uncomfortable agitation, in need of satisfaction.

Additionally, curiosity is a desire to close an information gap.[41] It is a discrepancy between what is known and what is unknown. It is a desire to resolve a cognitive deficit. When a predator sees an animal pass behind a bush, for example, but the animal fails to reemerge according to expectation, heightened state of attention and expectation arises. This is different from the category mismatch, such as when a previously benign animal suddenly delivers a painful bite, or when fire is encountered for the first time, or when conjoined twins are born. The desire to close an information gap is somewhat different. When an animal's food cache is surprisingly missing, the animal experiences an information gap (as well as possible frustration), and a search may ensue. According to this model, curiosity increases as information lacunae get nearer to closure.

In Goualougo Triangle, Republic of Congo, researchers encountered high degrees of curiosity-SEEKING in chimpanzees with no previous human contact. Unlike the fearful, wild chimpanzees who have encountered humans before—and the bored zoo chimps who no longer register human spectators—these chimps stared intensely at the human researchers, moved closer to get a better look, threw branches and slapped tree trunks to elicit responses from the humans, and engaged in intensive vocalizations.[42] These are the affective and cognitive precursors to our own culture of curiosity, but they remain byproducts of the SEEKING system (and the relevant cognitive activations and discrepancies) rather than opportunities (in the case of humans) for multigenerational investigation (and eventual cause-and-effect theorizing).

LUST

LUST behaviors are easily identifiable: genital arousal, pursuit of copulation mounting, an open mouth, bared teeth, neck biting, vocalization, submission and dominance displays, and so on. In the same way that our SEEKING system is rooted in a specific brain circuit, LUST is also a unique brain pathway in mammals, extending through the hypothalamus, ventral striatum, and insular cortex.[43] In LUST, norepinephrine and dopamine increase, serotonin drops, and androgens—like testosterone—fuel both male and female sexual drive. It is a different circuit than attachment bonding, or CARE.[44]

LUST is an affective system shared by many vertebrates, but it often operates in a mechanical way. Female estrus triggers chemical and physical changes that draw males to copulation. Like moths to flames, males perceive the chemical changes in females, then libido ignites and copulation follows in short order. When you filter this affective energy through different prisms of primate culture, however, you get unique sexual customs.

The genital area of female chimpanzees, which become sexually mature around eight years old, swells up and changes color. This new perceptual affordance initiates attention from many males, who attempt mating. Females copulate around eight times a day, often with different males, and this may be an unconscious strategy to keep males calm and non-aggressive (because male orgasms increase the quieting chemistry of oxytocin). But a pattern of possession quickly develops. Male chimpanzees use three strategies to establish ownership over the female and subsequent offspring. A male might start a dedicated bodyguard routine, fighting off competitors, or, if the number of males is too large, then two males will establish a coalition—sharing the copulations. Or finally, a male might sequester the female, taking her away from the group (by persuasion or force) to copulate privately for a period of days.[45] Chimpanzee paternity is important, and infanticide might become a default solution, but such aggression is usually averted by these various sequester techniques. And the fact that most males in groups are brothers may also reduce the paternity aggression considerably, leaving most in-group aggression for issues of mating access rather than paternity.

Obviously, humans have evolved a more byzantine sexual culture than chimpanzees, but fundamentally there are homologous affective brain and body systems at work. Presumably, early human sexuality operated along the lines of other primate strategies, but which one or even whether there was a combination of strategies is unclear. Were our early LUST adventures more like those of chimpanzees, bonobos, or gorillas? Data from most gorilla populations, for example, reveal that they have a single-male mating system (i.e., groups contain only one fully mature male that serves as silverback for many years). The gorilla's evolved anatomy reinforces the mating system (and vice versa), since the male is so much larger than the female (i.e., gorillas may be the most sexually dimorphic primate).[46]

Bonobos and chimpanzees are both members of the genus *Pan* and probably split around 1 MYA. Bonobos, which were not even discovered until 1929, are smaller than chimpanzees, matriarchal, less sexually dimorphic, and less aggressive; they also live in a rich diet environment and engage in almost constant sexual activity. Males copulate with females, but males also engage in genital manipulations (or penis fencing), and females pursue genital-genital rubbing.

The larger size of human men suggests that testosterone-fueled competition (like in chimpanzees) sculpted our sexual dimorphism. The larger size of *Homo* males makes a matriarchy somewhat doubtful—matriarchal female hyenas, for example, are 9 percent larger than males.[47] Still, an early human matriarchy is not unthinkable, since dominant bonobo females are still slightly smaller than males. So, it is presently unclear whether we originally had alpha-male harems (qua gorillas), roaming bands of brothers (qua chimpanzees), or some other primate system. Nonetheless, out of one or more of these procreative strategies emerged a form of pair-bonding cooperation.

Some paleoanthropologists have suggested that pair-bonding emerged as far back as 4 MYA with the australopithecines, on the grounds that sexual dimorphism reduced significantly then (revealing cooperative rather than competitive body types).[48] This argument seems undercut, however, by the fact that *Homo habilis* (2 MYA) was highly dimorphic (although the theory might be sustained if we remember that biocultural patterns—like monogamy—need only endure in populations, rather than in species, genera, or families). Pair-bonding may have been like fire-starting skill in the sense that it came and went many times before it spread to more universal proportions.

A recent genetic argument places the human shift from polygyny to monogamy as recent as 18 KYA.[49] Analysis of female mitochondrial and Y-chromosome DNA reveals that the number of reproductive females increased significantly in conjunction with *Homo sapiens* migration out of Africa, but diverse male contributions did not increase until much later (circa 18 KYA). One explanation is that prior to 18 KYA, many women would have been reproducing with the same few men.

As these data in areas from genetics to paleoanthropology increase, we will converge on a more detailed picture of *Homo* life, but for our purposes, the timeframe when strict monogamy emerged is not crucial (although see Chapter 7 on the role this social technology plays in civilization). The mechanism of interest is the feedback loop between emotional systems and biocultural innovations such as increased cooperation, apprenticeship social learning, alloparenting, and so on. It doesn't matter to the argument whether human social life was comprised of fewer men with multiple females and broods, or modern monogamy. What matters is domesticated (emotionally modulated) males, who are cooperatively provisioning females and needy offspring. For example, studies of gorilla social groups reveal that the level of silverback tolerance for and affiliation with infants and juveniles in his group strongly influences the degree to which females stay in his troop.[50]

This is unlike chimpanzee parents—where mothers do the rearing almost exclusively and human fathers contribute significantly. Just like other primates, women are trying to maximize their genetic investment and mate with the fittest candidates. Females try to "choose wisely" through sexual selection techniques (e.g., performance displays), but they also hedge their bets with optimized deceptive copulations (common in primates and humans alike). In some cases, men attempt various vigilance strategies on specific women—guarding them and provisioning them—and in other cases, males simply play the odds, broadly investing fertilization but not much else.

Since the Holocene period, options for humans have settled into relatively formalized monogamous and polygamous patterns of reciprocal long-term partnerships. Because kin expansions, through marriage partnerships, have benefited group survival, and such pairings are susceptible to break down from deception, sexual mores have become fairly conservative in most human societies. Sex and LUST have been channeled and transformed by the survival benefits of cultural affects like loyalty. For humans, sex has been decoupled technically from reproduction, and we've turned LUST into a connoisseur recreation. The pleasures of LUST, which motivate procreation, have been exapted by culture, but they remain tethered to their origins. Human

LUST, which doesn't need to wait for estrus triggers, has nonetheless been culturally constrained by the demands of survival partnerships, especially when we compare it with bonobo sexuality.

Among bonobos, the LUST system has exapted from pure procreation to other functions, but bonobo sex is not just fun and games. The amount of sexual activity increases whenever potential conflicts, like food-sharing, arise. In the same scenarios where chimpanzees and humans will fight and display aggression—namely, competition for resources—bonobos will mount each other and restore the peace with doses of sexual ecstasy. As Frans de Waal has noted, bonobos are the hippies of the primate kingdom.[51] It should be noted however that Gottfried Hohman's work with bonobos gives us a less sanguine picture and reveals that aggression is still alive and well in the supposed hippies.[52]

When primate LUST meets top-down, competition cultures, it produces chimpanzee-like social intelligence. The major contributing factor to these cultural structures is simply access to resources. If food is scarce, then males are competitive for females and LUST tends to be channeled into a hierarchical social system, with despotic sexual politics. If resources are plentiful, as in the case of bonobos, then competition reduces, and LUST can channel into sexual "egalitarianism." Since female bonobos are sexually receptive to all, males do not compete like chimpanzees.[53] In addition to keeping the peace, radical promiscuity solves the paternity problem shrewdly by confusing it beyond any possible tracking.[54] If every offspring could be yours, then you are less likely to harm any of them.

Primate mating cultures are complex, however, and one needs to be careful to avoid a simplistic cause-effect dynamic between resource availability and sexual politics. Yes, bonobo egalitarianism corresponds reasonably with easy food availability, but then again, procreative fecundity also rises in most primates as food supplies increase. And the increased population density of males increases competition for female access in other primates. So, for example, in orangutans, sexual coercion has been observed to increase along with food availability (e.g., high fruit abundance).[55] But generally speaking, mating cultures vary according to ecological differences rather than differences in endogenous endocrinology and neurochemistry (i.e., the LUST system).

The innate sexual drive is a relative constant in primate physiology / psychology, but the expression of that force is flexible. Free love and bohemian philosophies have long held out the hope for human sexual liberation, for example, but bonobos appear to have us beat. Humans have decoupled sex from procreation, but we look like amateurs next to the bonobos. Their LUST circuit has been untethered from one or even a few mates, and it has subsequently diversified and expanded into all manner of social grooming. So, homologous affective urges that drive procreation can become platforms for the emergence of unique cultural transformations.

CARE

Maternal CARE, like SEEKING and LUST, has brain-based signatures. Unlike other vertebrates, mammals care extensively for their young and other kin. As primates, we share important homologous attachment mechanisms in the brain. Mammal mothers have a distinctive circuit from the hypothalamus, through the *stria terminalis,* to the *ventral tegmental area* (VTA), in which the neurotransmitter oxytocin travels. Damage to this system destroys maternal feeling and behavior, and direct injection of oxytocin into the VTA (in rats) actually produces maternal behavior.[56]

Furthermore, researchers have known about the phenomenon of *imprinting* for many years. Behavioral scientists, working on animals, have described and successfully manipulated this simple form of bonding for decades. Researchers can get baby birds, for example, to imprint on the scientists themselves, on beach balls, and even on beer bottles. The imprinting occurs because a "window" of bonding opens right after birth and closes quickly, so whatever proximate thing is nearby becomes "mom." Mammals have the same, albeit more sophisticated, mechanisms for fastening together parents and offspring.[57]

Mother-baby bonding is an essential skill for any animal born into a hostile environment. Prey animals, especially herd animals, are born with generous physical adeptness. They can walk and even run within minutes of birth. This mobility is important in a predator-filled world, but it puts them at great risk of potential separation from their mothers. It's not surprising, then, that herd animals have very tight windows of opportunity for identifying their mothers and latching on. Failure to lock onto the mother (for *any* mammal species) usually means death for the offspring and compromise of gene-line transmission for the parents. Consequently, natural selection pressures for bonding are significant.

Nature has not left bonding up to chance; it also has not waited for rational deliberation or cognitive identification to evolve (i.e., many animals are great at bonding, despite a lack of intelligence). Instead, internal chemical changes spike during the window of opportunity in the brains and bodies of parents and offspring, cementing them together in ways that are incomparable with other conspecific relationships.

Specific neuropeptides—oxytocin, opiates like endorphins, and prolactin—all rise profoundly in the last days of a mother's pregnancy. Oxytocin regulates several aspects of maternal biology (facilitating labor and breastfeeding) but also plays a crucial role in nurturing behavior.[58]

Oxytocin bonding is a time-sensitive process. Sheep have a very short window for the mother to bond with offspring—only an hour or two. If a lamb is removed from its mother for two hours, the mother will not be bonded and will subsequently reject the lamb. Remarkably, after the bonding window has closed, it can be *reopened* again for a couple hours by injecting oxytocin into the mother's brain. Once oxytocin floods the system again, the mother can lock onto her offspring and engage in maternal behaviors.[59]

Oxytocin is more than a lever or switch for turning on motherhood. Found only in mammals, oxytocin was one of the first neuropeptides to be isolated and sequenced. Its presence in the breasts (letting down milk) is well known, but more recently, neuroscientists have been studying its role in the brain (it's made in the hypothalamus, stored in the posterior pituitary, and then released into circulation). Discovery of oxytocin receptors in the brain signify that the brain is also a target organ for oxytocin.

Oxytocin probably evolved from the ancient brain molecule vasotocin, which regulates sexual activity in reptiles. The evolution of oxytocin also reminds us of the way that natural selection conserves available resources, repurposing their original adaptive functions into new functions. The neurochemistry of mothering and nurturing seems to be a reconfiguration of sex chemistry (oxytocin plays a role in orgasm), rather than some unprecedented adaptational jump in brain chemistry.[60]

Oxytocin calms down aggression and dramatically reduces irritability—important mood alterations for new mammal parents. Male moods are equally transformed by oxytocin, which floods the male brain after sex.[61] Male mammals become more nurturing and less aggressive after sex. Very recently, vasopressin processing in the brain has also been linked with pair-bonding and prosocial behavior in mammals.[62] Interesting correlations have cropped up in the kinds of vasopressin receptors found in sociable, reciprocal species like bonobos, humans, and prairie voles. A similar mutation in the AVPR1A receptor has been found in sociable mammals but is lacking in the more detached chimpanzees and montane voles.[63] Recent research has indicated that oxytocin may also accentuate or heighten negative affect as well, so the causal picture is still unclear.[64] Oxytocin may be a chemical amplifier of whatever affect is dominant at a given time.

In chimpanzees, this CARE system is very limited in scope. Mothers and babies bond strongly for approximately seven years, but strong family bonds end there. Sarah B. Hrdy points out, "In roughly half the 300-odd species of living primates, including all four great apes and many of the best-known species of Old World monkeys, such as rhesus macaques and savanna baboons, mothers alone care for their infants."[65] Hrdy has persuasively argued that human cooperation was facilitated by unique cultural shifts in childrearing. Unlike chimpanzees, *Homo erectus* children were raised and provisioned by caregivers beyond mothers. Grandmothers, aunts, uncles, siblings, and fathers (alloparents) contributed to childrearing and constituted an expanded circle of empathic filial feelings.

Human offspring need extra work—a whole team of caregivers—because they are relatively helpless compared with most other primates. A complex mosaic of causal patterns leads to the unique human childhood (see Chapter 5). Our *australopithecine* ancestors had short childhoods and short lifespans. They also had wider hips, suggesting that their fetal brains probably developed more fully in utero, like chimpanzees. Their ontogenetic assortment of behavioral options was presumably more hardwired. But by the time of *Homo ergaster,* the hips had narrowed, and fetal brain development had to be postponed until after birth. Consequently, members of the genus *Homo,* including present humans, are born "prematurely," and our brains develop *ex utero.* The result is a much larger window of infant dependency that requires staggering amounts of parental and alloparental care. It also means, because of neuroplasticity, that our brains are still wiring as we take in information from our environment—including the rich social environment.

Our brains are slowly softwiring during our infancy, and our interaction with alloparents creates wider circles (beyond the mother) of affective bonding (see the Chapter 8 discussion of fictive kin). Humans bond with several caregivers, and the bonding window remains open almost indefinitely after we become independent. This feeling-based flexibility of attachments endows humans with unique powers of cooperation, compared with other primates. And these emotional developments glued together early families, creating (as a by-product) the stability of environments that enhanced social learning. Mothers have always been paragons of care, but early human fathers evolved impressive abilities of delayed food consumption in order to provision their families. It is unlikely that these cultural changes could have happened without emotional evolution.

Wismer Fries, Seth Pollak, and other psychologists at the University of Wisconsin, Madison, discovered that oxytocin is vital in *human* bonding.[66] Researchers wanted to know why some kids fail to bond with their parents. Some children suffer from "attachment disorders," failing to seek comfort in others, even their own families. Using a control group of non-adopted kids, the researchers collected baseline oxytocin (and vasopressin) levels in eighteen four-year-old kids who were adopted from Russian and Romanian orphanages. The children from the orphanages had a history of being neglected. They devised a test in which oxytocin levels were checked before and after comfort/play time with parents. The children were held on the laps of their mothers while they played a computer game together, engaging in intimate playtime that included whispering, tickling, petting, and so on. Immediately after this time, the pleasurable oxytocin levels spiked in the non-adopted children but remained the same in the adopted children. It appears that the anxiety-reducing, calming effects of oxytocin have been *primed* in us by our earliest nurturing experiences. If a child is neglected in an overpopulated orphanage or a cold family situation, they fail to form the normal attachment chemistry. We do not entirely understand the mechanics yet, but it looks as though early experience with a loving caregiver "wires" the brain to associate a specific person or people with pleasurable, happy states. This association, which is both chemical and psychological, appears to be the template for positive social bonding in later life.

We know that Neanderthals cared for extended kin, because paleoanthropological evidence shows they supported sick and elderly members who became dependent. Moreover, human funerals go back to the Neanderthals and provide suggestive evidence for filial attachment beyond the maternal template.[67] Once CARE is filtered through the cultural innovations of reproductive cooperation, alloparenting, social learning, and so on, we move beyond narrow, dedicated bonding to open-ended, flexible bonding (i.e., maternal bonding, nuclear kin bonding, extended kin bonding, fictive kin bonding, etc.).

Compared with chimpanzees, human cultures have exapted their emotional CARE systems well beyond their origins. However, it is unclear how to interpret bonobos. Is the bonobo culture of sexual pacification an exaptation of the LUST system or the oxytocin-based CARE system or both? Chimpanzees are much more xenophobic toward strangers, frequently killing newcomers to a group. Humans and bonobos are much more tolerant of, and cooperative with, strangers. Bonobos will quickly engage in sexual activity with strangers from another group, but bonobos seem to use sex to ameliorate aggression, and it's unclear if this is also tantamount to bonding or attachment. Recent research suggests that female bonobos create strong bonds with other females through sexual interaction and copulation calls.[68] Unlike chimpanzees, female bonobos do not migrate out of their groups for procreation and subsequently have stronger female–female social ties. The female migration model (e.g., chimpanzees) has been correlated with more despotic social systems and infanticide. It is unclear at this point whether bonobo males have relatively shallow social alliances—since they are ecologically contoured by a combination of promiscuity, ready resources, reduced hunting parties, and down-tuned competition.

Flexibility

The model we have been developing tries to acknowledge the homologous emotional features but also the diverse species-specific and tribal-specific manifestations that emerge from ecological constraints. Instead of positing a series of native genetic modules that cause filial affection or monogamy or information-seeking, we have suggested that the affective systems can take us a long way toward the modern human mind. This model requires that the affective systems can be decoupled (at least in part) from their dedicated targets and recruited for new functions, ultimately giving rise to cultural loops. These bio-cultural loops are made possible by associative emotional learning. Evidence that such processes are not only possible but probable is clear from neuroscientific studies of brain plasticity, especially as it regards limbic / behavioral conditioning.

It is not the purpose of this chapter to give a detailed articulation of emotional learning, but some sketch of the prerequisites and dynamics may be helpful. A pliable associational system is deeply homologous in mammals and it presumably allowed early humans, just as easily as contemporary primates, to be conditioned by positive and negative feedback. Antonio Damasio's well-known somatic marker hypothesis is one attempt to articulate such a plastic associational / deliberational system without significant appeal to conceptual inferences.[69] Our own model may depend upon the success of something like Damasio's somatic markers, but even subcortical enrichments may prove to be sufficient to extend the associational system. For example, Heyes makes a very compelling argument for how early visual perceptions wire together with motor associations to create default mirror neurons in humans.[70] The cognitive architecture of imitation connects a sensory representation of an action to a motor representation of the same action. So I see a hand grasping, and this matches with an inner motor sense or feeling of my own hand grasping—these are "matching vertical associations." Observational learning requires a conversion of visual patterns to bodily patterns (action and affect), and mirror neurons act as the requisite converters. This kind of mechanism would allow for cultural (and other ecological) changes to be incorporated (softwired) into the affective systems.

When recently observing Mountain gorillas (*Gorilla berengei*) in Rwanda, we observed a silverback male administer brutal punishment to a lower-ranking male who had been playing too rough with a juvenile. The kind of conditioning that results from this punitive experience is swift and relatively durable. The low-ranking male learns to avoid abusing his nephew by associating the behavior with a new set of affective experiences, namely pain and fear. The juvenile also learns that certain kinds of screams produce rescue. But the current social hierarchy is not eternal; power relations flip, alliances change, bodies grow bigger and older, and so on. Any system of adaptive associations must change to fit the changing social realities. When the alpha male is overthrown, for example, then the lower-ranking male and everyone else will need to have his social taxonomy re-encoded with affective associations.

The positive valence axis of this associational system is probably underlined by the ventral tegmental area (VTA) of the midbrain and its dopamine system, while the amygdala is probably the provenance of the negatively valenced axis. We don't need to leap to cognition per se to get the subtle biocultural changes we've been describing. All we really need are slight improvements in *Homo* mnemonic abilities.

We know that neurohormonal modulation can occur at four stages of memory: encoding, consolidation, storage, and retrieval.[71] All of these stages of memory have affective encoding, and such emotional modulation is a significant reinforcement mechanism for procedural learning, priming, and reflexive conditioning. The biocultural changes we've been describing (e.g., SEEKING transforming to social learning, LUST and CARE to family structure, etc.) require enrichment of only the non-declarative memory systems. Some of the brain systems that undergird procedural memory run between the amygdala and the striatum, while those underlying motor learning run between the amygdala and cerebellum, and those responsible for perceptual priming run between the amygdala and the sensory neocortex. These are separate from, and presumably evolved before, the semantic declarative circuits between the amygdala and the prefrontal cortex (PFC) and the medial temporal lobe.[72]

Emotions not only provide primates with salience and attentional focus in complex experiences, but they also help wire sequential motor skills into adaptive action patterns. If Robert Barton is correct, then cerebellar expansion (especially neuronal density) outpaced the neocortex during hominid evolution and gave humans the ability to order motor sequences in increasingly complex ways—setting the stage for technology, social ritual, and other syntactical processes like language (see Chapter 7).[73] Linking the mammalian associational system to modestly enriched memory processing (together with motor sequencing extrapolation) arguably produces all of the ingredients needed for apprentice learning and extended family care.

If a hunter's non-declarative pictorial memory of his offspring or partners can be retrieved (even imperfectly and intermittently) while he's away hunting, then he may also experience emotional CARE triggers that cement bonds and motivate provisioning behaviors. More memory creates more care, which creates more provisioning, which creates more social stability. This social stability loops back to create more information-rich childhoods, nutrition, affective domestication, and memory enhancement (among other cognitive improvements). And the cycle continues.

Humans are peculiar in the scale of their group devotions, but we do not need a set of brain innovations in order to explain our conviviality. Our views are sympathetic with cultural evolutionists Peter Richerson, Robert Boyd, and Joe Henrich in the sense that we similarly refuse to postulate a prosocial cognitive module to explain human congeniality. Like Boyd, Richerson, and Henrich, we think cultural evolution explains significant aspects of *Homo sapiens* success. Nonetheless, we think these culturalists have not sufficiently factored emotions into their model of cooperation and group commitment.

How do kin-devoted primates scale up to large-group altruism and even nation-state commitment? We'll have more to say on this topic in Chapter 8, but a few aspects can be addressed here. Richerson and Boyd argue that our original selfish or egoistic orientation is counteracted by the evolution of "tribal instincts" in humans.[74] These tribal instincts are, according to Boyd and Richerson, a combination of gene-culture coevolution, including selection for cognitive innovations like cheater-detection, improved social imitation, the desire to punish norm violators, in-group and out-group symbol markings, and other brain-culture novelties. Once humans attained these tribal instincts, Boyd and Richerson argue, it's a short step up to nationalism and other large-scale commitments.

While we agree that culture is a crucial force that scales up our prosocialty, we diverge from their specific line of argument. Richerson and Boyd make the common mistake of assuming an individualist rational-choice model to explain the original small-scale kin reciprocity of our early ancestors; they need to suggest a whole new level of altruism (tribal instincts) that evolves at the group level and eventually overcomes primordial individualism. But affective neuroscience shows that individual mammals already display deep group commitment from the very start, via the CARE system and imprinting. The idea of an original, autonomous, selfish agent is suspect, and makes unnecessary demands on an evolutionary model. The CARE system suggests that rudimentary altruism and empathy are already in place, and neither culture nor biology need to invent tribalism and nationalism de novo.

Our emotional niche approach also helps provide some of the phenomenological and psychological texture for these altruistic and caring behaviors. The agents themselves are not cost-benefit calculators but feeling-based creatures. Indeed, when we add together the feeling states (pleasure, reward, pain, lust, approach, avoid, etc.) with the tendency to imitate others in our social group, we have the main mechanisms of successful cultural transmission. As Joe Henrich's empirical work shows, humans acquire adaptive behaviors (e.g., detoxification of plants, hunting techniques, technology skills, etc.) without much logical conscious calculation.[75]

Many assume that humans became successful because they used logic and evidence to solve their environmental challenges, but, in fact, one can be very successful if accidentally imitating the right people. Cultural and psychological preparations evolved to make our imitation process limn advantageous behaviors (i.e., natural selection made us better learners), but the agents themselves are relatively unconscious to it. Henrich studied children in Fiji and found that they learn skills obliquely by following perceivable cues. They are drawn to imitate members in the community who have prestige; that is to say, they like and imitate the person that other people like. They also tend to imitate people who are older than they are, who are of the same ethnicity and sex as they are, and who are successful at a behavior (e.g., the hunter who came back with a bigger kill). We suggest that following these cues is affectively driven and not done by conscious, logical deliberation. What motivates a person to imitate another member of the group is not some logical assessment of the efficaciousness of the specific behavior, but feelings of attraction, admiration, care, lust, and so on. Affect is a powerful motivator for attention.

We agree that humans are tribal, but no additional set of "tribal instincts" needs to evolve in the brain in order to explain it. The affective systems, like SEEKING, CARE, and LUST, are capable of motivating the work of tribal and other large-scale commitments if cultural channels guide those feelings accordingly. Of course, cognition is relevant here too and inextricable from cultural learning, but we don't see the need to postulate modules or strictly genetic innovations.

Conclusion

We have now extruded three primary emotional systems—SEEKING, LUST, and CARE—through three different primate ecologies to see the resulting forms of biocultural intelligence. Obviously this is a conceptual model, and while some evidentiary data have been signposted, much more needs to be done.

We may be able to test some of these hypotheses by radically modifying the social and physical environment of primate groups and carefully observing affective/behavioral changes. Unfortunately, upsetting primate societies dramatically by depriving them of resources, or even overloading them with resources, may have unacceptable ethical implications because outcomes may be detrimental to individuals and groups. Some natural experiments, however, reveal compelling evidence that affective systems are channeled by cultural constraints into stable but revisable pathways. Sapolsky details a savanna baboon group, "Forest Troop," that was dominated by the usual aggressive males.[76] All the belligerent top-ranking males died suddenly after contracting tuberculosis while eating at a garbage dump. The result was that less aggressive males survived and were outnumbered by females, creating a completely different affective biocultural dynamic in the group. This more relaxed cultural style (e.g., more affiliative behavior, less dominance fighting, etc.) persisted for decades and even shaped the behavior of new generations and males who immigrated into the Forest Troop.

Such stable but revisable social intelligence systems are present in chimpanzees, bonobos, and humans as well. And we've tried to show that emotions, not rational cognition, are enough to explain many layers of this social complexity. Homologous primate emotions are channeled through the cultural folkways of each species, which are constrained by the matrix of ecology, technology, and reproductive strategy. Modest mental powers, like memory, conditioning, and social learning, are the only prerequisites for chimpanzees, bonobos, and early humans to have serviceable cultures. The emotional learning systems (e.g., affordances, somatic markers, and mirror neurons) are the bottom-up mechanisms that allow primate populations to scaffold up the cultural changes we've been discussing here. It was not until much later that *Homo sapiens* developed propositional problem-solving, so we should resist projecting that skill back into deep time. We also should not anthropomorphize—or ratio-morphize—current apes.[77]

Underneath the cost-benefit explanations of animal behavior and animal culture are the *feelings* that really motivate the animals. The research program that we are proposing will try to find empirical data to support plausible affect/emotion evolution scenarios, editing the possible scenarios down to the probable by the usual comparative methods in ontogenetic and phylogenetic sciences. Without these feelings and emotions, the animal is either much smarter than we thought (i.e., a cognitive inference-drawing agent) or much dumber (i.e., an empty behaviorist black box of inputs and outputs). Affect theory shows how pleasures, pains, anticipations, lust, care, and other gut-feelings can build up, in complex social ecologies, to produce sophisticated social intelligence. Epigenetics, also, may eventually shed more light on how these affective ingredients get expressed differently in primate groups over multiple generations.

Eventually, humans devised representational ways to communicate and improve the norms of our social contract. But before rules, morals, and laws, we had prosocial affective systems—like kin loyalty, empathy, and so on—that served to organize small-group cohesion. As Kim Sterelny puts it, "Prosocial and commitment emotions evolved before moral cognition; they made possible the cooperation and cultural learning that prepare the evolution of explicit normative thought."[78]

Finally, the mechanism of decoupling may be the most interesting implication from this story of emotional evolution (see our discussion in Chapter 6). Dedicated emotions can be decoupled from their original target functions and broadened into more plastic, open-ended suites of general responses. Primary emotions, like LUST and SEEKING, emerged early in vertebrates and served very specific adaptive action patterns. But as mammalian social life evolved, including high-investment parenting, feelings and emotions found new uses in driving groups toward resources and away from threats. Emotions that could be enlisted in the improvement of cooperation, especially for primates, would have been selected for. The means by which this decoupling (or opening up) happened for humans was probably a change in the rate of ontogenetic maturity—allowing a larger role for environmental social influence on the software coding of our brains. The neuroplastic brain generates and assigns affective values (and circumscribes default emotional tendencies) long before cognition learns to *represent* and *propositionally manipulate* the external world. Human helplessness and alloparenting would have put many hominin babies and caregivers in each other's arms, providing them with opportunities to experience important affective bonding triggers, such as touch, warmth, vocal soothing, and so on.

Emotional plasticity was obviously not just a human mutation, however, as mammals have been evolving in social groups for hundreds of millions of years. The bonobo, to name just one primate, has obviously decoupled and redistributed some dedicated LUST affects. So, there is no ladder of evolutionary development here, with human emotional plasticity at the top. Successful adaptations are always relative, and in a static environment where tiny nuclear kin groups survive best, it's easy to see how tighter and more exclusive social bonds would have been selected for (including the affective roots of loyalty). That was not, however, the case for humans. Of course, our emotional cultures are still evolving, but the primate precedents are simply too diverse to leverage any predictions or wagers about our own social future.

5

The Ontogeny
of Social Intelligence

IN THIS CHAPTER, we describe the ontogeny of social intelligence through the infant-primary caregiver relationship.[1] Our argument draws from research on the adaptive co-evolutionary nature of cultural learning and the developmental impact of early experience. The infant-caregiver relationship is a critical aspect of the ontogeny of social intelligence in humans. It plays a constitutive role in defining the capacities necessary for appropriate social interaction; for example, the accurate interpretation of emotional information transmitted through nonverbal social cues. In this way, the infant-primary caregiver relationship is a crucial process of enculturation—of how affective behaviors are calibrated and developed to serve socially adaptive ends.

We also consider the relationship between childhood development and the evolution of emotional plasticity and sensitivity. It has become commonplace to think of plasticity as an intrinsic virtue of the mind-brain, but we need to locate it in the larger context of adaptive strategies, including in non-plastic emotional and cognitive styles.

Finally, we employ recent findings regarding the ontogeny of social emotions to throw light on a longstanding problem in the philosophy and psychology of emotions: How do emotions come to have

122

the objects they have? Our discussion of the ontogeny of social emotions provides a prototype for the intentionality of emotions.

Reframing the cause of behavioral and cultural modernity toward painting a picture of the affective roots of culture and cognition in humans requires that we integrate our knowledge of how social structures and culture evolved with an appropriately pitched interpretation of the social emotions and changes in neuroanatomical networks that facilitated greater cultural learning. We believe archaeological data from modern humans has been overly interpreted in terms of cognitive explanations, while the record itself also supports the possibility of a gradual build of social and cultural adaptations.[2]

In an interpretation of the archaeological data that favors social and cultural causality, one would expect to see unique aspects of human development that integrate greater volumes of sociocultural information into fundamental affective motivational systems. We believe recent research on the ontogeny of emotions, namely apprentice relationships, mimicry, and the evolution of cooperation provide guidance for rethinking a causal story of what makes humans and human culture unique.[3]

Important elements of our social nature, like trust, were probably assembled in the late Pliocene or early Pleistocene period. If "prosocial, affiliative emotions generated by joint success" are an integral part of trust in hominin cooperation, then the cognitive element of trust began "as an exaptation of an ancient psychological mechanism."[4] Indeed, "the evolution of prosocial emotions in hominins did not begin from a zero baseline" but rather, "[r]elevant changes in motivation and affect were built incrementally."[5]

In this chapter, we describe how cultural input into the developing mind influences an individual's subsequent socio-emotional capabilities. Below we focus on one example of a prosocial mechanism: how social and emotional information from a primary caregiver (PCG) informs the response thresholds and regulatory control parameters of emotional systems in the developing infant brain to produce an adaptive form of social intelligence. The infant-PCG relationship is a paradigmatic process in the extrapolation of social intelligence to human cultural norms because such a relationship is necessary for any human infant's survival. Furthermore, this relationship forms a locus where basic affective and motivational systems ratchet up toward systems for cultural learning. In this form of intelligence, perception, affect, and prior experience interact with social affordances to produce context-appropriate motivations and behaviors.[6] This type of social intelligence is present not only in humans but also in our primate relatives.[7] *Homo sapiens* emerged with some of this social intelligence already pre-adapted from our ancestors.

Phylogenetic Prerequisites for Social Intelligence in Humans

Several neuroanatomical changes paved the way for the unique ontogeny of social intelligence in humans. Neurological evidence suggests affective abilities develop and "come online" before frontal cognitive abilities, suggesting the crucial nature of developing affective motivational circuits within the socializing process. The human brain undergoes a massive expansion in the first two years after birth wherein the total volume of the brain increases by 101 percent in the first year, and a further 15 percent in the second.[8] That amounts to 40,000 new synaptic connections per second during the postnatal period. Within this expansion, the volume of the limbic brain and brain stem increase by 130 percent in the first year and 14 percent in the second.[9] Changes in the limbic regions are pertinent to our argument as they house the neurological systems of affective motivation in mammals.[10] Notably, the human brain growth spurt is right-hemisphere lateralized, and the right hemisphere is dominant in prelinguistic (i.e., non-verbal) communication abilities.[11] The right hemisphere disproportionately stores and processes emotional experiences during early development.[12]

Although we share basic emotional processes with other mammals, significant postnatal brain expansion—a rapid increase in volume coupled with dynamic changes in cerebral tissues—appears to be uniquely human.[13] Crucially, these changes do not happen in our closest living relative, chimpanzees. Indeed, "a dynamic reorganization of cerebral tissues of the brain during early infancy, driven mainly by enhancement of neuronal connectivity, is likely to have emerged in the human lineage after the split between humans and chimpanzees and to have promoted the increase in brain volume in humans."[14]

In addition, human brains are comparatively less developed at birth than those of our primate cousins; in relative terms, humans are born *earlier*.[15] By one estimate, humans would have to undergo an eighteen-to twenty-one-month gestation period (as opposed to the usual nine) to have the neurological and cognitive development of chimpanzees at birth.[16] According to "exterogestation," a human infant functions as a fetus for a significant amount of time after it is born.[17]

These phylogenetic neurological factors, neoteny, and an expanding brain, constitute the endogenous factors in the development of social intelligence. Our brains, greatly underdeveloped at birth, must undergo significant development *after* parturition. The bulk of neurological development occurs first in limbic fields, notably the hippocampus, yet complete and utter dependence at birth makes humans particularly sensitive to social and environmental information and events.[18] Because we are born into a sociocultural environment, significant brain developments—such as the burgeoning of affective systems in the limbic brain—happen under the direct influence of a sociocultural environment. In fact, Portmann claims this dependence and sensitivity is a prerequisite for social learning.[19] While early birth and extended childhood in a cultural environment are necessary, they are not sufficient for the development of social intelligence. We argue that affective processes are essential in the socialization that allows the collection and reception of cultural information.[20]

Hominin culture harnesses the affective foundations of the primate brain toward malleable, socially receptive behaviors, communicative abilities, and motivations.[21] The "information-rich, expertise-dependent forager lifestyles" of humans and their ancestors "depended on (an) organized learning environment and specific adaptations for social learning."[22] Accumulating cognitive capital in our expanding social niche ratcheted up social-learning capacities. Ultimately, cross-generational cultural learning expanded and became "a core cause of the increasing phenotypic differences between humans (ancient and modern) and great apes."[23] Changes in hominid brain morphology co-evolved with a changing social niche—including changes in reproductive cooperation, kinship structures, length of childhood, and resource cooperation in the Pleistocene era.[24] Some argue that brain growth occurred because of selection pressures exerted in an increasingly complex social niche.[25]

Such complex fundamentals of social dynamics as dominance hierarchies and coalition formation certainly influenced the development of the mind; but in addition to so-called Machiavellian motivations (i.e., the ability to deceive), we must extend our discussion to adaptive species-specific mechanisms for enculturation, such as clear communication and the ability to learn. For example, Herrmann et al. describes how the infant mind is intensely attuned to social stimuli; the authors explain that a series of tests administered to 2.5-year-old humans, chimpanzees, and orangutans demonstrates that humans perform at approximately the same level as chimps in tasks that test for physical cognition, but humans surpass both in tasks of social cognition.[26]

But how might a changing social niche have enabled an accentuated attentional capacity for social information? Our answer builds upon arguments developed in the Chapter 4 on how innate affective systems serve in symbiotic changes between primate brains and cultural evolution. Through the concepts of social intelligence and biological intentionality, we have emphasized how social forces channel affect to produce adaptations like social learning without a reliance on Machiavellian intelligence, modularity, or language. We continue this argument by introducing the concept of emotional intentionality as an integration of prelinguistic intelligence within social enculturation and the eliciting lubricant of affective processes.

Importantly, it is an expanding *emotional* brain in a social environment that materializes the unique adaptations of human social learning, as evidenced by our heavy reliance on emotions to interpret nonverbal interpersonal interaction, such as facial expression and bodily movement.[27] Changes in social niche made it possible for caregivers to give infants a greater level and higher quality of attention during a developmental window in which foundational parts of the emotional brain undergo expansion and neuroplasticity programming. Obtaining food and providing safety for the infant became less time-consuming as complex social cooperation strategies increased. Such strategies included reliable pair-bonding, shared food-processing, enriched environments, and longer, safer childhoods, all which provided more fruitful opportunities for social learning through parenting, play behaviors, and apprenticing.[28] Increasingly, complex social relationships and cooperative modes of resource provisioning have allowed cultural input to actualize the latent socio-emotional capacities of the human brain that are necessary for adaptive social and cultural participation. To accomplish such a feat, cultural processes and brain expansion converged to actualize social intelligence by integrating high loads of social and cultural information into fundamental emotional systems in the brain. The resulting emotional domestication (e.g., reduced aggression, increased patience, directed salience, and impulse control) eventually facilitated cognitive processing, when the quality and quantity of information was increased by the symbol revolution of the Upper Paleolithic. The examples discussed below illustrate how the cultural dynamics of the infant-caregiver relationship guided the development of social intelligence necessary for adaptive enculturation.

Research presented in *Evolution, Early Experience and Human Development* emphasizes that a positive, nurturing, attentive, and prosocial relationship between an infant and its caregiver in early life plays a crucial role in optimizing not just emotion and stress systems but also foundational capacities for nonlinguistic social intelligence more specifically.[29] There is a wide overlap between fundamental motivational systems and social intelligence (i.e., the expression, experience, and self-regulation of emotion and stress as components of optimal social functioning), as distinct from more cognitive forms of social intelligence like IQ, linguistic ability, logic and reasoning skills, theory of mind (TOM), conscious self-reflection, and abstract and conceptual thinking.

Clinical data suggests orbitofrontal cortical deficits and insufficient infant-caregiver attachment impact the content amplitude and saliency of the social information available to a person. To be more specific, the skills adversely affected include pairing facial expressions with emotions, gaining social information through speech prosody and social gesturing, being adept at navigating interpersonal interactions, and maintaining social networks—all important skills for navigating the social world. This is not to say that persons with adverse infant-caregiver experiences, right orbitofrontal cortical deficits, or mental illness are not intelligent. To the contrary, persons in all three categories can, and often do, portray typical capacities for conceptual intelligence. But context-appropriate behavior, the ability to maintain and participate in broad social networks, and the capacity to interpret and identify subtle, context-dependent social cues consistent with the expectations of others are integral to social intelligence while not being dependent on language or higher-order cognition.[30]

The ontogenetic sketch presented here is not reviving misguided theories like the *refrigerator parent*—the widely discredited theory that autism is caused by cold mothering.[31] Rather, research into both abnormal psychology and the ontogeny of emotional systems paints a picture in which fundamental deficits of mental illness may be characterized as "atypical socio-emotional reactions and compromised social behaviors."[32] These compromises are linked to vulnerabilities during important periods in the development of stress and emotional systems, including, in one important example, the infant-caregiver attachment relationship and, as elaborated above, its influence on the broad multifunctional network of the right orbitofrontal cortex and its affective feeder circuits. In this way, infant-caregiver attachment is a critical part of a human ontogeny of social intelligence and cultural learning.

The Infant-Caregiver Relationship

Recent research demonstrates the fundamental impact of early life experiences in optimizing the development of emotion and stress-response systems in the human brain. We know, for example, that nonlinguistic positive social interactions (e.g., touch) are translated into "psychological experience related to a sense of safety and a physiological response strategy related to an improved capacity to cope with uncertainty."[33] In this context, the quality of caregiving rendered to an infant child is presented as both a bellwether and a behavioral manifestation of the above descriptions of affective and social neuroscience, developmental neurobiology, and genetics. The optimal development of affective motivational systems is a part of an evolutionary context of "evolved, expected care": a series of experiences anticipated by the brain that are necessary in order to actualize a prosocial orientation.[34] Natural childbirth, breastfeeding, physical closeness, affective touch, social bonding, prompt and caring response, play, and multiple alloparents are optimizing forces in the development of social and emotional systems in human brains.

In the late 1950s, John Bowlby began articulating attachment theory.[35] Bowlby argued that infants are born with a psychobiological need to turn to their caregiver when experiencing fear or stress. We are born primed for intimacy, and that bond is forged with touch, eye contact, vocal soothing, and general attentiveness of the caregiver to the needs and emotional states of the child. Infants continue to explore their environment when they sense that their caregiver is nearby and vigilant. When the mother or father provides this safe haven, adaptive attachment occurs, otherwise known as *secure base* attachment. However, this secure bonding is often disrupted or fails to develop, and this leads to less successful social navigation in later life.

Mary Ainsworth and colleagues turned Bowlby's model into an empirical research program, observing infants and mothers as they related to each other in the Strange Situation experiments. In Strange Situation, a baby and mother are introduced to an unfamiliar woman. The two women talk for a while and then the mother leaves the room, returning after a three-minute interval. Over many trials, and with subjects from diverse racial, cultural, and class backgrounds, results confirmed Bowlby's model of secure attachment behaviors and feelings. Secure children seek closeness with their caregiver and show distress when separated from her. The two other patterns that clearly emerged were that some children showed extreme stress behaviors when separated and were difficult to console even when the caregiver returned. These children, who required constant attention, were designated to have an *anxious style* of attachment. Another group of children showed remarkable detachment during the Strange Situation— they were not distressed by separation from the caregiver and treated the stranger and the caregiver similarly. This third group was designated to have *avoidant style*.[36]

It is clear that the infant-caregiver attachment relationship has a crucial impact on developing stress- and emotional-response systems and should be viewed as part of the environment of evolutionary adaptedness.[37] The primary caregiver unconsciously sculpts or regulates the infant's behaviors and, subsequently, her developing limbic system, making way for the child to successfully navigate the rapidly changing social environment.[38]

In the decades since the work of Bowlby and Ainsworth, the theory of attachment styles has been corroborated in a wide range of studies, and as we discuss below, neuroscience has begun to elucidate and localize the proximate chemical processes beneath the attachment styles.[39]

The presence of generally positive, nurturing, attuned, and prosocial caregivers plays an optimizing role in infant development by triggering endogenous opioids and oxytocin during prosocial experiences. For example, the infant-caregiver attachment relationship shapes the sensitivity of stress response in the hypothalamic-pituitary-adrenal axis through developmental processes.[40] The hormones oxytocin and prolactin mediate maternal care and infant-bonding in a brain-based system that is "foundational for key forms of human prosocial attitudes, from the establishment of a secure base for emotional maturation, to the capacity and enthusiasm for sustained playful-friendly interactions, to expertise with and refinement of all the prosocial capacities of our species."[41] Indeed, there is evidence that orphans with a history of neglect during early experience lack brain-based oxytocin spikes typically associated with comforting parenting behaviors.[42]

As we discussed in Chapter 4, oxytocin has been correlated with many prosocial behaviors. For instance, intranasally administered oxytocin increases a subject's ratings of faces as attractive and trustworthy.[43] Administering oxytocin also increases trust when subjects are playing social trust games.[44] And the ability to read emotional information from faces is increased or enhanced by oxytocin.[45]

However, clinical work reveals that oxytocin is not a sufficient cause of prosocial feeling or behavior. In subjects with Borderline Personality Disorder (BPD), for example, an increase in oxytocin actually intensifies negative feelings and decreases trust and cooperation.[46] One explanation for this seeming inconsistency is that BPD and healthy individuals respond differently to the feelings provoked by oxytocin. "For healthy individuals, feelings of closeness and intimacy associated with oxytocin are generally seen as positive. However, individuals with BPD may view the same feelings of closeness as threatening, thus experiencing decrease in trust and collaboration after receiving oxytocin."[47] Unsurprising, perhaps, is the fact that *disorganized* infant attachment (a result of unpredictable sequencing in parent rewards and punishments) is a reliable predictor of later BPD symptoms.[48]

A review of neurobiological profiles underlying the different attachment styles shows that subcortical limbic areas are heavily involved in social *approach* behavior, while cortical limbic areas are more dominant in social *aversion* behavior and attitude.[49] Moreover, the amygdala is hypersensitive in anxiously attached subjects, while it is hyposensitive in avoidant-attached subjects—suggesting that anxiously attached individuals have heightened reactivity to emotionally salient social cues. Avoidant-attached individuals, on the other hand, may have an intrinsically decreased activation of prosocial affect, or they may have developed strategies for down-regulating their own emotional experience (possibly as a self-protective buffer against socially initiated negative affect).

A striking indication of the power of healthy caregiver activity can be seen in the ability of good parenting to counteract genetic sources of personality disorder. A disorder of emotional regulation (like BPD) has been correlated with a genetic variation (viz. a polymorphism in the serotonin transporter gene 5-HTTLPR). But research reveals that adolescents with secure attachment styles did not experience the disorder, despite having the genetic basis for it, leading to the inference that healthy bonding and secure attachment prevented the expression of this form of emotional disorder; whereas, when the genetic trait was correlated with insecure attachment temperaments, personality disorders followed.[50]

The close relationship between endogenous oxytocin levels and the early social environment of the child has led many researchers to recognize the oxytocin system as a plastic and adaptive interface of nature and nurture.[51] "Oxytocin plays an important role in priming mammals to form social bonds, but in turn, the early social environment may also be able to shape the development of the oxytocin system."[52] Research by Larry Young on monogamous and family-devoted prairie voles has shed light on the evolutionarily conserved system of bonding through oxytocin. Prairie voles, compared with their cousins the meadow voles, have a higher density of oxytocin receptors (in associational regions of the brain) and this correlates with high degrees of social bonding. It also may explain caring empathy, like consolation behaviors toward injured or stressed partners, and so on.[53] The connection between secure attachment in mammals and later-life social functioning is enough to motivate more inquiry into the chemical origins of bonding, but we are also interested in how this adaptive brain–body loop gives meaning, color, and texture to the human mind.

The infant-caregiver relationship regulates the human brain's growth spurt in a way that impacts the baseline expression, experience, and self-regulation of emotion and stress for adult life.[54] Expansion of the right hemisphere in the first two years after birth typically happens under the direct regulation of a caregiver. Lacking language, the communication between infant and caregiver during this time is primarily physical and behavioral. Infants absorb and process social information from the body language, voice, eye movement, and touch of the caregiver.[55] Significant right-hemisphere attachment episodes that are related to visual facial, auditory-prosodic, and tactile interactions with the caregiver occur during this time. This physical and emotionally laden information is modulated, regulated by, and processed through the affective systems of the infant's burgeoning right brain.[56] These social schemas can take the form of the social intelligence complex discussed in Chapter 3 and may therefore rely on conditioned secondary-process affective mechanisms. Neuroimaging data reveal that parent-child emotional communication involves participation of right-hemispheric cortical and subcortical systems.[57] Moreover, stimulating the plastic system properly guides the definition and actualization of baseline socio-emotional action tendencies and psychological potentials in the brain. The brain wires into prosocial pathways, like attachment styles, under the nurturing influence of caregiver stimulation.

Foundational socio-emotional action tendencies and psychological potentials result from a basic system of emotional and stress responses set by three integrated circuits of the brain that come online in the first two years of life under the influence of the caregiver. A complex of specific emotional circuits in the limbic brain develops connectivity and activates during the first year. First, lower circuits of this limbic complex harness the major motivational systems that "generate autonomic expressions and arousal intensities for all emotional states."[58] Second, the higher orbitofrontal-levels of this limbic complex are responsible for generating "a conscious emotional state that expresses the affective output of these motivational systems."[59] And finally, at the end of the second year, the orbitofrontal cortex matures on top of this system in the right hemisphere, forming the highest executive-control mechanism for emotional behavior.[60] The location of the orbital prefrontal cortex facilitates quick connections with the insular cortex, anterior cingulate, and amygdala—in short, forming a convergence zone between cortical and limbic activity. The wiring of this brain area undergoes major maturational growth during the second year of the infant's life—the same time that the attachment control system (activated earlier by external social interaction with caregivers) is also maturing and solidifying.[61] For the remainder of adult life, the right lateralized prefrontal region is largely responsible for regulating affect and stress.[62]

Orbitofrontal cortex maturation, under positive, prosocial, nurturing, and attentive care from the caregiver, is crucial for optimizing social functioning, as seen in the following discussion of dysfunctional relationships.

Dysfunctions in Social Intelligence as a Result of Inappropriate Infant-Caregiver Relations

If the regulating system for prosocial emotional management is the orbitofrontal cortex, and if that regulation system is optimally wired by nurturing, then we should expect to see altered wiring or gross anatomical change in the orbitofrontal cortex of abused or neglected mammals. A recent study of male rats performed by Marquez reveals this correlation: young rats were exposed to a variety of stressful situations, and these rats were subsequently more aggressive as adults.[63] Brain scans further revealed reduced activity in the orbitofrontal cortex. Behavioral geneticist Carmen Sandi explains, "In a challenging social situation, the orbitofrontal cortex of a healthy individual is activated in order to inhibit aggressive impulses and to maintain normal interactions, but in the rats we studied, we noticed that there was very little activation of the orbitofrontal cortex. This, in turn, reduces their ability to moderate their negative impulses."[64] In conjunction with changes in orbitofrontal cortex functionality, overactivation of the amygdala, which suggests intensified emotional activation, was observed. Of course, violent adult humans have been known to have experienced abusive childhoods; less well known is that violent offenders demonstrate reduced orbitofrontal activity and / or mass.[65]

Studies on adult patients with right orbitofrontal cortex deficits demonstrate that this region of the brain influences aspects of social behavior.[66] The right orbitofrontal cortex is critically involved in what has been termed *adaptive interpersonal behavior,* and damage to this area has been linked to behavior with strangers that is more appropriate for non-strangers.[67] Orbitofrontal damage causes inappropriate, gregarious behavior, which is thought to be a deficit in adaptive interpersonal behavior rather than an overactivation of social communicative systems.[68] For example, damage has been linked to the tendency to greet strangers in an overly familiar manner.[69] Furthermore, impaired interpersonal behavior stemming from orbitofrontal cortical damage is correlated with emotional deficits.[70] Lastly, impairment in social functioning is among the hallmark criteria for schizophrenia, and the schizophrenic brain exhibits decreased volume in the orbitofrontal cortex.[71] Such dysfunctions in social decision-making are demonstrated in both schizophrenics and orbitofrontal-damaged patients.[72] As an illustration of the consequences of these social and emotional disabilities, consider a 2011 study of schizophrenic and schizoaffective disorder patients, which found that the average number of friends per patient was only 1.57 and that approximately 48 percent of the patients reported having no friends at all.[73]

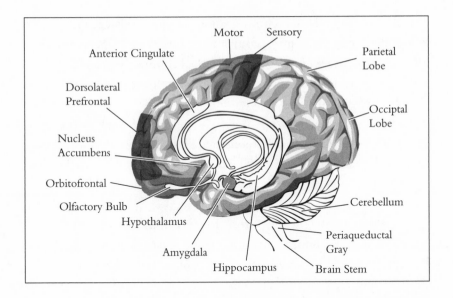

Both the quality of the infant-caregiver relationship and orbito-frontal cortical deficits have been linked to impaired nonverbal social communication in an infant's first two years of life. As reported by Carbone, O'Brien, Sweeney-Kerwin, and Albert, eye contact serves an important social function for young children even before vocal responding begins to develop.[74] In early development, it serves to regulate face-to-face social interactions and contributes communicatively to social interactions.[75] Later, eye contact responses coordinate the visual attention between another individual and an object of interest and have been found to be an influencing variable in language acquisition.[76]

Decreased visual interaction from a caregiver during infancy has been linked to impaired abilities for visual-social communication in children raised in institutional environments; for example, orphans are less adept at pairing facial expressions with emotions.[77] It is known that "[t]he infant's early maturing right hemisphere . . . is dominant for the child's processing of visual emotional information, the infant's recognition of the mother's face, and the perception of arousal inducing maternal facial expressions."[78]

Parker and Nelson support the view that compromised infant-caregiver interaction during infancy serves to limit the information available through visual communication for a person later in life.[79] The signal of eye contact requires complex interpretation by perceptual and affective systems that manage interpersonal communications and allow a person to navigate the social world. Failure to recognize the emotional content of such eye contact stems from a viewer's limited ability to grasp a social interaction (i.e., a social affordance) in the most comprehensive manner, as demonstrated by those on the autism spectrum and, morphologically, by the decreased white matter in the right orbitofrontal cortex, which is associated with the social deficits of autism.[80]

The DSM-IV diagnostic criteria for autism includes a set of symptoms that arises from impairment in the use of *nonverbal communication*—including eye-to-eye gaze, facial expression and recognition, body posture, and regulatory social gesturing—as well as from deficits in social reciprocity, which includes things like poor speech prosody and difficulty expressing emotions.[81] Indeed, morphological and functional development of the orbitofrontal cortex has been implicated in the following social disorders: affective psychosis, bipolarity, personality disorder, aggression and violence, alcohol and drug addiction, panic disorder, and depression.[82]

Social Affordances and Enculturation

The emotional and stress systems of the early developing brain are forged under the influence of early experiences, and they play a fundamental role in creating or inhibiting prosocial orientations in the individual. A "prosocial orientation" includes, in part, an ability to experience and interpret behavioral social cues in a way that is continuous with the customary understanding of others and in a way that is continuous with the normative environment of cooperation. Negative or inappropriate attention in the infant-caregiver relationship, as well as deficits in the orbitofrontal cortex, can cause deviation from registering social cues like eye contact, speech prosody, and interpersonal interaction. Consider Temple Grandin, the acclaimed author who is autistic, describing typical and atypical experiences of eye contact:

What a neurotypical person feels when someone won't make eye contact might be what a person with autism feels when someone *does* make eye contact. . . . For a person with autism who is trying to navigate a social situation, welcoming cues from a neurotypical might be interpreted as aversive cues. Up is down, and down is up.[83]

The three parts of our social intelligence model, articulated in Chapter 3, are exemplified in Grandin's scenario, as she explains how emotional information gleaned through perception informs possibilities for social action. First, physical perception of a social affordance occurs when an individual perceives a social cue—eye contact is the social cue in this example, but such a cue could also include social affordances like observing a bodily gesture or hearing a vocal intonation. Second, this perception is processed in the emotional and affective networks of the brain in some form analogous to the somatic-marker hypothesis (generally localized in the orbitofrontal cortex) or the "vertical association" development of mirror neurons.[84] These plastic networks have certain parameters, thresholds, and sensitivities forged under the influences of early experience via conditioned associative mnemonic processes. The perception will be colored with an affective hue to orient a cue to a behavioral response in the context of conative drives. The result, as Grandin describes, is what the person "feels" in the broader context of "trying to navigate a social situation" (i.e., the implicit and explicit affective tugs on the options of actions available to the perceiver). The affective valence of the somatic marker generated by the affordance percept of eye contact pushes or pulls the viewer in relation to one or more action-response tendencies. In this way, active perception of nonverbal social cues has an affective dimension; it provides possibilities for motivated social action as social affordances.

As Grandin states, eye contact may be interpreted as either a signal of "welcoming" or one of "aversion." What she is describing, of course, functions at implicit levels in individuals without compromised social intelligence. The developmental impact of early experience and the infant-caregiver relationship play a crucial role in orienting an individual's social behavior. In this example, these developmental processes are guiding factors in determining whether a person will interpret eye contact in a way that is consistent with the meaning of the conspecific's gesture. Social interaction thus requires a set of skills acquired and calibrated during our social and emotional development; namely, eye contact, social pain, play behaviors, etc. Acquiring these social skills is our earliest appropriation of norms and values. In the broader evolutionary story, the ontogeny of human social intelligence through the infant–caregiver relationship is an example of the way that cultural forces adaptively shape affective potentials.

The cultural adaptation of affective potentials is not uniquely human; as we discussed in Chapter 4, it is present across the great apes as well and should also be seen as a contributing factor in their social interaction. The type of social intelligence produced under the ontogenetic process of the infant–caregiver relationship is only a case-specific example of how emotional and affective capacities co-evolve with manifold varieties of sociocultural scaffolding, along reciprocal adaptive trajectories.[85] It may be that our affective capacities are elaborated through specific cultural innovations, resulting in unique sociocultural architectures that subsequently feed back into phylogenetic differentiation.

The infant–caregiver dynamic is an adaptive enculturation process because it serves as a means for realizing the social and emotional intelligence potentials of the human brain and for mapping such potential onto the stream of cultural information. It does this mapping in a way that tends to coordinate social interaction and confirm consistency in the meaning of nonlinguistic communication. Cooperative cultural changes in the human lineage have allowed for an increased intensity of attention between the infant and caregiver and a more enriched environment for the infant to explore. In that space, the unique relationship between infant and caregiver has now *itself* become a cultural process serving adaptive ends.

Plastic and Non-Plastic Survival Strategies

Now we shall connect our ideas concerning social intelligence and the infant–PCG relationship with neo-Darwinist principles through the concept of plasticity. We know that populations that adapt well to environmental changes survive. In the case that adapted trait options come via heritable genetic mutations only, populations will adapt slowly over many generations. If the organism has the ability to forward alternative trait options using a means beyond simple genetic mutation (say, via learning, or conditioning), then the population will increase its ability to respond to new environments. Conditioning brings significant flexibility to mammals, and cognitive sophistication and memory add to the palette of responsiveness for primate clades. In general, these advantages underscore the importance of plasticity for evolutionary success.

Recent understanding of phenotypic plasticity (genetic flexibility in response to environmental change), neuroplasticity (axonal sprouting, synaptic pruning, cross-modal reassignment, etc.), and epigenetics (heritable gene-expression switching) have restored developmental biology to a place of pride after a long twentieth-century romance with molecular biology. The important role of the environment (pre- and postpartum) in shaping the body and brain of the human ape is seemingly beyond doubt. The ability of a flexible operating system that can learn from experience and recalibrate deep physiological pathways (such as metabolism rates) is evidenced by well-known neuroscience findings of post-injury brain compensation, the epigenetics of obesity, and so on.[86]

But, paradoxically, it is not always good to be flexible and changeable. Sometimes the best evolutionary strategy is stubborn inflexibility. In the *Origin of Species* (1859), Darwin noted how the brachiopod genus *Lingula* has remained largely unchanged for 500 million years.[87] Its remarkable sameness reflects the stable conditions of the warm and shallow seas it inhabits. If an animal's environment is very stable, the generational mutations are stabilized and the population preserves average traits, rather than extreme differences. There are, for example, many orders of Crustacea for which phylogeny can be traced back 400 million years, with little or no discernable change.

Humans, of course, are not crustaceans, and our ancestors evolved in a fluctuating environment, with radical climate change, food-source variations, and social instability. Humans became very flexible responders to novel environmental challenges. We became good improvisers. As Daniel Dennett suggests, one of the main functions of the representational mind is to "produce future" by developing an internal model of the world in order to make predictions about the external world—the "tower of generate and test."[88] We have been arguing that emotional plasticity preceded representational mind and gave us behavioral options that replaced more limited hardwired instincts, and, as such, that plasticity is a large element in our control over adaptive opportunities. But there are no absolute or intrinsic advantages in evolution, and if plasticity is a property of systems, then it too is susceptible to selection.[89] Even in humans, we should expect to find some variation in the degrees of mental and behavioral plasticity.

If we consider the context of childhood development, then we can appreciate the pros and cons of early plasticity. By analogy, just as the *Lingula* remained phylogenetically the same—to its great advantage— we can consider the benefits to a human organism that stays the same under environmental changes. Can a child, for example, have too much receptive plasticity?

Applying an evolutionary perspective to childhood development, Jay Belsky proposed that individuals should vary in their susceptibility to environmental influences, especially in the "critical period" of the rearing environment.[90] Our exploration of attachment styles above revealed that infants come into the social world primed for bonding and for the acquisition of social and emotional intelligence. Humans are not so determined as birds and ungulates, which imprint mechanically, but we are still governed by chemical triggers and requisite experiences. In humans, the juvenile must acquire its caregiver target, but it must also acquire substantial amounts of information about its environment. A seemingly simple environmental mismatch between the organism's body and the external world could be ruinous; so, too, an environmental mismatch between the mind and the world can be deleterious. The developing mind must be very sensitive to the social environment—parental influence, in particular—in order to limn it properly, but there are costs to such sensitivity (i.e., plasticity).

There are energy costs associated with possessing sensory sophistication, regulatory mechanisms, and processing power; indeed, having a highly plastic system that is fine-tuned for receiving and simulating environmental complexity costs neural energy. In contrast, less flexible individuals have less of this energy expenditure and may benefit in other ways from lower receptivity to parental influence. That is, plasticity is not an unmitigated good but rather "a trade-off that confers both costs and benefits to the organism."[91]

Some people are more susceptible to and thus more apt to be shaped by their original environment ("plastic strategists," following Belsky), while others are less affected by their early experiences ("fixed strategists"). This difference has historically been interpreted in purely clinical terms. The clinical interpretation suggests that some children have a dispositional vulnerability to childhood stress and adverse experience, while others display resilience to adversity. Reexamining this distinction in light of adaptation and evolution suggests a more interesting theory than the usual dichotomy of normal versus pathological adversity response.

The ideal plasticity of a developing mind would be highly sensitive to the emotional information available in its original environment, allowing the child to master the relevant features of the proximate milieu. But this only works well if the original environment is positive and supportive. When parental or caregiver discord is violent, or the child is maltreated, or the general environment is abusive, then the more plastic or sensitive the child is, the more damaging the effects on her subsequent personality.

Resilience is an affective style that can be measured by the rapidity of recovery time from adversity.[92] According to fMRI data, the resilient person shows a more rapid down-regulation of negative affect (observable in the interaction of the amygdala and ventromedial prefrontal cortex) after seeing painful images, whereas the non-resilient person has much slower restoration of positive affect (and cortisol homeostasis) after negative stimuli.

Different affective styles or emotional-intelligence strategies may act as natural variations upon which natural selection can do its work. The clinical literature underscores a small normative palette of prosocial personalities on the one hand and many disorders on the other. But some temperaments (e.g., fixed insensitive strategy) may be better suited for some early rearing environments (e.g., non-supportive, volatile households).[93] Children with more sensitive or plastic developmental temperaments can benefit greatly from supportive caregiver environments, but they will also be more adversely affected by negative or abusive caregiver experiences. They are disproportionately advantaged because they can absorb high volumes of social information with high fidelity, compared with the less-penetrable style of the robust but fixed strategy.

Highly sensitive adults make up approximately 20 percent of the population, as measured by their nervous-system sensitivity, awareness of environmental subtleties, and tendency to be overwhelmed in high-stimulus surroundings. Adults with highly sensitive personalities who also had high-stress childhoods, experience high degrees of shyness and negative affect as adults, whereas sensitive adults who had nurturing, prosocial childhoods reported no such tendency toward shyness and negative affect.[94]

Interestingly, the calibration of the individual's affective style and social intelligence is a combination of caregiver experiences and genetic predispositions. Two genetic markers are currently linked with emotional sensitivity: one is the serotonin-transporter-linked polymorphic region 5-HTTLPR, and the other is the dopamine receptor DRD4. The serotonin transporter gene helps regulate the effects of serotonin in the brain by adjusting the reuptake of serotonin from the synaptic cleft. Studies found that a mother carrying at least one copy of the short allele of HTTLPR was more susceptible and sensitive to her marriage dynamic. If the marriage was without much conflict, the mother was a mild-mannered, non-punitive parent. But if the marriage contained conflict, then the short-allele mother was much more harsh and punitive as a parent. By contrast, mothers without the short allele were not particularly susceptible or sensitive to the emotional environment of the marriage (suggesting a fixed, resilient temperament). The short allele of HTTLPR thus appears to be best characterized as a *sensitivity amplifier*.[95]

In addition, a specific variant of the dopamine receptor DRD4 has been associated with the high-novelty-seeking behavior demonstrated in ADHD patients. "As it turns out, a number of studies indicate that children carrying this putative-risk allele are not only more adversely affected by poorer quality parenting than other children, but also benefit more than others from good-quality rearing."[96]

There are several takeaways from this work. One important lesson regards the complex interaction between genetic and environmental causal pathways that results in the emotionally calibrated child. The affective system is not just turned toward relevant objects, say alloparents, but is also dialed in or tuned in to different levels of sensitivity. The original family environment acts as an emotional amplifier for positive or negative social affects. But some infants are biologically predisposed to more sensitive plasticity and, therefore, to more sensitive acquisition of social information and affect—for good and for ill.

Recent empirical research on childhood stress adaptation confirms the strong influence of adversity on a child's value system and behavioral strategies. But the resulting strategies are not all impairments. Some of them are highly adaptive in certain kinds of social environments. Gillian Pepper and Daniel Nettle have argued that a behavioral constellation of deprivations in childhood (e.g., poverty, malnutrition, abuse, stress, etc.) creates a value system and a behavioral style that pursue short-term goals and satisfaction, because of the learned affect that delayed rewards will never come due.[97] Nonetheless, Willem Frankenhuis and Carolina de Weerth discovered certain cognitive enhancements in people who developed in stressful situations. Against the common wisdom that children from violent households are cognitively impaired, Frankenhuis and Weerth found that "these people may exhibit improved detection, learning, and memory on tasks involving stimuli that are ecologically relevant to them (e.g., dangers)."[98] It may be the case a child raised in a stressful environment develops specialized knowledge and predictive powers regarding that specific environment.

We can see that the ontogeny of social intelligence does not unroll along a single successful path (e.g., secure attachment) or a forest of divergent paths (e.g., insecure attachments and personality disorders). The evolutionary framework complicates this normative map and reminds us that adaptive emotions (and adaptive affective styles) may not always correspond to modern human-value expectations. Even plasticity itself, which is by-and-large a great boon to the primate mind, might, on occasion, be less preferable than a robust and obstinate emotional style. Our species has benefited dramatically from the level of emotional sensitivity that produced the likes of novelist Marcel Proust, but such sensitivity can be a deleterious burden under other environmental conditions. Fixity and plasticity in emotional development combine with other factors to produce the prosocial or antisocial person.

Emotional Intentionality Revisited

It is important, now, to revisit the issue of emotional intentionality in the context of childhood development. As we have already argued, emotions are more than just mechanical responses to stimuli, but they are not intellectual appraisals either. Human emotions occupy a unique middle ground between instinctual reactions and deliberative judgments, but they share territory in each of those domains. We argued in Chapter 2 that the biological creature has a broadly intentional structure, which is frequently misconstrued; here we will develop the notion that emotional intentionality is a subset of that biological intentionality. In Chapter 3, we argued that mammalian sensorimotor systems, homeostasis, and affective mechanisms, along with communication abilities, converge into body-world loops. These loops thus comprise a kind of social intelligence that is adaptive, despite the fact that it is not in the animal's head, so to speak. Now, before we start to consider how the linguistic mind interacts with our sketch of the affective roots of the mind in the coming chapters, we can articulate our vision of the prelinguistic intentionality of mammal life. The key concept is that our intentionality orientation, emerging in childhood development, is the formal structure that also codes the content, via salience-marking, of our meaningful world. This early form of emotional intentionality is similar in kind with the way animals direct their affective experiences or even, as we'll discuss in Chapter 6, populate their spatial maps with affective content.

Emotional intentionality is twofold. First, an emotion usually takes an object. I am angry at Huck, for example. Or I am afraid of this approaching dog. Or I am lusting after this particular person. This aspect of emotion—the ability to be about a specific object or target—distinguishes emotions from moods. Secondly, emotions are intentional because they are usually forms of evaluation. An emotion does not simply pick out an object, as visual perception might pick out a character from a background visual field. An emotion takes an evaluative stance toward the object in question.[99] When I am afraid of this dog, I am evaluating this animal as a threat. When I am drawn toward a friend via the affective care system, I am implicitly evaluating the friend as a person to approach rather than to avoid—someone to have contact with rather than evade. The warm relation I have with the friend is, in part, an evaluation of them. And my enemy is also constituted, in part, by the negative emotional evaluation I have of them.

The two aspects of emotional intentionality show that they are in between bodily changes and judgment, while physiological bodily changes are constitutive of the basic emotions.[100] For example, fear is constituted in part when my amygdala triggers the sympathetic nervous system, epinephrine increases my heart rate, norepinephrine increases blood flow to skeletal muscles and releases glucose into the system, and cortisol increases blood sugar and calcium. On the other hand, fear is constituted in the evaluation that this slithering shape, for example, is a threat. There is a discriminatory aspect to emotions.

This has led many philosophers to ask, rightly: How do emotions come to have the objects that they have? We agree that the question is important, but the traditional answers have been misleading and unhelpful. Traditional philosophical approaches to emotions have lacked biological perspective. Deonna and Teroni's *The Emotions: A Philosophical Introduction* is a good example of the problem, but Martha Nussbaum, Colin McGinn, Robert Solomon, and most recently Lisa Feldman Barrett have all suffered from a similar confusion.[101]

It is common for these philosophers to start their analysis of an emotion, say fear or joy, by examining the way humans speak about emotions. A frequent strategy is to start with some oft-heard sentence construction like "Bernard fears the lion" or "Bernard has good reason to be elated if he has just heard from a reliable witness that his wife is in better health."[102] Philosophers turn to language as their principle database for gathering intuitions about emotions. There are historical reasons why they proceed in this fashion, because for a hundred years, philosophers have made this same ordinary-language move for every other topic (e.g., What do we mean by God, by soul, by justice, by science, by art?). Philosophers look at how people use a term like "anger" or "God" in natural language and try to decode the deeper meaning. The linguistic turn in philosophy still generally holds sway in both analytic and Continental traditions.

In the case of emotions, however, this is the wrong starting place. It is wrong for several reasons. First, it guarantees that we will not learn anything about mammalian emotions generally—and that seems misguided since human emotions are a homologous subset of those ancient adaptive resources. Two, it presumes that language will be a mirror of emotional reality, but there's no reason to think this. In fact, there's plenty of reason to think that self-report and indicative linguistic descriptions of conscious life are already highly corrupted by the cognitive biases and agendas of the agents involved. A person using emotion terms like "fear" and "anger" has no privileged access into what fear is, and a person describing their own fear is not necessarily the best source of reliable information about fear either.[103]

The chief reason why language is a bad starting place for understanding emotions is that philosophers immediately become distracted by the search for criteria to designate when emotions are "correct or incorrect." When we begin our analysis of emotions by tracking their appearance in language, it is a but another short step to the place where we ask when emotions are "justified" or "warranted," which can lead to a variety of other questions that treat emotions as a subset of belief or judgment. Deonna and Teroni are symptomatic of this tendency when they move immediately from recognizing the evaluative intentionality of emotions to asking about an emotion's "correctness." "Fear of a dog is an experience of a dog as dangerous, precisely because it consists in feeling the body's readiness to act so as to diminish the dog's likely impact on it (flight, preemptive attack, etc.), and this felt attitude is correct if and only if the dog is dangerous."[104] We agree with the first part of this claim—fear as action-readiness—but we take issue with the verificationist turn that animates so much philosophy of the emotions. Likewise, on their account, "shame" is the feeling of the body as it tries to disappear or conceal itself when it believes it is being degraded, but they assert that this emotion is only *correct* when the emotional person is *really* degraded.

Trying to figure out when an emotion is "correct," without any appeal to evolutionary adaptation, is a dubious approach. We suspect that it stems from a category mistake that (1) incorrectly sees an emotion as a propositional form of evaluation (when other forms of evaluation are, in fact, more primordial; viz., as a value point upon a general approach-or-avoid spectrum); and (2) incorrectly searches for the meaning of such evaluation in the correspondence relation to an objective referent (or referential state of affairs). This latter approach hasn't even paid dividends in the realm of indicative descriptive speech, so why it haunts the philosophy of emotions is hard to fathom— except as a byproduct of an incorrect linguistic model of meaning in the mind.[105]

With no appeal to evolution, these discussions treat emotions as accurate or inaccurate pictures of reality or as appropriate behavioral responses to agents—but with the protocol for appropriateness undefined. How, for example, am I to grasp whether my shame is a correct mark of my degradation, but only when I am *really* degraded? It's hard to know what this could mean, and more importantly, why we should care about the correctness of an emotion, when this seems like the least interesting question about the emotions.

The fascination with the veridical status of emotions betrays the linguistic and even positivist residue in the investigation. This approach looks for a correspondence theory of emotion, wherein the meaning of anger is referenced back to its accuracy in evaluating a perception. And the correctness of the emotion is not just in its correspondence but also its rationality—it's optimal calculation of appropriate response. The more suitable question for the emotions, however, is not their accurate representation of objective traits in the environment and optimal rational responses, but whether they are adaptive or maladaptive. To assume that the most adaptive response is the most rational is to misunderstand the nature of biology. A biological approach to the adaptive nature of emotions has no illusions about the rational optimality of the behavior. Almost nothing in biology (e.g., behaviors, traits, etc.) is "optimal," but is instead "serviceable" given the options available at the time.

Implied in the traditional philosophical approach to emotions is the requirement that emotions be conscious. When the evaluative aspect of emotional intentionality is modeled on language or linguistic thought, then fear or sadness must be raised from being merely bodily to the level of conscious reflection. Appraising a dog as fearsome, on this propositional model, is akin to subsuming a *token* under a conceptual *type*—an intellectual judgment that somehow produces fear as an ensuing secretion of proper categorizing. But there are unconscious or preconscious forms of appraisal too.[106]

Our view is that behavior and action tendencies (potentiality) are much better places to start than language. We suggest that emotions are evaluative to a large degree, but we should not confuse evaluation with representation (and thereby equate it with belief and judgment). Another model of valuation has already emerged out of our discussion of Gibsonian affordances and Millikan-style pushmi-pullyu imperatives. The traditional philosophy of emotions is confusing evaluation with representation and then, along the primrose path, worrying about whether the representation corresponds with objective reality.

We recommend taking the question of the "correctness" of emotions off the table entirely, at least as it is currently articulated. In its place remains the more interesting issue of the meaning of emotions. And here we see that the semantic content of our emotional lives is dictated originally by the unique associational pairings of our mammalian affects and our earliest social environment of caregivers. In a nurturing family environment, positive affect (underwritten by endogenous opioids, oxytocin, dopamine, etc.) will be associated with parents, siblings, and alloparents as "objects" (i.e., recognizable individuals) and with behavioral activities (i.e., touch, feeding, play, etc.). In this way, emotions have intentional objects—specific aboutness. Another way of saying this is that objects, in our world, have an emotional aspect right from the start. The correctness of these intentional objects of the emotions is then difficult to appraise, since (1) context determines how much plasticity is appropriate for the formation of the infant's social-emotional palette and (2) the emotional meaning of a

given intentional object shifts across situations and through the dialectical process of acquiring new associations.

Our biological approach is consistent with that of DeLancey, who
recognized that the intellectualizers of emotions turned emotions into
judgments because they had no other way to frame intentionality.[107]
The intellectualizers argue that anger, for example, is a judgment by the
agent that someone or something has demeaned or disrespected the
agent (and / or the agent's family, etc.), and they assert that this judgment is accompanied by the familiar bodily changes. But if this kind
of judgment is the constitutive aboutness of emotions, then how
could infants and apes have anger?[108] We are led to the dubious conclusion that infants and animals either have secret propositional minds
or they don't have emotions at all. In forwarding a truly "biological
aboutness," we reject this false dichotomy and, subsequently, the traditional judgment model that follows therefrom.

While it is true that human emotions are heavily contoured by judgment ("I am angry because his sarcasm demeaned my wife!"), other
animals, human children, and our *Homo* ancestors aim their basic emotions at specific objects directly, not at representations or elements of
propositions. DeLancey recognizes that biological creatures intentionally direct their anger or fear at a "concretum"—another specific animal,
or thing, or state of affairs.[109] I do not need a propositional judgment
about snakes, or a folk theory about their dangers, combined with a
sense of my own vulnerability, in order to be afraid of the snake. I just
need the feelings of fear directed at the actual approaching snake.
This is the kind of emotion (directed at the snake—a concretum) that
we share with other mammals. Our additional form of emotional intentionality (directed at propositions) is a unique human form of
aboutness; that is to say, an outlier and certainly not the paradigm case.

Our view of emotional intentionality is truly embodied because the
body and its mindbrain mark the objects of the world with constitutive feelings (cf. somatic markers and / or vertical associations). Ontogeny is a fundamental case study because, long before propositional
thinking, the baby associates perceptions of caregivers with positive
or negative affect (including disorganized patterns for inconsistent parenting), stores those percept-affect chunks and schemas in memory,

and begins building (unconsciously and then consciously) behavioral responses to those meaningful experiences. These prelinguistic forms of meaning allow the infant to make implicit predictions or anticipations about future experience and generally allow for an adaptive navigation of the idiosyncratic conditions of early social life. The "emotional lexicon" of this simple social intelligence is admittedly small at first—approach, avoid, fear, anger, care, and so on, but a great deal of prediction and manipulation of alloparents can result when infants and toddlers begin to recognize feeling patterns and associations. As executive control slowly comes online, the emotional management of self and other becomes more sophisticated. We examine this issue more carefully in the coming chapters—namely, how does a small palette of basic emotions or affective systems become the rich diversity of familiar, complex human emotions? Is it through fresh alignments between basic emotions and acquired associations, or some kind of mixing of emotions, or a linguistic / conceptual interaction with emotions, or a cultural managing of emotions, or something else?[110]

A recent study by Steven Frankland and Joshua Greene reveals an interesting feature of emotional meaning, and it is a feature that combines well with our thesis of emotional semantics.[111] The investigators recorded subjects' brains in an fMRI, carefully tracking the region of the brain that correlates with recognition of meaning. In this case the meaning was rudimentary and consisted of being able to identify who or what was an "agent" or actor, and who was a "patient" or receiver of action. For example, if I say, "the car hit the truck," the agent term is "car" and the patient term is "truck." And this meaning is preserved even when I switch word order, as in "the truck was hit by the car."

The experiment suggests that the left mid-superior temporal cortex (lmSTC) is the precise region that encodes "who did what to whom." As a foundational building block of linguistic meaning, we must understand who is doing an action and who / what is receiving that action, and this precise brain area appears to toggle according to this distinction. The brain appears to have a simple dichotomy functor system that receives data (written sentences in this case) and slots the variables into a specific position of "agent" or "patient." This immediately brings some understanding to the sentence, without having an infinite database of sentences to check against. "The dog bit the man" and "the car hit the truck" have the same logical format regarding agent and patient, and both can be computed by this brain architecture of thought. The meaning changes when you simply reverse the variables.[112]

In fact, the investigators themselves, Frankland and Greene, consider this experiment to be a kind of corroboration of the computational model of thought. Our own reading of the experiment is different. One of the interesting aspects of the experiment was the introduction of affectively charged sentences, where changing the variable positions produces different emotional meanings. So for example, subjects were given the sentence "the baby kicked the grandfather" or the sentence "the grandfather kicked the baby."

Obviously these two sentences have very different meanings, and the second sentence triggered a strong reaction in the amygdala, while the first one did not. The pattern of activity appears to flow quickly from the perception of the written sentence, to lmSTC categorization, to the amygdala. The reverse causal pathway—often postulated by proponents of a ventral route for hot cognition (i.e., system 1)—was not the order observed in the experiment.[113] Nonetheless, we want to seize on this amygdala evidence as relevant to our own thesis that meaning is significantly constituted by affect. Even if we concede that a rudimentary categorization of agent / patient occurs like a Boolean algebra computation at some tertiary level of cognition, we argue that this represents only the most basic level of meaning—an important formal structure, but without much content. This linguistic categorization is only the base level of the *indicative mood* that we discussed in previous chapters (cf. Millikan's PPRs). The *imperative* meaning—in this case outrage, worry, etc.—of "the grandfather kicked the baby" is constituted by the affective system as it meets the relational coding and the representational, associational, ontogenetically-derived memory bank of secondary-level affective processes.

The limbic system is implicated in the meaning of the experimental affectively-charged sentences. Simple computation of the agent and patient role needs salience added for it to be actionable. Our claim is that childhood development is where that affective salience is first coded and where nontrivial meaning is established. Ontogenetic development is where each of us acquires an affective style; the world itself (the semantic environs) is thus also shaped (via somatic markers, vertical associations, schemas, scripts, and affordances) in this early development.[114] These meanings are stable enough that we all have a more or less shared affective world, which remains changeable and revisable, too (i.e., adaptive to idiosyncrasies of ontogenesis and later experiential learning).

Categorizing experience via indicative computations is clearly important for basic meaning. But the linguistic brain is not the best model for thinking about how animals (or our hominid ancestors) engage with meaning. Imperative meaning comes when the amygdala and other affective regions get involved. Parallel to the trivial categorizing computation, it is childhood development that sets the amygdala associations that give content to the simple slotting system. Slotting which thing is the agent and which the patient is a primordial valuation of experience, giving structure, coherence, and base-level predictive power. This slotting is accomplished by the conditioning of associations in secondary-level affective processes. But such a formal structure is a kind of abstraction from the action-based life of the animal, for whom such distinctions are attenuations of real-life enactive relations—involving predators, prey, mating, feeding, nursing, fighting, and so on.[115] A wash of limbic emotions without some categorical organization (e.g., agent or patient) will be fairly incoherent, as will agent/patient computation without affective meaning (e.g., "This is a threat to avoid," "This is a mate to approach," etc.). The two together make experience meaningful. Just as Immanuel Kant synthesized the rationalist and empiricist traditions with his oft-quoted "concepts without percepts are empty, percepts without concepts are blind," we submit a synthesis: categories without affects are empty; affects without categories are blind.

6

Representation and Imagination

THE STUDY OF AFFECT can be the basis for organizing a research program concerning how the embodied, associational mind developed executive and linguistic dimensions. Beyond the minimal embodied representations of action and social intelligence that we share with other animals, there is a story to tell about how affect came to be decoupled (from its primary- and secondary-level functions) and how this led to our unique cognitive realm of symbols and concepts. In Chapter 5 we described the ontogenesis of emotional intentionality, in this chapter, we sketch two related transitions: (1) from the behavioral affordances of direct perception to the mental simulation of action and perception in spatial navigation and (2) from the affective reconsolidation of memory in the involuntary imagination of dreams to the conceptual and linguistic symbol systems of voluntary imagination. These transformations describe a transition from a behaviorist notion of mind to a mind capable of simulation and cognition.[1]

Representational abilities were decoupled from perceptual tasks, expanding possibilities for simulation and executive cognitive abilities. While qualitatively different, these representational abilities remain nested in the affective brain systems we have been describing in this book. They make possible new potentials for mental functions that, through the pressures of expanding social complexity and cognitive load in response to ecological conditions, lead to the human mind as we know it now. The key shift was from real-time perceptual behavioral affordances to valence-tagged simulations decoupled from here-and-now demands for the purposes of planning and decision-making. As we describe below, spatial navigation, the dream state, and body-task grammar served as the crucibles for this evolutionary transformation. And in Chapter 7, we turn to the decoupled simulators and executive processes of imagination that constitute cognition and representational mind, which made possible the evolution of concepts. Dialectical cycles between these mental abilities and complex social processes led to our infinitely iterative natural language.

If the human mind is in fact an evolutionary kluge, it must, in some respects, still function in an embodied-extended-embedded nonconceptual manner.[2] The missing theoretical middle ground between percept and concept is ripe for analysis as its evolution must have passed through several stages. For example, before you have a modern eye, you need a simpler optical predecessor, and before that you need a responsive light-sensitive tissue. Evolution scales up from the ground, so to speak. Initially, evolution built a crude imaginative faculty that was then slowly refined into a sophisticated set of processes. The crude system, made up of affects and affordances, is still alive and well in the substrata of the mindbrain, despite the user illusion we experience of more explicit symbolic material.

In previous chapters, we focused on how mind may function with minimal cognitive activity; we did this to make space for understudied non-linguistic affective capacities. In Chapters 7 and 8, we describe a causal story concerning how cognition and representation arose from the affective mind sketched so far.

Nonconceptual Content

There are many shortcomings of the traditional representational view, which include a failure to produce good behavior-based robots, examples of creatures able to navigate in plan-like behavior without the need for representations, and the ineradicability of the role of the body and environment in cognition and perception-action.[3] In contrast to representational information-processing views of perception, which hold that a creature must have the ability to think thoughts that have truth-values before they plan a course of action, we endorse the existence of nonconceptual content and the idea that perception is best understood as a form of action or a set of abilities to act in the world.

Our suggestion is that representations are constructed in part by an automatic, unconscious bundling process. This process is the imagination, but there are several layers of imaginative complexity (starting in involuntary dreams and scaling up to voluntary waking processes). In pre-sapiens, babies, and nonhuman animals, the imagination works with sense-perceptions, affects, memory, and the feedback loops of brain-body-environment. In linguistic *sapiens,* the bundling is upgraded by a syntactical system that allows for more complex sorting and recursion.

While philosophers of mind have spent decades working out theories on propositional content, there has been far less written about nonpropositional and nonconceptual content and its relation to perception and affect. Herein, we craft an evolutionary account of imperative, perceptual, and affordance-based subpersonal content in nonlinguistic mindbrains.[4]

Nonconceptual content can be attributed to a mental state that represents the world in creatures who lack mastery of the concepts required to specify the content.[5] The epistemic abilities necessary for nonlinguistic creatures to complete actions cannot depend on the conceptual machinery observed in *Homo sapiens;* therefore, some Anglo-American philosophers of mind suggest there is a range of mental states within the developmental arc of the evolution of representational systems.

Christopher Peacocke's autonomy thesis says that the fine-grainness of perceptual information outstrips its conceptual content insofar as the content of a percept—for example, the shades of blue on a faded wall—are too fine-grained to be described by concepts.[6] In other words, nonconceptual content suggests that we can have more perceptual experiences than we have concepts of perceptual experience. Perceptual content is also dependent on context, complicating the notion that creatures have concepts for every one of their experiences.[7] For example, in color constancy—the perceptual phenomenon whereby one sees the same color despite the wide range of differences in how much light is being refracted by a surface—perception does not depend on the creature having the concept of the context within which perception is taking place.[8] Nonconceptual content speaks not only to the richness of perception but also to its situatedness in affective and action possibilities.[9] We suggest that there is a range of nonconceptual content that does causal work in the mindbrain, and that it consists of affective goads and perceptual affordances that motivate behavior.[10]

It is at the subpersonal level that we believe one can make ascriptions of thoughts to non-linguistic and prelinguistic creatures, not to mention a wide range of our own sensory and motor behaviors and responses. Affect in this scheme is a nonconceptual goad that promotes adaptive control through the selection of affordances in relation to valence appraisals of the environment.[11] There are conative connections between homeostatic processes, the SEEKING system, and contextualized perception-action sequences functioning largely at subpersonal levels. The "direct perception" approach we espouse leaves space for the developmental evolutionary continuum of nonconceptual mental states.[12] Eventually, representational processes and intentionality evolved atop and interleaved with the affective mind, and the whole nature of the equation became transformed by new dialectical interactions between newer and older parts of the mind.

Launching from this brief overview of nonconceptual content, we argue that a nonconceptual dorsal stream of perception and action is where affect can be decoupled in simulations for planning (e.g., spatial navigation) and organization of social hierarchy in dreams.[13]

Affordances and Perception

The study of perception reveals underlying metaphysical commitments; the percept has alternately been considered to be raw nerve signals, the perceptual given, concepts, and digital data.[14] Cognitive science betrays a dualist opposition of world and mind, with perception characterized as a series of mechanical operations upon private mental representations of information picked up by the five senses.[15] Direct perception develops a space for sensorimotor links between perception and behavior that do not necessitate mediating representations.[16] This active direct perception approach we espouse describes how machines and creatures can use sensory data to answer specific queries, without having to build comprehensive inner models.[17] Let's take perceptual search as an example. In this paradigm, the body engages in continuous subpersonal perceptual interrogation of what the environment affords as well as its effectivities relative to the homeostatic needs and dispositions of the organism. There is no percept or mental image; rather, there is the activity of perceiving, which is a "constructive act of fitting a perceptual hypothesis to its sensory source."[18] Drawing from the phenomenology of philosophers Merleau-Ponty and Husserl in the last several decades, the direct perception approach has matured into a robust picture of how perception works, claiming that animals are active in the world, moving their heads and bodies to make information available to their senses and to direct their actions.[19] This mutual dependence between body and environment is part and parcel with the relational nature of meaning we have been portraying as central to the emotional mind.

J. J. Gibson's ecological psychology describes lawlike correlations between environmental stimuli and behavioral responses. That is not to say that every percept is linked to a response, but that there are co-evolved relations between mind and niche that seem teleological, or as we wrote in Chapter 2, that seem to possess biological intentionality.[20] In the case that such lawlike correlations hold, perception and action consist of nonconceptual first-person contents.[21]

Affordances and effectivities demonstrate the unity of perceptual and action systems. Affordances are dispositions given by features of the perceived environment to support behaviors, and effectivities are a given animal's dispositions to undertake afforded behaviors in the appropriate circumstances. Effectivities complement affordances in an informational coupling between perceiver and perceived. Similarly, in this approach, proprioception and exteroception imply one another.[22]

Affordances exist in the environment itself; they carry information that suggests actions (e.g., surfaces reflect light in a way that tells the perceiver what substance the surface is made of).[23] Here, light is a relational feature of the environment; it tells one about the refractive surface as well as the viewer's relation to the light and relevant features of the environment. Some behavioral outputs that the perceptual features of light can indicate include one's height relative to other objects, the texture and composition of objects in the environment (and thus how they may be interacted with), and so on.

W. H. Warren's rigorous empirical research on the relationship between leg length and body scale illustrates a stair-climbing affordance.[24] This research found that a subject's body scale had the same optimal ratio of distance from step to riser height; that is to say, a subject would use this directly perceived ratio to afford how high to raise his leg and climb the stairs. Both short subjects (people with a mean height of five feet and four inches) and tall subjects (with a mean height of six feet, two inches) visually perceived optimal stair height to be one-fourth their respective leg size.

Another research program in direct perception is optic flow and the variable Tau, in which the flow of textural elements radiating out from the center of one's visual field (i.e., the *looming* of visual elements) directly affords information about one's locomotion, pursuit, collision avoidance, direction, and velocity of movement.[25] Looming is used not only by hummingbirds and pigeons, but also by humans in activities such as somersaulting, putting a golf ball, steering a car, and hitting a ball with a bat. In optic flow, it is self-willed movement that changes the nature of incoming percepts such that information states co-vary between environmental features and the body of the perceiver. Gaining pertinent information in these paradigms is a matter of movement and direct perception rather than the consultation of an inner representation of the environment.

Another pertinent research program is work on visual entropy (i.e., the amount of information in a visual percept that allows an animal to decide upon the sameness or differentness of features of the environment). The density of a visual scene / percept here serves as a form of analogical reasoning (same or different) in direct perception.[26] For example, a pigeon can make accurate decisions as to whether two displays are the same or different, depending on the amount of information in the visual percept of each display.

Lastly, haptic perception in dynamic touch scenarios (which usually center on the wrist), where a creature manipulates objects in the environment to gather particular informational states, is also relevant.[27] The way we move our limbs to animate our touch receptors and manipulate objects in the world is a form of affordance and effectivity creation. Furthermore, when considering direct perception, it is important to consider how smaller-scale abilities are nested into more complex abilities.[28]

This embodied-extended-embedded-enactive (4E) model provides us a basis for understanding nonconceptual content in perception-action. Below we describe a possible evolutionary story for how such affective, imperative content was the space for decoupling of affective tags and imagination, and, in consequence, made possible the evolutionarily later representational processes of symbolic thought.

Affect and Decoupling

We view primary-process homeostasis, secondary-level learning and memory, and direct perception as neural systems that co-evolved and provided the necessary space for the decoupling of affect from its here-and-now functions to encapsulate complex social and imaginative representational capabilities. Furthermore, perceptual affordances and the dream state discussed below may have been the mental space where executive and inhibitory processes came online.

The property of decoupleability refers to offline processing of information. It indicates thinking of a concept when it is not part of the creature's present action, target of action, or perceptual context, like daydreaming of the summer heat while sitting in a freezing apartment in February in Chicago.[29] Decoupling is the process that cleaves present-tense perceptual indicative percepts from instrumental proto-beliefs.[30] Affect as conative motivational drive is amenable to being decoupleable because it predates—and remains functional—through all evolutionarily later cognitive abilities; that is, its primacy ensures that it has a use within any mental context. And, unlike other mental functions, affect can filter through any mental operation, infusing pertinent elements with salience; affect dyes our thoughts with value and meaning. Accordingly, we have described several roles played by affect including, as a mode of presentation, as an intentional arrow, and as motivation for locking onto appropriate affordances.

Perception and action depend upon the imperative forms of informational transfer between creature and environment, which we described as affordances. Salience within the perceptual world occurs via affective goads that dynamically covary with homeostatic needs and lead to action patterns, such as further information-seeking behaviors.[31] The function of affect in perception is as a mode of experience, and specifically, as a subjective motivational force. Affect functions as an approach / avoid value in affordance space, whether it be social space, as discussed in the Chapters 3–5, or the spatial navigation landmarks described below.

A core concept in the cognitive sciences, representation, has the functional role of acting as a decoupleable surrogate for specifiable *extra-neural* states of affairs.[32] While full-blown decoupleability is certainly a paradigmatic quality of representational mind, our thrust here is that affect is susceptible to a more simple decoupling process that could function via mnemonic processes and direct perception of landmarks in a minimal or pre-representational mind. We believe the varied role of affect has long been overlooked in this regard and claim that, in both spatial navigation and dream states, affect is a decoupleable mental quality that, in the evolution of mind, may have served as a bridge to tertiary-level representational processes. As an attractor in spatial navigation and as a somatic marker in dreams and mnemonic processes, affect could be said to specify *inner* neural states of affairs in a feedback relation to *extra-neural* states of affairs, thus establishing the decoupling of primary-level affective states from their origins in the here-and-now of homeostatic functioning.

Our focus shall rest on nonconceptual content as minimal modes of mental representation. We claim the initial instances of functional decoupleability arose within a perception-action core, with affective valence-coding as goads in spatial navigation, and within the dream state wherein affective valence tags are unconsciously manipulated for the purpose of organizing social space. In the involuntary imagination of dreams, the landmarks in cognitive maps, and the burgeoning imaginative faculty, affective tags may be decoupled from their initial objects, creating a fluid space for the abstraction of value and meaning. It is only after these formats are in place that symbol systems and the subsequent compositional explosion of the linguistic mind becomes possible.

Affect and Spatial Navigation

Mediating between mind, behavior, and world, cognitive maps are mental structures that offer a substantive example of how a structural relation between observer and environment can afford movement, simulation, and opportunities for decision-making.[33] Such spatial cognition necessitates fine-tuning of motor movement, executive planning, and even inhibition-related delayed gratification. The most primitive body structure in the brain is the somatosensory and proprioceptive body image; next are spatial egocentric and allocentric maps.[34] This layering of maps of the bodily self suggests a functioning isomorphism between a set of homeostatic and movement processes in the brain and behaviorally important aspects of the world, such that elements of the map can be considered as symbols, or proto-symbols. As the inner model develops from an egocentric to an allocentric map, it eventually allows the capability for simulation (i.e., simulating different possible routes) and structured mnemonic storage of information. Allocentric maps differ from evolutionarily later forms of symbolic representation, like language, in several ways; most notably, they are iconic, lack predication, are analogue, and are not compositional.[35] These allocentric maps are good candidates for nonconceptual content, specifically as landmark affordance simulation spaces.

Cognitive maps as mnemonic packages necessitate acquisition, coding, storage, recall, and decoding of information about the relative locations and attributes of phenomena in a spatial environment.[36] One school of thought considers the cognitive map to be a memory of landmarks that allows novel short-cutting; it is a representation of the environment that indicates the routes, paths, and environmental relationships that an animal may use in making decisions about where to move.[37] Another school considers a cognitive map to be an allocentric representation of space; specifically, a record in the central nervous system of macroscopic geometric relations among surfaces in the environment, which is then used to plan movements through the environment.[38]

Let's look at two examples of how cognitive maps are used: *localization* and *piloting*. In both, one observes landmarks relative to oneself using observations and prior knowledge to figure out one's position.[39] An example of the former is when ants find their way back to the nest via dead reckoning, and an example of the latter is when a particular landmark, like an old tree, orients an animal to its relative vicinity.[40]

For the piloting function of cognitive maps, the question of how we store information about the landmarks is pertinent to our discussion. For example, how does an animal store the tag that a given landmark was close to a food source in a past instance? Remembering a large number of landmarks will require holding (and pausing) homeostatic goals as well as making decisions as to which landmarks to remember—and subsequently which geometrical properties of space are specified by the remembered landmarks. We suggest that landmark information is initially built up by enactive and direct perception, and then integrated by cognitive mapping processes like dead reckoning and piloting to create bearing and sketch elements of the map. This process brings forth egocentric maps that afford action–oriented representations (AORs), such as movement in a particular direction. In some animals, this process also eventuates in allocentric maps.

We contend that the value or meaning of the landmark is an affective valence tag that summarizes mnemonic information regarding the landmark as a motivational vector.[41] The affective tag will simply average, via conditioning, past affective states at the landmark—for example, "Food was in this spot" equals positive valence—to indicate "approach" or "avoid" as a simplified attractor goad on a cognitive map. In more complex maps, or layers of maps, a landmark can indicate relational geometrical information pertaining to destination, relative to goal states. Goal states may be generated by primary-level homeostatic needs and then associated with landmarks by secondary-level learning and mnemonic processes like operant conditioning. A recent ethological example of such a process is the so-called episodic memory in scrub-jay food caching, which demonstrates how elements of a spatial map can have values determined by the length of time a food article remains edible.[42] In these studies, it was determined that scrub-jays seem to keep track of which articles of food are stored in which areas. Our interpretation of the data is that spatial cognition and valence of map elements (e.g., the affective urgency of retrieval) is responsible for the bird's behavior.

Within perception, affect provides direction as a motivating internal context; viz. the biological intention or conative drive. Indeed, across developmental processes intentionality locks onto relevant aspects of the environment in a way that shapes our subpersonal processes, which then serve the intentional aspects of act-planning.[43]

Experimental work reveals that affect plays an important role in concentrating or diffusing attention and retaining downstream memory of salient information.[44] Negative, visually arousing stimulus causes humans to remember less peripheral details from the environment. Subjects shown pictures of snakes in the river and also squirrels in the forest remembered less spatial peripheral detail about the snakes image than squirrel image. Presumably, slight affective fear focuses attention on the potential threat agent (snake), diminishing perceptual detail from the surrounding environment. This is another example of how affective tagging effects navigation of the environment.

The highly complex cognitive maps that humans seem to possess include landmarks that provide cues to declarative memories; for example, the "memory palaces" discussed by Luria.[45] Even early texts on the art of memorization, like the Roman *Rhetorica ad Herennium* (circa 80 BCE), suggest that memory of facts is improved when associated with strong emotional pairings in virtual space. Elaborating on our earlier discussion of the mind as an embodied-embedded-extended process, we suggest that affect in spatial navigation plays an affordance-like role; it suggests basic behaviors—viz. approach / avoid—toward landmarks.

. . .

One characteristic of affect in cognitive maps that we would like to emphasize is how affect can be moved around and is thus decouple-able from the map, depending on its internal salience within the homeostatic economy of the body and its external context (including past experience and environmental factors). The affective tag is thus a reflection not only of the landmark that is being tracked in relation to its surroundings but also of bodily needs that covary with the internal motivational homeostatic needs of the creature. This variability in the affective code suggests an early form of decoupling from internally generated embodied needs to embedded-extended physical space. Furthermore, this internal decoupleability of affect implies levels of control over motivational variables and perception-action in the context of the environment. Control enters the equation when inner compulsions meet external compulsions, and decision-making processes must decide between body states and behavioral priorities. How this is accomplished and the nature of its functioning on the continuum of nonconscious-conscious processes is a worthwhile question, though we will not attempt to answer it here. Suffice it to say, this incipient-level control may be an early form of executive functioning.

Affect and Dreams

The areas in the brain responsible for the REM dream state are more ancient in brain evolution than areas necessary for waking consciousness. Emotion-mediating areas of the brain, like the amygdala, as well as mnemonic storing areas, like the hippocampus—which, in dreams, is in an EEG theta rhythm state common to the state in spatial exploratory behavior—are active during dreaming, while planning areas in the prefrontal cortex are quiescent.[46] With inspiration from Panksepp's original thesis that REM states are a space for ancient emotional circuits to be integrated with the newer cognitive skills of the waking brain systems, we describe here how dreams are a form of involuntary imagination that serve as a mediating state between perception and judgment.[47]

In dreams, when atonia, REM, and changes in neurotransmitter levels occur, the dreamer is largely absent from the perceptual here-and-now and the mind enters a (re-)calibration zone of emotional training. At this level of consciousness, the dreamer indeed still experiences affective states; we know this because motivational mechanisms and dopaminergic activity are essential for the generation of dreams.[48] In the dream state, affect does not always line up with its appropriate objects; for example, standing on a cliff may not produce fear, while "random" inanimate objects may generate great anxiety. Since perceptual information is not entering the mind and triggering affective circuits, affect can be said to be decoupled from both the here-and-now and its task of adaptive functional appraisal of percepts. But why would that be? In being decoupled from agency and here-and-now perceptual-motor tasks, synthetic mental processes are able to organize / reorganize affect relative to mnemonic content. Imagination in the dream state, where the mind unconsciously and involuntarily mediates between perception, memory, and judgment, is a good candidate for predecessor-to-conscious forms of explicit decoupling, simulation, and voluntary imagination.

The adaptive function of dream decoupling in our story is as a form of mnemonic-affect schema consolidation; this could occur for spatial landmarks, as well as in social animals, for the purposes of generation and refinement of appropriate social behaviors. An example of the latter would be the stabilizing of dominance hierarchies, an activity notably undermined by REM deprivation.[49] The hippocampus, a subcortical limbic structure, is known for its role in mediating cognitive maps, receiving affective components of incoming sensory stimuli, creating declarative memories, computing spatial orientation, and providing information about the context in which conditioning has taken place.[50] Behavioral neuroscience evidence about the role of the hippocampus dovetails well with our hypothesis that both spatial landmark goads and dream memory reconsolidation occur when affect is decoupleable. These two forms of decoupling may take place in non-human animals with minimally representational minds.

It is in the dream state that, we hypothesize, a type of memory reconsolidation occurs, wherein malleable emotional memories like dominance hierarchies can be modified and recalibrated.[51] Experiments suggest that fear conditioning, as well as procedural memories and some memory distortion in episodic memory, can be modified in the reconsolidation process—in or outside of the dream state.[52] We suggest a dynamic reconsolidation process where affective tags (or landmarks) can be shifted and learned in an implicit, unconscious format at the level of secondary-level affective processes, in both spatial navigation and in the involuntary imagination of the dream state.

Dreams not only demonstrate how affect can be decoupled from mental functions, but also teach us how affect colors mental functions. In fact, it is possible that affect can color any mental function— something we have known since Damasio's *Descartes' Error*, with its memorable description of a man with limbic damage who is rendered unable to make even the simplest of decisions.[53] We can thus imagine how subjectively experienced affect can be decoupled via associational processes in emotional memories during the dream state. Likewise, such emotional tagging is used within the conditioning paradigms of behaviorism (cf. Thorndike's law of effect) and, subsequently, in the development of language. What we observe in dreams is a form of simulation of affective pairings, with sundry mnemonic objects. This process is seemingly run on associationist principles, and it may be the dawn of imagination.[54]

The role of memory and its evolutionary trajectory is pertinent at this point, because we are claiming that implicit, or nondeclarative, memory processes are at the core of affective decoupling in both spatial cognition and dream states. The standard story is that the earliest form of memory is classical conditioning (including fear conditioning), followed subsequently by operational conditioning, and then forms of procedural memory, early declarative semantic memories, and finally episodic and prospective memory processes.[55] It is prior to declarative memory, at the secondary level of associative conditioning processes that we believe memory reconsolidation through affective decoupling occurs. Indeed, an affective center, the amygdala, seems to modulate the encoding process of memory consolidation.[56] This may be how memory is marked with valence within a somatic-marker process.[57]

Memory storage is an ongoing dynamic process, wherein old memories become malleable when retrieved in new contexts.[58] If the function of dreaming is organizing mental space, then one important category for social animals is coping with anxiety through the simulation of manipulating affective valence tags onto social actors. Rehearsing social situations and giving value to members of your tribe's dominance hierarchy is crucial to appropriate social behavior. But where is this reflective or concatenative process undertaken? We suggest it is in dreams that, amongst other tasks, social position and the broadcasting of unconscious suggestions—regarding future motor, social, and problem-solving plans—is worked out. We speculate that the function of dreams is the affective organization of objects, situations, and social actors.

In the dream state, affective tags may be decoupled from their initial objects, creating a fluid space for the abstraction of value and meaning, and perhaps demonstrating the first springs of conscious imaginative faculties. Our example of such a process is that when an animal dreams of members of its community, each member has affective values, akin to somatic markers, that determine their position relative to the dreamer and in the broader dominance hierarchy. Humans continue this tradition of dreaming of conspecifics in their community, in the common dreams of sexualized scenarios and violent domination schemes, or in dreams of friends or strangers who shift between familiar and unfamiliar. These manifest dream contents suggest a latent dance of affective tags and mnemonic social objects, which have the purpose of quelling our social anxieties and hierarchizing our social worlds, much the same as the dreams of our animal cousins.

A parallel story may be told about the development of concepts, which themselves must have developed from cross-modal affect-perception units within a mnemonic system like procedural memory or in the space of working memory. Through evolutionary pressures, these concepts become the toolbox for simulation according to associationist principles.

Imagination

Imagination is a faculty of mind that once enjoyed philosophical and psychological attention but has been largely ignored for the last century or so. In part, this is because the imagination has been broken into constitutive parts (e.g., perception, memory, binding, etc.), and researchers have focused on these sub-faculties—tacitly denying the existence of an independent power called imagination (relegating it to a folk category of the mind). The disappearance of the imagination from most explicit research programs (and faculty psychology in particular) should not be taken as evidence that its career is over. In this section, we want to suggest that the imagination forms an important locus that is still vexing for cognitive scientists, evolutionary psychologists, philosophers, and neuroscientists. The imagination may not be a separate faculty of the mind—indeed, it may be a convenient way for picking out a family of processes where perception already has noetic content—but it forms a useful placeholder for the significant territory of image-based thinking. We argue that the imagination is a bridge from the older sensory-dominated, real-time world of our pre-sapiens ancestors to the modern mind—a rich linguistic world of internal deliberation and inferential judgment. Our exploration of dream decoupling above already paves the way for this argument, as does new neuroimaging evidence that mind-wandering (the "default mode network") engages similar brain mechanisms as dreaming.[59] Imagination amplifies the decoupling process found in dreaming and makes novel representations available. Such mental work appears to happen in the default mode network, but eventually, goal-directed waking thought (the "task positive network") is able to harvest and make use of these novel resources.

Plato introduces the word "imagination," *phantasia,* as a kind of sense-perception *(aesthesis),* but in the *Sophist,* he adjusts the function of imagination to include judgment.[60] He suggests that imagination contains sense-perceptions but also has true and false qualities—a mixture of sense-perception and opinion.[61] Plato draws a distinction between real-time, present sense-perceptions (aesthesis) and past sense-perceptions that are retained in the mind. These past sense impressions are memory images, and he suggests a metaphor of a wax tablet in the mind that receives a sense impression and stores an imprint copy (just as wax receives a pressed form). Some scholars take this wax metaphor of memory images to be an attempt to describe the imagination *(phantasia)*. In some places (e.g., *Theatetus*), Plato treats cognition as a kind of matching system, where sensory images are understood to the extent that they match the internal images of the knower.[62] In this view, images have a crucial role in the formation of knowledge, since the rational intellect makes use of them in cognition (or re-cognition). Elsewhere (e.g., *Republic,* book VI) Plato denigrates images and the imagination as mere sensory simulations of the real—distracting and alienating us from the rational realm of knowledge (modeled now on the Pythagorean paradigm of mathematics and propositions).

Aristotle expands on Plato's view of the imagination in *De Anima*.[63] He distinguishes it from sense-perception on the grounds that imagination produces images when there is no perception, as in the case of dreams. Moreover, animals seem to have powerful sense-perception but no imagination (suggesting that imagination is not reducible to sense-perception). Finally, he points out that perception is always true to the subject (i.e., phenomenologically manifest), while imagination can be false (e.g., chimera, etc.).

Aristotle suggests that the imagination is a middle faculty between our sense-perceptions (colors, sounds, tastes, etc.) and our mind (the realm of concepts and judgments). Like sight, the imagined form has sensual properties and shares some aspects of the bodily. The imagination, which he says is "that in virtue of which an image occurs in us,"[64] provides us pictures that have particular shapes, sizes, colors, and so on. But sense impressions are never really false, according to Aristotle. They are like raw data. When the drunk person perceives the walls moving, he truly perceives the walls moving. Imagination, however, joins perceptions to additional mental data and sometimes forms judgments. In this way, imagination is more like mind *(nous)*, which abstracts out particular sensual data and considers the universal defining features of a thing. Mind is a "form of forms"—able to ignore the material aspects of natural things and record and process their formal aspects. Mind can run *code* versions of experience, where mental processes transform elements of representations of perceptual and mnemonic content.

Although there are many differences between Aristotle's and Kant's philosophies, Kant seems to agree that the imagination is an unconscious synthesizing faculty that pulls together sense perceptions and binds them together in coherent representations that have universal conceptual dimensions.[65] For example, I see a fluffy brown shape moving in the field and quickly judge it to be a rabbit—a creature that fits into a formal conceptual category (of the family Leporidae, inside another category Mammalia, inside the subphylum Vertebrata, etc., or the folk-taxonomy equivalents). The imagination (an inscrutable black box) plays some role, according to Kant, in subsuming particulars (percepts) under universals (concepts).

According to contemporary philosophers, our image-making faculty helps package our experience into manageable units that can be plugged into cognitive judgment faculties. These cognitive judgments are propositional in the sense that they have subject/predicate attribution structure ("The rabbit is brown") and that the judgments are categorial ("This brown creature is a mammal").

However, a little reflection reveals the strangeness of this model. It is extremely rare to see a moving shape in a field and suddenly turn into Linnaeus, cataloging and classifying my experience into a cognitive system. We rarely engage with the world by explicitly categorizing it and using essentialist definitions. In rarefied endeavors, like science, we try to relate our experiences to abstract models and form judgments and predictions accordingly, but most of our imaginative work is well below this erudite, indicative level.

This model "jumps the gun," racing straight from perception to propositional conceptualization and missing the huge middle ground. We recognize the rabbit more by automatically *associating* it with memory images, not by subsuming the percepts under a formal abstract concept. It seems more likely that we associate this brown creature to a prototype memory—a learned and stored master image, or schema of a rabbit, and this helps put the experience into an overall context of meaning. We manage to *judge* the experience in many ways that are not like logical inferences. In the act of recognizing the rabbit, for example, I am affectively or emotionally drawn ("Oh, isn't he cute?" or "I'll get that varmint!") These positive and negative affective judgments are very tightly conjoined with our perceptions and slip into the psychological mix well ahead of the conceptual processing. These affective aspects are a part of the "seeing as" core of perception.

According to Darwin, "the *Imagination* is one of the highest prerogatives of man. By this faculty he unites former images and ideas, independently of the will, and thus creates brilliant and novel results. *Dreaming* gives us the best notion of this power; as poet Jean Paul says, 'The dream is an involuntary art of poetry.'"[66] Moreover, contrary to Aristotle (and Descartes, among others), Darwin recognized the imagination as a faculty in nonhuman animals. "As dogs, cats, horses, and probably all the higher animals, even birds, have vivid dreams, and this is shewn by their movements and the sounds uttered, we must admit that they possess some power of imagination."

As we have already discussed, dreams furnish us with an exemplary case of involuntary imagination, where imagination is understood as a novel combinatorial synthesis of decoupled stored images, memories, feelings, and representations. But imagination proper is the same combinatorial activity under some level of executive control or agency. The dreaming dance of images and narratives that happens automatically is commandeered in waking life and given direction and guidance. The adaptive function of waking imagination is different in degree, but not in kind, from the social-emotional management we attributed to dreaming. If an animal can imagine a social or environmental scenario (in a low-stakes mental space), then he can anticipate adaptive moves in the high-stakes real world. While human imagination has its roots in the automatic associations of mammalian dreaming, it can also be (1) constrained by understanding (rule-governed cognition) or (2) in free play (unconstrained by teleology, and rationality).

The imagination, in this view, is a prelinguistic or pre-propositional meaning system.[67] How does it produce meaning? The imagination has a cognitive architecture that includes the following: (1) it has a middle position—part sense-perception, part judgment or appraisal; (2) it allows for a novel combination of representations or properties; (3) it allows counterfactual representations; (4) it allows domain-crossing (i.e., decoupling a tool from one function and repurposing it for another, etc.); (5) it has aboutness or intentionality (in its components and its overarching projects); (6) it can subsume tokens under types (more on this shortly); (7) it allows inferences (e.g., nonlinguistic image-based inferences); (8) it recruits the affective / emotional systems; and (10) it has two modes: voluntary (agent directed) and involuntary (e.g., dreams, mind-wandering).

Body Grammar

We will now elaborate on an intermediate imagination process, body grammar, that decouples movement from the basic nonconceptual sensorimotor function of spatial cognition. A symbol system needs both a semantic generator and syntax. It is common to think that only language can provide such meaning structure, but closer consideration of nonhuman animal communication and pre-sapiens cultural achievements suggests that images and motor-control sequencing (a simulation system) can provide a communication framework that plays a role in cultural learning.

The computational model of mind considers representations as information packets that insert into algorithms, like words into grammar, or zeros and ones into Boolean algebra functor sequences. Our contemporary minds can do such information-processing (and AI has been successful modeling along these lines), but we forget the fundamental "action" of the mind. As anthropologist Robert Barton suggests, we need to think of mind first and foremost as "internalized movement," not as a spectator or recorder of data bits.[68]

Our ability to coordinate our bodies into sophisticated action sequences, such as in rhythmic entrainment or tool use, stems in large part from the cerebellum. A dancer on the ancient savanna and a good jazz drummer or knitter today are sequencing motor patterns in a sophisticated manner, but the roots of motor-sequencing evolution can be seen in our primate cousins. We are beginning to understand the evolution of the cerebellum, in particular, and its underappreciated link to the cortex.

So much research has emphasized the expansion of the neocortex that it has gone relatively unnoticed that the cerebellum grew even more rapidly in size and complexity. Primate cerebella, especially ours, are not just relatively larger than that in other mammals but also extremely dense in neural connections. In humans, the cerebellum has around 70 billion neurons, but we have only a vague understanding of their functions. Using a comparative study of monkeys and apes, Barton discovered that cerebellum evolution happened six times faster in apes than in other primates.[69] Gorillas, chimpanzees, and humans had a rapid cerebellum expansion that might be uniquely important for explaining our unique mental and cultural advances.

The cerebellum is important in modeling, predicting, and organizing behavioral sequences. It is crucial in correcting spatial and temporal relations, like those necessary in dance and tool-use sequencing. It is also important in fine visual-motor dexterity—the kind that apes excel at when engaged in foraging and food preparation. Gorillas, for example, engage in complex food processing that turns painful stinging nettles into an important source of nutrition. Stinging nettles are flowering plants that have burning stinger thorns all over their leaves and stems. Gorillas carefully strip the leaves off the stems and then gingerly fold them in a way that keeps the stingers from piercing their mouths as they eat. They also have similar seriatim steps for processing bamboo.

The ability to string together such behavioral steps is facilitated by the cerebellum (not higher cognition), and it makes social learning possible, but is also improved by social learning. We consider the elements of body grammar mediated by the cerebellum to be an important element in the manifestation of the action-oriented representations discussed above.[70]

In the domain of foraging and food processing, apes are particularly good at building up complex sequences from subroutines, called "behavioral parsing." Presumably, early action chains of behavior would have been small, irreducible sequences that had to be performed in an exact order or the result would not be achieved. As memory and motor coordination became more robust, the behavior sequences could extend out—by replication of subroutines. Decomposing larger routines into parts also creates the possibility of rearranging the sequences, not just repeating the parts. Thus, a kind of *grammar* begins to emerge. All the while, innovations of sequence must be edited by natural selection— if a novel routine fails to produce food or shelter or protection, then it will fail the innovator, and the troop mimicry is unlikely to replicate the innovation.

In foraging, apes are too big and heavy to jump around on trees like monkeys, so, Barton suggests, they had to carefully plan and predict their pathways through their physical environments.[71] This meant there was selective pressure on embodied simulation—a cortico-cerebellar loop that gave apes better coordination of learned sequences and better predictive power about bodies in space and time.

Subtle changes in the ape brain increased sensorimotor *task grammar,* or action syntax abilities. Barton suggests that this bodily skill probably scaled up, in early humans, to help organize, predict, and model social domains as well.[72] Such motor coordination of steps is prerequisite for tool use and may be the beginning of a kind of gestural syntax system that could give rise to gestural communication and eventually language in humans. Recent studies of apes reveal that some of the seriatim food-processing techniques, like gorilla nettle preparation, are culturally transmitted to members of the same band or troop.[73]

Most bodily motor sequencing may be simple stimulus and response, but it can also be decoupled from immediate stimuli. Its status, as de-coupled, is unclear, but sequences must "reside" in the loop of muscle memory, ecological trigger, and affective intentionality. We might think of these motor sequences as "premodern concepts" because they are not linguistically grounded, but they have the potential for organizing kinds of experience. Procedural memory, for example, is a form of implicit (often unconscious) memory that consolidates motor responses in long-term memory. The potential complexity of such sequencing is attested to by the way musicians and dancers store and recall elaborate patterns (most of which entered the system via rote routines). An embodied task grammar that allows for manipulation of such premodern concepts provides a mechanism for increased tool use and gestural communication.

Image Grammar

In addition to thinking with our bodies, we also think with images. Psychologist Lawrence Barsalou, for example, has argued that images are forms of thinking too.[74] Cognitive activity, according to Barsalou, is a kind of simulation of external reality, via perceptual systems of representation rather than via code system. This hypothesis bears similarity to our discussion of affordances and AORs as embedded in percepts and cognitive maps.[75] When we have a visual experience of a dog, for example, the brain extracts an image symbol from the perceptual experience. The specific details of a German shepherd, for example, are abstracted out of the image symbol—and we retain a morphological or shape simulation of the shepherd, without the specific color and so on (powers of selective attention are enough to create attenuated symbols from the data). This more generic symbol of a dog is not a transduction of the image properties into code properties. Rather, the structure of the percept is simulated in the symbol that is stored in the brain. The symbol recreates or maps the relations of the parts as it captures (imperfectly) the whole entity. In other words, the legs of the dog are ventral, and the head is anterior, while the tail is posterior, and so on. These image symbols, derived from the visual system, are then stored in long-term memory to function as concepts. The generic symbol of "dog" now acts as a prototype concept that helps us to organize other experiences. For example, comparing a new creature we've encountered to the extracted and symbolic dog prototype leads us to identify it as a member of a generic (type) category. Moreover, the symbol prototype of "dog" gives us inferential abilities regarding features or properties of the creature that may or may not be visible in the moment (e.g., I am unable to see the lower half of this creature, but since its head matches the dog prototype, I can infer some sense of its leg structure and function). This is all unconscious processing, for the most part, and we are not aware that our mind is forming these image concepts. Consciousness rarely reveals the machinery of knowledge itself, and we're forced to catch it obliquely.[76] Of course there is also significant conscious thinking with images, as when we use Venn diagraming techniques to derive insights about sets and members of sets. Here we see a modern manifestation of a much older way of thinking about class membership as physical containment.

An image-cognition system is imperfect but serviceable. Organizing the world with image prototypes is sometimes crude. I might incorrectly identify a hyena as a dog, for example, because it has strong morphological similarities with my generic dog prototype. From a genetic perspective, hyenas are closer to cats than to dogs, but then again, such a sophisticated taxonomy would have meant little to our ancestors, who needed a rough-and-ready system.[77]

There may be good evolutionary reasons why humans can form generic mental images. In a constantly changing environment (like the one our ancestors evolved in), it does no good to store very precise visual memories that fail (by their specificity) to map onto the diverse problems of life. Generating an abstract prototype out of specific perceptions would be more adaptive because it would be more capacious for the taxonomic work and more fruitful for the predictive work of a mind trying to limn nature.[78] It might be objected to that precise memories are necessary for infant imprinting on the primary caregiver—indeed, we argued as much in Chapter 5. However, our case study of a prosopagnosia subject (Chapter 1) reveals how a lack of precision in visual representation can be compensated for by robust redundant systems of identification (i.e., affective markers, olfactory markers, etc.).

Our suggestion is that image-based thinking (like the decoupled motor sequences of a body / task grammar) may have dominated our prehistory and formed another domain of "premodern concepts," but such a modality is still with us, albeit obscured by the propositional dominance of modern mind. Indeed, contemporary psychology research on images may corroborate or at least lend credence to our conjecture about the Pleistocene mind. Recall, for example, Stephen Kosslyn's empirical evidence for where and how mental imagery is managed in the brain.

Kosslyn showed that mental rotation of an object takes time because our mind actually seems to inspect our internal mental imagery.[79] This productive reinspection of mental imagery seems to confirm that the medium of the image is not digital code or algorithmic encryption, but something pictorial—albeit in some neural substrate format.

Kosslyn presented a theory of cognitive imagery that allowed for perception-based pictures in short-term memory only. He suggested that perceptual images exist temporarily in working memory but probably transduce or translate into amodal code in long-term memory. Barsalou, in contrast, goes further and argues that even long-term memory is storage of symbolic derivatives of visual perception. But both Kosslyn and Barsalou remind us of a nonlinguistic realm of representations that allows for inspection and inference.[80]

Historically, philosophers considered thinking to be comprised, in large part, of imagery, and only in the positivist and cognitive science of the twentieth century was this view replaced by a paradigm of *thinking as code*—in part, because meaning was reduced (by schools like logical positivism) to linguistic representation. David Hume typifies the comfort with visual thinking, when he says, "[I]f we see the limbs of a human body, we conclude that it is also attended with a human head, though hid from us. Thus, if we see, through a chink in a wall, a small part of the sun, we conclude that, were the wall removed, we should see the whole body. In short, this method of reasoning is so obvious and familiar, that no scruple can ever be made with regard to its solidity."[81]

We want to bring back the pre-computational paradigm of embodied mind, because we now have tools and data that can fill in many of the earlier gaps.[82] Consider, for example, the dynamic narrative structure that results when we add together the subcortical embodied brain loops, like Robert Barton's cerebellar-cortical system, and the image gestalts of perceptual symbols. The same *task grammar* of behaviors that can organize subroutines into complex sequences can serve as a prelinguistic syntax system that governs images too. This would give the mind the power of "scenario creation."

The upshot of all this is that mental images and artistic images are not simply recordings, like camera snaps. They are modes of human appraisal. They are early forms of judgment. This model of knowledge appears controversial because many philosophers, from Aristotle to Wilfrid Sellars, claimed that only propositional language enables the thinker to subsume individuals (e.g., Julien) under universals (e.g., *Homo sapiens*), as well as to attribute properties (is hungry), and draw inferences (is now seeking food). In this model, knowledge is declarative. But another way of thinking about knowledge reveals its imaginative core. We are suggesting that a person who simulates a thing to a high degree of detail (either with body gesture, or drawing, or mimicry) can be said to understand that thing—to have substantial knowledge of it.

This view of meaning or understanding as simulation is ancient and can be seen in the Stoic appreciation of virtue as a craft, or *techne* (e.g., Cicero's *On Ends,* book V), and in Daoist philosophy's appreciation of *wu-wei* simulation of nature.[83] Wisdom, according to this tradition, is an accurate simulation of nature (which, in turn, makes one natural and not just a copier of nature).

These prescientific glimpses of simulation epistemology can now be augmented with new data from neuroscience and linguistics.[84] Meaning occurs when we recreate a relevant virtual reality, composed out of remembered and constructed perceptions and actions. This is not as simple as reaching inward for the abstract symbol or concept that matches our experience. The animal body itself has intentionality, and so the embodied mind is caught up in those projects.

Even when mature language does give us a rich symbol system for easy manipulation, those abstract symbols have their semantic roots in bodily activity. As Merleau-Ponty puts it, "The visible world and the world of my motor projects are both total parts of the same Being."[85] And again, "my movement is not a decision made by the mind, an absolute doing which would decree, from the depths of a subjective retreat, some change of place miraculously executed in extended space. It is the natural sequel to, and maturation of, vision."[86]

Even when seemingly neutral, lexical terms are processed by our brains, we betray the deeper simulation system. When we hear the word "cup," for example, our neural motor and tactile systems are engaged because we understand language by "simulating in our minds what it would be like to experience the things that the language describes."[87]

We can now suggest, given our story so far, that the *imagination* is just this *simulation system.*[88] Aristotle and Kant glimpsed some of this system in the gap between percept and concept, Darwin said it was the faculty that created brilliant and novel results by uniting former images and ideas, and Freud saw it at work in the dark poetics of dreams. But our discussion in this chapter reveals that the imagination is a type of knowing that has two distinct modes: a creative mode that is *involuntary, instinctive, spontaneous, and unplanned,* and a creative mode that is *voluntary, deliberate, and planned.*

Voluntary and Involuntary Imagination

The move from involuntary to voluntary imagination seems crucial for the rapid success of *Homo sapiens.* And it is not just language that ushers in the new era of imaginative, inner headspace. The rise of voluntary simulation or imagination is the result of (1) a new social world, with greater cooperation and emotional regulation; (2) enrichment of the recursion / embedding systems that allow elements to be accumulated and assembled (of which language is but one case); (3) greater brain-based executive control (more on this when we get to the self below); and (4) prestige social status for creative innovators (where skillful imaginative behavior becomes a phenotype that can be selected for by natural and sexual selection).

It might be tempting to think of the two simulation modes as *weak* or *strong,* and to think that Pleistocene *Homo* had weak simulation while Upper Paleolithic and Holocene *Homo* had strong simulation faculties. But this characterization is somewhat misleading, since the strength of Pleistocene simulation must have been very intense, phenomenologically speaking. Pleistocene simulation activity probably included biomusical entrainment (social rhythm), conditioned gestural mimicry, and dreams, among other things, while Upper Paleolithic simulation added visual representations like painting and carving to these precursors.[89]

Evolution of Imagination

Upper Paleolithic (50 KYA)	Sophisticated tools Cave art Musical instruments Jewelry		
	Funerals		Cold Cognition
		Symbols	
Homo sapiens (200 KYA)	Language		Voluntary Imagination
Homo erectus (2 MYA)	Self consciousness		
Australopithecus (4 MYA)	Tools (e.g., fire)		
Apes (15 MYA)	Task grammar Social complexity		
	Memory enrichment		Hot Cognition
Primates (55 MYA)	Representations (image) Decoupling		
			Involuntary Imagination
Mammals (250 MYA)	Simulation	Dreams	
Vertebrata (500 MYA)	Perception, motor and spatial coordination		

The earlier Pleistocene forms of imitation are anything but weak forms of simulation. Indeed, this prehistoric involuntary simulation may have been overwhelming, especially if cognitive decoupling came later and allowed psychological distance as well as increased control over affect–induced states.

Evolution does not produce a new mental faculty and then "fire" or eliminate the old one. Instead, natural selection slowly repurposes forms and functions—modifying, adapting, and exapting what went before, and generally conserving previous forms. This makes the mind, just like the body, a highly redundant kluge of mechanisms built on top of mechanisms.[90] The voluntary mode of imagination does not *replace* (phylogenetically or ontogenetically) the involuntary mode, but rather *subsumes* it. When the simulations can be stored, accessed, and manipulated by voluntary control, then the system rises to the authorial level that we readily recognize as imagination. But the unconscious, involuntary, raw elements are alive and well inside this more deliberate creativity.

Affect or emotion forms a large part of those raw elements. For example, the image of the body of one's lover (or any sexually attractive body), whether remembered or depicted, has arousing affective impact built into it. What deliberative imagination can do with such an element is almost infinite (as attested to by everything from the Venus of Willendorf, to Egon Schiele paintings, to the latest pixelated pornography), but the affective tinting will already be in the anatomical element under consideration. Likewise, a carved or painted emblem on a forehead, shield, or shell, has powerful emotional implication if it is the image that your tribe has always used to designate membership. But even trivial images, or those without obvious adaptive tether, have low-grade affective tone, acquired through individual developmental history. The affective systems mark everything automatically through conditioning, but the values of such markings are not a priori or always significant.

Darwin isolates some of the more adaptive components of imagination when he says, "[T]he value of the products of our imagination depends of course on the number, accuracy, and clearness of our impressions, on our judgment and taste in selecting or rejecting the involuntary combinations, and to a certain extent on our power of voluntarily combining them."[91] If I am imagining a way of taking down a large animal, say a giraffe, with a small hunting party, then I need relatively accurate impressions about possible weapon materials, giraffe behaviors, distraction techniques, landscape potentials and constraints, colleague skill levels, and so on. In the beginning of the process, I probably have a series of realistic and cartoonish scenarios rush into my conscious field—for example, we could create poison spearheads. But also I have the maladaptive impression (quite involuntarily) that I might be able run alongside the giraffe and tackle him around his legs. As Darwin points out, the use of imagination is not simply in its creative associational phase but also in the editorial phase of judgment. I need to reject, for example, my involuntary scheme to tackle a giraffe. Ruling out this disastrous option may come from my own conditioned experience of playing rough as a kid, or witnessing and remembering disasters of other agents, or never having witnessed such an extraordinary event, or just having a gut feeling or affective repulsion to giraffe-tackling. Alternatively, I could just be reading affordances and effectivities when the actual situation occurs, rather than prediction-processing in the imaginative realm of simulation. But, as Darwin points out, I attain the highest level of adaptive imagination when I have voluntary control over the uniting of impressions and the testing of scenarios—when I can conduct internal simulations of possible outcomes (using impressions, folk physics, gut feelings, and variable conditions).

Two advances are happening in this stage of the evolution of mind. It is unclear which of them comes first or whether they are contemporary innovations. First, we have the development of more information-rich representations (moving from action-oriented representations, to associational gestalts, to prototypes, to concepts). As we'll see shortly, concepts—whether classical or prototypical—have additional structural properties that allow increased information load, as compared with perceptual memories and contingent associations. The offline representations are decoupled from immediate sensory experience, but they also bear more information (affective and factual). Second, increasing executive control makes the emerging symbol system into a non-random, organizable, toolbox that is accessible on demand. Having a volitional centralization in the brain allows for monitoring and managing the previously subpersonal elements of mind.

Recently, the theory that the mind-brain is a "prediction processor" or prediction machine has become increasingly popular.[92] On this view, the brain responds to its environment by creating ever more accurate predictive models of the world, in which error signals are minimized. The models are thought to be representational networks that help the brain extract salient signals from ubiquitous informational noise. In one sense, we are sympathetic to this approach, as it treats the mind as an active, practical navigator rather than a "spectator" or "mirror" of nature. On the other hand, this view of the mind is not particularly novel, as it simply rephrases in computational language the adaptive activity of mind, first formulated by Darwin and further articulated by John Dewey. Our own view is compatible with the prediction processing model and may be said to provide needed affective components. Not only is the cognitive information in any given prediction going to have affective coding, but the emotional systems themselves can be considered the earliest forms of prediction processing. Perceptual and memory gestalts of affectively coded experience act as feeling models for predicting salient social behavior and for navigating the physical environment.

In Chapter 7, we look at the underappreciated entanglement of emotion with language and concepts. We have finally arrived at the propositional symbol system of human cognition—the tertiary level of mind where most traditional philosophers begin their analysis. Our work so far on the evolution of affective mind will have important implications for language and concepts.

7

Language and Concepts

IN THIS CHAPTER, we add two more decoupling processes to our list that includes spatial maps, dreams, and body and task grammar. First, we discuss the evolution of language and then we discuss the evolution of concepts, raising the notion of the decoupling force of analogical thinking. We emphasize the affective function of both meaning and linguistic communication.

The Evolution of Language

Currently, there are an estimated 7,000 living languages. The origin of language, however, is shrouded in mystery. Of course, components of language exist in many other species, but these are properly speaking *communication systems,* not true language. Bird songs can be very elaborate and changeable, dog barking can indicate threats, and whale songs can signify navigational information. But animal communication systems do not come close to the prolific semantic possibilities of human speech.[1]

For nativists, language is a cognitive module that evolved in the shadowy past and suddenly gave us linguistic capacity.[2] This solves certain problems—like the mystery of how merely associational child-

184

hood experiences with speech become increasingly grammar-bound and how linguistic ability seems to come online suddenly as a developmental benchmark (with a critical period for activation). But it raises other problems, like how do we empirically test and verify the existence of such an a priori innate module? Is universal grammar a properly scientific theory or an *ad hoc* way of avoiding an evolutionary puzzle?[3]

Against the universal grammar thesis, psychologist Paul Bloom offers a line of critique, showing that nonlinguistic perceptual capacities create the preconditions for word recognition and use.[4] His empirical work with infants shows that humans have a tendency to see the world as comprised of objects, because our perceptual equipment keys in on things that have continuity, cohesion, and solidity.[5] For Bloom, the all-important skill of "naming" in language emerges out of this preexisting perceptual bias for attending to objects in experience. When babies hear recurrent speech stimuli, they are already primed to see objects as principal candidates for naming. Waving a banana in front of a baby and saying "banana" repeatedly will naturally forge a connection between the word and object, not the motion or color. Bloom argues that developmental studies bear this out, since children generally acquire substantive nouns first and verbs second, relatively speaking.[6]

We are somewhat skeptical of Bloom's suggestion that language follows naming, which, in turn, follows an early bias for object ontology. It is worth pointing out that the earliest enactive minds of infants and mammals generally are as concerned with activities, or projects, as they are objects. Babies and primates *use* the world before they ponder or examine it. Inhabiting a world is an active process of exploration. And it may be that objects, per se, only emerge slowly from the animal's various interactive projects or *teloi*. As Heidegger pointed out, our primary mode of encountering the world is as ready-to-hand *(zuhandunheit)* rather than present-at-hand *(vorhandenheit)*.[7] In other words, I use a tool as an extension of my hand (ready-to-hand), and only notice it as an object per se (present-to-hand) when it breaks or malfunctions. Careful observation of prelinguistic children playing and of primates interacting with their environment suggests that

object-awareness may be secondary, not primary. Our view that embodied mind is embedded in a functional environment (getting food,
social soothing, etc.), before it is a Cartesian spectator, leads us to some
doubt about an object-based epistemology. Embodied experience is
made of unified situations, or holistic *umwelten*—and language helps
the young human to discriminate the world into discrete objects (not
vice versa). But bracketing this issue, for now, we acknowledge the
various empirical approaches—like Bloom's—that nicely counter the
oft-quoted poverty-of-stimulus argument for universal grammar.

The evolution of language must be understood in the context of its
affective social value. An evolutionary biological approach to language
recognizes its ability to manage prosocial emotions. Describing the
world is an idiosyncratic latecomer activity when compared with surviving the world, and for social mammals, survival depends on
grooming the other members of one's group. Food-sharing and picking
parasites off of others are serious work in a state of nature. Primates
spend many hours each day grooming each other in order to create
and strengthen social bonds.[8] The utilitarian functions of grooming,
like hygiene, are matched and possibly outweighed by the positive
emotional effects of touch. And just as touch itself is a communication system (e.g., for purposes of calming, agitating, seducing, etc.),
speech sounds also communicate quite a lot, even when the sounds
are not symbolically referential.[9] Our own proto-language was certainly a wash of emotionally salient nonsense.[10]

Anthropologist Robin Dunbar argues that human language emerged
as a kind of verbal "grooming at a distance."[11] When a social group
gets too large for an individual to physically groom all the important
members, other remote kinds of grooming are selected for, and language owes its origin to this change. Social reciprocity is based on the
fact that I scratch your back and you scratch mine. If our social group
gets into the tens and hundreds, however, it is too costly and ineffi-
cient for me to spend my whole day removing parasites from everyone.
Vocal grooming, however, is relatively energy-inexpensive, and I can
service many relationships at once. The earliest kinds of true language
storytelling, according to Dunbar, were probably forms of gossip.
Gossip not only feels good and bonds people but also shares important

information about possible social opportunities (procreative and af-filiative) and liabilities (regarding freeloaders, and defectors). Our pre-sapiens ancestors picked parasites off of one another using their hands, but our sapiens ancestors picked social parasites off of the body politic using linguistic gossip.

Recursion and the Development of Communication

One of the major arguments in the evolution and philosophy of language concerns how humans arrive at the grammar recursion and embedding ability of our syntax. It is hard to see how even extreme social learning dispositions can give humans this syntactical ability. However, instead of postulating a hypothetical "language acquisition device" in the brain (Chomsky's solution), we should consider the role of the embodied "task grammar" system that we described in Chapter 6. Remember that the cortico-cerebellar loop in apes is important in modeling, predicting, and organizing behavioral sequences like pre-paring food and collecting water. This loop is crucial in correcting spatial and temporal relations, like those necessary in dance and tool-use sequencing. Such motor coordination of steps is prerequisite for tool use and may be the kind of gestural syntax system that acted as a platform for signal communication, and eventually language, in humans.

In what way could such a system be a platform for subsequent lan-guage evolution? The two key features of language—recursion and hierarchical embedding of sub-clauses—could derive from serial motor sequences and the incorporation of subroutines into larger routines.[12] Music, for example, has these same properties. Did the properties de-rive from language or did they precede it? Consider, for example, the American Song form. Most songs of this genus (like "Blue Moon," "Moonlight in Vermont," "Exactly Like You," etc.), contain the well-known AABA structure: eight bars of the same chord pattern (A1), followed by a slightly modified repeat pattern for eight bars (A2), and then eight bars of the bridge pattern of chords (B), and finally a re-turn to the original eight bars (A3). Music is built on such repetition, and it's reasonable to suppose that even Stone Age flute music had "parts" in repeated sequence and parts embedded within other melodic passages (like dropping a B sequence or clause in between the recursive A patterning). As Steven Mithen puts it, "recursion is one of the most critical features of music."[13]

Even before structured music was rhythmic dance, an even more foundational example of simulation and sequencing. The dancing body is another example of prelinguistic "grammar," because it has infinite recursion and "step" embedding. A good dancer can subsume many subroutines inside a larger frame of movement repetitions. The basic Tango is a five-step pattern (slow, slow, quick, quick, slow), using eight musical counts. Our pre-sapiens and sapiens ancestors may not have done the Tango, but they may have done something equally complicated. The nativists may be correct in thinking there is something special and uniquely human about recursion and embedding, but they would be wrong to think this is only a feature of language. Berwick and Chomsky claim the minimal meaning-bearing elements of human language are different from anything known in animal communication.[14] But their cognitive science approach is head-centric, not allowing for the considerations we discuss concerning the role of body grammar and task grammar as embodied-embedded forms of recursive activity.

When did human language capacity evolve? Some scholars suggest that human language evolved quite recently, dating back to the time of the cave paintings, approximately 40 KYA.[15] The underlying logic of this dating is that language is a function of a mental development—specifically, the ability to represent symbolically. Language thus evolved as "meaning with sound."[16] Art, according to this theory, is evidence that this ability came online in our species in the Upper Paleolithic, and language is a crucial aspect of this wider cognitive revolution. This logic is reasonable, but we should be wary about chronologically correlating visual thinking (which has archaeological data) and linguistic thinking (which has no direct data and must be inferred indirectly from archaeological data).

In contrast to this theory of recent language origin (40 KYA), some theorists mark the FOXP2 gene (discovered in 2001) as the potential spark of language. Humans have a unique FOXP2 mutation (occurring sometime in the last 1 million years). This particular gene controls for oral-facial muscles and movements but also for little understood cognitive differences. At first, some scholars like Steven Pinker leapt to the idea that FOXP2 was a "grammar gene" and that our particular mutation marks our capacity for language (Neanderthals, it turns out, also share our particular version of FOXP2).[17] More recent study of the gene, however, reveals that it is part of a complex suite of anatomical and cognitive conditions for speech, not a smoking gun that shot language into the species. Whenever we date the origin, we all recognize that language is a game-changer because it ramps up the representational powers of the species.[18]

Some philosophers have argued that the mind is so shaped by language that a nonlinguistic mind is not really a mind at all.[19] This hard-line position claims that linguistic words create definitional concepts in the mind (via some form of abstraction) and that the manipulation of these abstract concepts is *thinking*. Thought is internal language, and language is thus the organ of meaning. And our thoughts become knowledge when the internal symbols and propositions match the external entities and relations. This classical objectivist view holds that mind is a mirror of nature.[20] The world has a kind of structure (i.e., *entities* have *properties* and stand in *relation* to each other), and our symbols (in this case, words) get their meaning from their correspondence to reality's structures. In this way, objectivists argue, our words represent or mirror reality. Good thinking, in this model, is when our concepts (linguistically derived and expressed) accurately describe the world. When I say, "The small cat plays with his toy," I reflect an objective reality of an entity (a cat) having a property (small) in a process relation (play) with another entity (a toy). While this nexus between language, thought, and meaning is a crucial consideration in the question of what makes us human and is therefore worthy of further study, we would like to point out that our argument thus far has strongly suggested that meaning as an affective relation to the world, and the organism's past, existed prior to language.

Linguistic Imagination

Language is an imagination pump. We saw in earlier chapters that body simulations and images can be stored in the mind as well as retrieved and reformatted with new associations and emotional markers. We even saw how simulations could be used to draw inferences and make predictions about our environments. But language is an extremely powerful instigator in our headspace. This is because language, according to linguist Daniel Dor, is an "imagination-instructing system." When I say, "The black cat is on the fence" to my friend, her perceptual processors take in the auditory stimulation, but her mind then decodes the signal, such that she "sees" in her headspace a black cat sitting on a fence. In this way, language allows each of us to have a parallel mental reality.[21] My sentence actually instructs the imagination of my friend to call up the relevant images. According to this view of language, "the communicator produces a code, a plan, a skeletal list of the basic coordinates of an experience, which the interlocutor uses as a scaffold for the construction of a parallel experience in his or her mind."[22]

This view of language is helpful because it focuses on the oft-neglected "mind conducting" aspect of language communication. So many evolutionary psychologists have focused their research on Theory of Mind Mechanism or "mind-reading" and have not sufficiently explored this step of mind conducting. We are not just reading the conspecific's mind. We are manipulating it. The imagination, in this view, is a crucial aspect of how language fundamentally works. The phrase "black cat" is just noise until it triggers the memory image of a cat that has all the emotional or affective associations that chunk together in a prototype concept. The speech code or signal instructs the hearer's memories, placing perceptual symbols in motion—simulating the complex meaning that the speech blueprint has triggered. Without imagination, the code might stimulate behavioral response but not produce a parallel subjective experience in the listener. The imagination is necessary for the symbol to *be a symbol* rather than just a stimulus. Affect and imagination are the other side of the coin of meaning.

As we have suggested throughout this book, the propositional mind emerges out of the affective mind. Psychology increasingly recognizes that the mind has two mental pathways—dorsal and ventral, cold and hot, indicative and imperative.[23] When we consider experiences like fear of a predator, we see that the two pathways are coupled—to put it simply, one is cognitive and the other affective. The emotion / cognition complex in predator fear is a dual experience, partly *imperative* (e.g., "I should run away") and partly *indicative* (e.g., "That creature is a snake") This dual-aspect representation (cf. PPR) is strongly coupled together in lower animals—mice, for example, simultaneously recognize cats as a kind of thing (in a category) *and* as dangerous (the affect of fear). In contrast, through symbolic thought, humans can decouple these two pathways (indicative and imperative), and fear can be reattached to alternative kinds of creatures / perceptions.

Language helps humans do two different kinds of "decoupling." A word like "cat" is a symbol, and it allows me to decouple and manipulate the concept from the immediate perceptual experience of a flesh-and-blood cat. I can invoke the idea of a cat (e.g., a big-cat predator) in my Pleistocene friend by giving the verbal signal. But if he has also had terrible experiences with lions and leopards (and he might well have), then I have also just frightened the daylights out of him. This is because the indicative representation of a cat also has imperative baggage (e.g., "Run away!") But language also has a way of decoupling the emotional imperative content from the indicative content of the representation. My friend does not become so startled and frightened that he runs around the campfire or cave when I say "cat." Maybe there's a frisson of fear, but he inhibits any desire to bolt.[24]

This separation of the emotional response (hot cognition) from the factual information (cold cognition) was probably fostered or furthered by the increasing abstraction of words. Some words, like onomatopoeias, have close connection with the things they describe, like "plop" or "fizz." Still other words could be closely tied to their referents through an imitation of the relevant sounds, such as when we talk to toddlers about a "moo-moo" for a cow or a "meow-meow" for a cat. Early natural languages may have started in this more tightly coupled way, where a hissing "*sssss*" could represent a snake.[25] Word sounds clearly move toward arbitrariness. There is nothing about "cat" that intuitively represents a cat in the way that "*sssss*" might do for a snake.

As language is increasingly untethered from its referent, humans gain a style of thinking that is more and more indicative and less emotionally imperative. As the symbols are neutered, they no longer subject the communicators to the emotional drama that is part of our mammalian inheritance. Language insinuates some emotional domestication because we need to inhibit our selfish and aggressive tendencies in order to engage in cooperative communication.[26] But language itself also helped *Homo sapiens* increase such emotional domestication because it allowed us to use arbitrary, agreed-upon signals that have no inherent emotional push or pull.

Once language helps decouple cold from hot cognition, we now have mental headroom to play with concepts, memories, and ideas, without triggering the life-and-death emotional systems that evolved to save our lives.[27] Our Pleistocene or Upper Paleolithic ancestors benefitted greatly from this abstraction of signals. Certain kinds of daydreaming, mind-wandering, and more structured imaginings could thrive and develop unchained from the emotional storms that usually accompany involuntary memories and dreams.

Additionally, as linguistic symbols grew more abstract and arbitrary, they need not trigger or interact with stored mental imagery—or at least the interaction could be glancing. This may be why humans can easily inhibit emotional responses to spoken communication; namely, the signals (words) have less emotional content. Images, being more directly sensual, have more emotional content than words.[28] Additionally, another mechanism helps explain the expansion of such a decoupled disinterested headspace; namely, the increase in brain-based executive control and self-awareness. Language lacks the emotional punch of hot cognition, but it can still have emotive motivational power. However, the newly evolved, top-down, executive-control system (via the frontal lobes) may continually frame the linguistic mental events as virtual. In other words, our linguistic minds seem to have a "staging" function, reminding us that our imaginings are not real. Moreover, we have a hazy sense that even language itself is but a player on that stage.[29]

The decoupling of semantic content from immediate experience, such that it can be recalled, manipulated, and projected into the future, is due to several evolutionary trajectories (some of which may remain obscured in the contingent history of deep time). Cultural changes in pre-sapiens social life (e.g., longer and safer childhoods), better storage, recall, and recursion of embodied action-loops (facilitated by the cortico-cerebellar system[30]), better executive control over default associational systems, vastly improved episodic memory,[31] and intensified social learning and then teaching[32] are but a few of the trajectories.

When these routes come together and produce a viable natural language (which probably happened in the Lower or Middle Paleolithic era), we arrive at the unique linguistic aspects of the human mind. We have been arguing that this is a slow decoupling of the indicative functions from the older imperative functions of verbal behavior. With natural language, the primate mind of action-oriented representations (AORs) receives a new layer of *delayed-action-orientated* representations and then *reflection-oriented* representations. What are some of the defining properties of this top layer of indicative linguistic meaning?

Analytic philosophy, in the days of positivism, searched to find the smallest unit of meaning, the supposed element of signification. Many problems arose with the positivist attempt to tether every elemental word to a discrete sense datum, including the problem of meaningful non-observables (e.g., quantum-level entities, causation, etc.) and meaningful language use by sensory-impaired speakers (e.g., blind people use "blue" meaningfully without sense-data correspondence). The subsequent holist movement noticed that any given element, like a word or phrase, takes its meaning from the whole language or at least from some connection to other terms, statements, and / or contexts.[33]

Philosopher Hilary Putnam provides us with a helpful vector for thinking about linguistic meaning. The meaning of a term like "water" can be found in its "(1) syntactic markers; (2) semantic markers; (3) a description of the additional features of the stereotype; (4) a description of the extension."[34] In other words "water" is syntactically a noun; semantically a liquid; stereotypically a transparent, hydrating fluid that fills our bodies, streams, oceans, and other things; and extensionally the set or thing described by the chemical formula H_2O.

Presumably, changes in the meanings of words are attributable to changes in their "stereotypes," which are altered by the context of the specific linguistic community. Unfortunately, analytic philosophers have never been very interested in the evolution of language per se, and the relationship between vector and real language development received little attention. Perhaps more disappointing, the analytic philosophy of language—even in its more holistic forms—has not sufficiently explored the affective or emotional content of such a vector, despite Wittgenstein's dictum that "meaning is use" (and affect surely plays a role in our use of language).

From our vantage point we can ask, what, if any, affective marking should be added to a Putnam-style vector? We have already argued that affect and feeling generally play a great role in the semantic structure of many representations, just as it does in their affordance predecessors. So, too, Putnam's "stereotypes" for terms as seemingly neutral as "water" would also have a significant affective component. Water, for example, also can take on such imperative meanings as "something I thirst for," "something I cool with," "something I cleanse with," "something dangerous," and so on. Affect is an important ingredient in the connotative and associational matrix that seems altogether absent in analytic approaches to language and concepts. Analytic philosophy of language may continue to help us parse the complexities of contemporary linguistics, but it is relatively unhelpful in tracking the evolution of mind. Moreover, its traditionally disembodied approach has failed to penetrate the adaptive functions of language.

The Evolution of Concepts

A successful theory of concepts will properly articulate the structural aspects of concepts (i.e., the relation of simple concepts to complex wholes, the relation of sense and referent, etc.) and the functions of concepts (i.e., how they categorize and entail). Additionally, a full theory will explain how concepts are acquired (i.e., synthesized from isolated sensory experience, analytically derived, connotatively associated, abstracted from simulations, etc.). We cannot resolve all of these questions here.[35]

It is our view that a given theory of concepts will win over its competitors in proportion to its ability (among other things) to incorporate affective precursors into conceptual structure. Each theory will eventually have to reconcile with an evolutionary account of affectively encoded affordances. Our project here will only suggest the way forward for prototype concepts (and perceptual symbols), since the analog nature of such non-classical concepts makes their affective structure more obvious. Our preference for prototypes, perceptual symbols, and enactive schema also results from the consilience of these prelinguistic concepts with biology generally (i.e., animal minds).[36] We leave it to others to annex their preferred theory of concepts with the affective semantics we have been detailing in this book.

In this section, we criticize the (dominant) philosophical tradition that pursues the propositional content of concepts and emotions, in some cases, going so far as to suggest that propositional content is not only necessary for meaning but also requisite for consciousness itself.

We start with a brief tour of the age-old propositional view of emotions. Next, we offer a clearer picture of how prototype concepts better interface with the more embodied forms of meaning that preceded our modern minds. And finally, we suggest a future research program into analogical thinking as a pre-definitional and prepropositional core of embodied cognition.

Classical Concepts and Prototypes

Imagination and language are ways in which humans gain greater control over their internal affective storms and dreamlike image streams. Based on the work of Stoic philosophers, from Epictetus and Seneca to contemporary psychologists like Jaak Panksepp and Joseph LeDoux, we have recognized that anger or fear is not just the instinctual physiological response to stimuli (e.g., fight or freeze) but also the cognitive judgment that "I have been injured" or "I am disrespected" and so on. Aristotle defines anger as "an impulse, accompanied by pain, to a conspicuous revenge for a conspicuous slight directed without justification towards what concerns oneself or towards what concerns one's friends." And he adds, "It must be felt because the other has done or intended to do something to him or one of his friends."[37]

The uniquely human form of anger, in this view, requires the angry person to judge a willful malevolent intention on the part of the offender. That is an assessment that requires TOM, or at least some distinction between accidental and intentional behavior. Human emotion is so intimately tied with higher cognition that Seneca thought it was impossible for animals to feel anger or other emotions. "Without speech, animals are without human emotions, though they have certain impulses that are similar to them."[38] As we suggested earlier, however, this level of conceptually entangled emotion is what affective neuroscience calls secondary-level affect, which involves interaction between sentient feelings and cognitive processes, often producing noetic content.[39]

Contemporary versions of this cognitive approach abound. Robert Lurz, for example, suggests that what makes one's beliefs and desires conscious is that one is conscious of their propositional content. Lurz goes so far as to suggest that the desire to drink water is unconscious until it causes you to have an awareness of the propositional content of the desire to drink water. That awareness of content is just a conscious desire to drink water. Conscious thirst, according to Lurz, is not a state of mind but a proposition or a packet of information.[40]

This view seems incorrect. We might be willing to concede that such a propositional approach maps onto beliefs—since some conscious beliefs can have a syntactic structure that derives from linguistic competence (e.g., I believe that my upstairs neighbors are dancing in high-heeled shoes). But desire and emotion are different from beliefs and cannot be grouped together in a facile way that bestows propositional content on them. Moreover, according to our view, adding propositional content to a desire to drink water, is, as the Chinese idiom goes, like "drawing feet on a snake" *(hua she tian zu)*—it adds properties that are superfluous. Thirst is the motivational primary-level feeling state of a homeostatic imbalance and a need to correct it—not an awareness of a proposition. Finally, we are at a loss to imagine what on earth the propositional content / information would be for thirst. This is not to deny that thirst can be redescribed in a propositional manner via a rational reconstruction. But this weaker claim is uncontroversial, and it is the stronger claim—that awareness of propositional content is constitutive of desire—that strikes us as a thoroughly nonbiological approach to conative and affective forces.

This nonbiological approach begins by asking: How much propositional content is in an emotion? Up until now, in this book, we have been resisting the familiar assumption that propositional cognition gives affect its intentionality. The modern human mind lends a certain phenomenological credibility to this familiar assumption, because we can indeed direct affect within the propositional framework of thought. Our linguistic system can recruit and organize affect. But premodern mind, we have suggested, is directed more by the affective appraisals contained in perception, navigation, and inchoate "signs" from waking and dreaming associations.

We have reached the point now, presumably sometime in the Upper Paleolithic, where language can be said to be constitutive of mind. But the familiar question—How much propositional content is in emotion?—is the wrong question. A deeper question, or the more accurate starting place is: How much conceptual content is in an emotion?

This is the accurate starting place because only some concepts are propositional, while others are prototypes. A popular version of conceptually constituted emotions can be found in Lisa Feldman Barrett's "conceptual act theory."[41] Barrett's theory is so tied to the conceptual aspect of emotions that she has to deny that emotions like fear, or anger, or lust could be "natural kinds."[42] Against neuroscientists like Panksepp (and emotional taxonomists like Paul Ekman), Barrett claims that there are no discrete emotional systems. She opts instead for a view of emotions wherein human agents mentally construct emotions from memories, perceptions, concepts, culture, and connotations. According to Barrett, this rapid emotional processing is largely invisible to the subject in the same way that perception is constructive but seems naïve and unmediated.

Unable to fully deny the embodied nature of an emotional experience, Barrett and other constructivists posit the existence of a core affect—a kind of physio-feeling tone, neutral in itself but pliable enough to take any conceptual packaging. The core affect becomes anger this morning and love this evening, depending on my own cognitive labeling. The core affect can have either positive or negative valence, and high or low intensity, and we apply a wide range of emotional labels to the same state. But this seems strangely circular, because we want to know why this advancing, barking dog always makes my affect "intensely negative" and mobilizes escape behaviors. Even if Barrett is correct about core affect, and we do not think she is, then the reliable and environmentally salient changes of core affect cut against her view.

While we appreciate the need to track the conceptual aspects of human emotion, we cannot side with a cultural and conceptual construction argument for several reasons. First, the fact that my anger draws upon a rich network of previous experiences, connotations, observations, and folk theories about the way the world works is trivially true, and it serves poorly as counterevidence for the existence of innate affective systems. After all, those of us who think the affects are innate also argue for levels of increased cognitive, linguistic, and even cultural top-down influence on those systems (i.e., primary, secondary, and tertiary levels of consciousness).[43] Yes, there is cultural and individual diversity and variation in the way anger or lust is felt and expressed, but these variations are explained sufficiently by tertiary-level complexity, not through a denial of emotional natural kinds. Moreover, a sophisticated natural-kinds view is supported by years of neuroscience that isolates specific subcortical regions (e.g., the amygdala, the periaqueductal gray, the thalamus, the ventral tegmental area, etc.) across phylogenetic clades. And perhaps most importantly, we cannot accept Barrett's extreme form of conceptually constitutive emotions because, as presently stated, it fails entirely to explain animal emotions. This may be the result of a poorly worked out theory of concepts, but at the present time at least, the "conceptual act theory" is more like a complaint than a theory.[44]

We have already claimed that minds can make very adaptive moves without representations (see Chapters 3–5), and here we want to suggest that when representations eventually come online, they need not have a language-like syntax. The "language of thought" hypothesis is assumed by most contemporary cognitive scientists and maintains that concepts are word-like units of meaning that have subject / predicate structure and submit to logical functors like quantification and conditional relation.[45] These are classical concepts. However, our discussion of images and visual thinking—together with the fact that ancient and Enlightenment philosophers took "concepts" to be mental *images*—remind us of how recent (and provincial) this reliance on the primacy of propositions for thought comes off in this era of the computer metaphor.

Recall our discussion of prototype thinking that emerged in the section on imagination in Chapter 6. We recognize a rabbit, for example, more by automatically *associating* it with memory images, not by subsuming the percepts under a formal abstract concept. We associate this brown creature with a prototype memory—a learned and stored master image of a rabbit—but this process also enables the kind of classification activity that traditional "classical concepts" furnish. I do not look for the necessary and sufficient definitional conditions for calling this approaching animal a rabbit, but instead perform a perceptual morphological comparison against a mental prototype.

The prototype concept is not a necessary and sufficient definitional structure, but it is a probabilistic category, including entities that satisfy a sufficient number (none necessary or exhaustive) of properties associated with that exemplar.[46] Prototypes are conceptual cores that can be enlisted in compositional thinking; for example, "The wild bird is hungry" has no corresponding prototype but is composed of the elemental prototypes "wild," "bird," and "hungry." In the prelinguistic mind, such composition can happen in internal simulations that have image and narrative structure, rather than definitional structure.

In cognitive science circles, the debate is usually framed such that one either supports classical (propositional) concepts or prototype concepts. But biology rarely works with such neat dichotomies, and there is a high probability that the mind employs both classical and prototype concepts. This can be understood through the model of parallel systems, which solve mental-processing challenges in digital and analog formats. Or it can be understood through the developmental, evolutionary model, which places nonlinguistic prototypes as pre-modern concepts (similar to perceptual symbols) and propositional concepts as later forms of decoupled linguistic representations. But this evolutionary view is not meant to suggest that propositional concepts replace earlier primitive prototypes in some ladder of progress. If prototypes are older—and animal minds suggest this—then they are still with us, operating both independently from language (in visual, motor problem-solving) and sublimated inside linguistic thinking. The mind is filled with such adaptive redundancies.

We are sympathetic to a developmental rapprochement between propositional and prototype concepts. But is there an intermediate species of cognition that acts as a missing link between prototype and propositional concepts? Or, is there a bridging cognition, shared by both modes of thinking? Here we make only a tentative suggestion, as a fuller exploration is beyond the purposes of this book. We suggest that *analogical cognition* may well turn out to be the homological capacity we are after.

Others have suggested that cognitive templates or schemata form a foundation for both language-based concepts and prototype concepts.[47] And our discussion of spatial navigation above may contribute to an understanding of how schemata first emerge. If we confine schemas to body maps or spatiotemporal extrapolations from sensorimotor experience, then they seem both possible and helpful as a bridge to concepts.[48] But most theoreticians describe complex schemata—erroneously, in our view—that are already composed of many concepts, leaving the priority question confused. For example, Pascal Boyer claims that my concept of a *walrus* has certain specific features (e.g., lives in the sea, bears live young, has tusks, etc.) but this concept is derivative of a template (or schema) for *animals* generally—containing variable slots for size, shape, local living environment, reproduction method, and so on.[49] According to Boyer, the learner builds the concept *walrus* quickly and easily by plugging information into the *animal* template. We do not take issue with this theory per se, especially as a learning theory for modern minds, but eventually, one needs to explain where the templates come from. And this is even harder to explain when the template, in this case "animal," is so obviously composed of complex conceptual content.[50] The egg-chicken question of priority is further reinforced when we consider that many "chunking" mechanisms of learning, like conscious goal-oriented chunking and automatic perceptual chunking, reveal that some templates are chunked together from more primitive elements (informational, conceptual, and perceptual).[51]

Analogical Modeling

For the moment, we set aside these higher cognitive maneuvers in favor of a more rudimentary mental activity—analogy thinking. Analogical cognition discovers similarities between behaviors, forms, sequences, and patterns generally. Discovery of similarities may be automatically achieved through association, or actively constructed. Analogies are useful and adaptive because they help the agent see a novel event as similar to an already experienced event, opening up response maneuvers and capacities. Douglas Hofstadter argues that analogy is the very heart of cognition.[52] And analogical cognition works best when its elemental terms are fuzzy enough to admit meaningful comparison.

Analogical cognition is the perception of common elements between two things. The sophistication of such perception is contingent upon other faculties, including memory, conditioning, representational power (e.g., motor or linguistically based, etc.), affective memory, and so on.[53] Analogy-making underlies a very wide range of animal skills, including the way we plan and map our movements through spatial environments and the sophisticated way Plato describes learning as climbing out of a dark cave. Linguistic metaphors and similes are just the tip of the iceberg once we realize that perception itself can have analogical aspects (e.g., "seeing as") and associative mechanisms.

The Relational Matching to Sample test, or RMTS, is a visual test that asks the participant to first see a visual sample (e.g., two circles) and then see subsequent sets (e.g., a square and a triangle, a rectangle and a circle, and a square and square), making an analogical connection between the first and the subsequent sets (e.g., the two squares). This ability to see things as similar or different can be taken into entirely new contexts where the perceiver has no previous experience. Apes have been tested using the RMTS and show significant analogical abilities, and more recently, monkeys have also been shown to possess analogical cognition.[54] We suspect that such tests are capturing the kind of fleeting and ubiquitous analogical connections that primate minds are regularly making.[55]

Analogies can be said to have fundamental structures and mechanisms. Consider the metaphors "Love is a journey," "The melody is rising," and "He is cold as ice." What Johnson and Lakoff describe regarding metaphors is true of analogies more generally—they have a source domain and a target domain.[56] The analogy maps a relatively known pattern (e.g., peanuts taste good) onto a relatively unknown pattern (e.g., newly discovered pistachios) and thereby gains better traction (pardon the analogy) on the environment. Arguably, *analogy is an early form of decoupling,* since it triggers and recruits a mental state (source pattern) that reveals information about the immediately perceptually-present pattern (target pattern).

Additional mechanisms of analogy have been explored in Hofstadter and Sander.[57] For example, it is common for single individual things to be "pluralized" into a multimember set. Here we see the prototype concept being used as a class concept, from simplistic levels of taxonomy (e.g., primate herpetophobia) to complex cultural categories (e.g., "This class is filled with little Einsteins"). Or consider the cognition of chunking, wherein primordial elements are synthesized into a new unity, now contained in a virtual membrane as one concept.[58] Diverse people are chunked together as one family, for example, and then these families can be chunked together into a community, and many communities can be chunked together into a town, and so on. The components inside the chunks are largely invisible as we use the downstream concepts. The analogical mechanisms that underlie the abstract connections in chemistry and physics, for example, are not different in kind from the simple play of a child who sees his building blocks as a "castle."

Important for our purposes is the crucial role that affect and emotion play in analogical thinking. Language obviously contains a large lexicon of phrased styles that communicate as much, if not more, than the lexicon of mere words. Indeed, we have suggested that the origin of language is in the affective manipulation of our social ancestors (e.g., grooming, soothing, etc.). The affective roots remain in our contemporary language. The same word signal can be conveyed aggressively, affectionately, fearfully, indifferently, joyfully, and so on.

From the most basic analogies to the most sophisticated, source patterns will have affective tone and thus affective content. This affective content will map onto the target pattern of the analogy, and the degree of this affect transfer will depend on the level of executive control of emotional editing the animal can do. Likewise, most target patterns—like a newly encountered animal or environment pattern—will trigger affective and conative content automatically. Indeed, in some cases, the affective experience is the largest part of the analogy because different percepts may stimulate a similar unconscious emotion (or set of valences), which drags a specific memory pattern to the surface of consciousness.

Our suggestion is that analogical cognition underlies both prototypical concepts and class concepts, the latter taking on increasing abstraction and decoupling via language. Affective systems may have greater penetration in prototype concepts, since those are more like attenuated percepts or action patterns, but emotion is also a large part of analogical language and propositional thought as well.

So, now we return to the question that began this section: How much concept is in emotion? In truth, the question should be reversed, and we can acknowledge that, except for math and other self-referential symbol systems (e.g., logic), a substantial degree of affect colors our concepts. Our concepts are not disembodied Platonic forms, but embedded representations that help us navigate the environment. Decoupling is the process by which representations are shorn from their birthplace in direct experience, and language is probably the most effective decoupler we have. When language emerges, *Homo sapiens* acquire a whole new system for analogically triggering adaptive affect, in addition to communicating needful information. A whole new offline lexicon of counterfactuals can conduct emotion and action between speaker and listener. But our propositional aboutness of language (indicative referential content) is already embedded in the emotional aboutness of our social interaction with other humans, who we are trying to assuage, impress, attract, or destroy.

8

Affect in Cultural Evolution

THE SOCIAL STRUCTURE
OF CIVILIZATION

AS SOCIAL INSTITUTIONS become a part of our lived environment, culture serves as a secondary niche for the species. This unique ability we have to learn from others and to transmit and reproduce knowledge opened a world beyond genes. While sociocultural arrangements are responses to ecological pressures, with each solution making survival possible in the given ecological circumstances, the most telling part of the anthropological record demonstrates how the ecological pressure of population growth called forth new technology (like hand axes), new forms of social organization (like the nuclear family), and political regulation (like the chiefdom). But how do sociocultural arrangements remain bound to, and grow out of, our affective needs and motivations? We will argue that social norms—from reciprocity to ritual restrictions—are ultimately forms of affect management.

Recent interpretations of the archaeological evidence emphasize rational adaptation as the mechanism for changes in social organization. For example, Bogucki states, "[T]he overall sweep of prehistoric society was the cumulative result of decisions made by self-interested individuals," while Samir Okasha and Ken Binmore have similarly sought to join evolutionary biology's notion of *fitness* with the economic understanding of *utility*.[1] We argue that a different interpretation of the anthropological record, as well as the findings of recent behavioral economics, including the work of Daniel Kahneman, Amos Tversky, and Kim Sterelny on cooperation, challenge these rational choice models of cultural evolution.[2]

While rationality surely played a role, we must account for the fact that the proximal causes of our interaction are based on social norms, which themselves are the result of social behavior underwritten by affective forces in the human mind, which, in turn, originated well before the rise of rationality and continued to develop through their synthesis. As a counternarrative to the rational and self-interested model of cultural evolution, we emphasize the role of affect in both orienting behavior and in shaping various forms of social organization. We argue that affective adaptation to the specific ecological and social topography of human groups is a causal factor in the creation, maintenance, and eventual transformation of the social norms that constitute culture.

Social Stages for the Evolution of Affective Norms

Society is the result of the interaction between population size, ecology (i.e., food sources, waterways, weather patterns, etc.), and technological innovations, which is aimed at bringing the needs of the population in line with the ecological possibilities for exploiting environmental resources.[3] Population growth seems to be the single most important factor in calling forth technological innovation, the social organization of production, and the political regulation of social hierarchy. The three interlocking evolutionary processes in the transformation of society are thus: (1) the intensification of subsistence efforts on available ecology, (2) the amount of political integration necessary to secure subsistence internally and in relation to bordering communities, and (3) the social stratification that dictates power and the sharing of responsibilities.

A standard way to describe the size of human groups in relation to environmental resources is as follows: *Family-level group* refers to a foraging society of up to thirty-five people that re-forms each season. In this hamlet community, leadership is not permanent; rather, it is specific to the task at hand, whether that be hunting a stag or making a net to catch fish. For example, the !Kung of the Kalihari Desert in Namibia make decisions by informal consensus reached through long discussion.[4] If a hunter demonstrates remarkable skill, he acts as the informal leader of the hunting party but does not hold leadership in other corporate enterprises.[5]

The *Local-level group* is five to ten times the size of the family-level group and, being less nomadic, includes organization for common defense and surplus food storage. The technology employed in the local-level group includes methods and spaces for the domestication of animals and plants, as well as tools for the exploitation of maritime resources. These actions require group cohesion and cooperation, and thus in the local-level group, we see further development of the kinship bonding that started in the family level. Humans, at some point (presumably at the family-level), become domesticated, and domesticated humans are bound to one another in a new way. Kinship calcifies one's social and emotional world through blood bonding and creates the conditions for a sense of empathy, norms of reciprocity, and impulse control.[6] This social bond is institutionalized in the family-level group (and more formally in the local-level group, etc.) via the defining of corporate kin groups in daily life and through ceremonial associations that present social hierarchy to the larger group within the context of status and its attendant rivalry. It is in this group that we observe the advent of ceremony, a political regulatory mechanism that publicly defines subgroups and, consequently, social hierarchy. In groups larger than the family level, one's affective system has to be formatted for kinship structures, which includes *fictive kin* bonding, like the village, state, or local sports team. Aimed at bringing individuals together into groups for purposes of social cohesion, these fictive kinship structures are perpetuated by ceremonies and adornment practices. For example, in the Trobriand Islands, when a daughter or sister is married, a substantial bag of yams must be given to the new son-in-law or brother-in-law.[7] We also observe the *kula* ceremony in these islands, off of eastern New Guinea, where gifts are publicly presented and received with a wide range of competitive display. Such ceremonies constitute political maneuvering and build reliable coalitions.[8]

The *Big Man* group is a type of local-level group of at least 350 and at most 800 people, featuring a leader who decides for the group how to manage risk in corporate enterprises. For example, in trade relations with other tribes, the Big Man is the go-between, or, in arbitrating internal disputes, the Big Man decides who is right, who is wrong, and what is to be done. The Big Man in central Enga, with its public dance ground and ancestral-cult house, also serves as a master of ceremonies in ritual dances and ancestor worship ceremonies.[9] While in the Indian societies of the Northwest coast of North America, the Big Man maintains great stores of provisions and a retinue of warriors in preparation for holding sumptuous feasts that display his wealth (and, therefore, that of his community) at interregional potlatch ceremonies.[10] These communal events display group hierarchy by formalizing intergroup relations (for example, who sits beside whom at the banquet table) and forging kinship and practical interfamily bonds (for example, in the wedding ceremony). Ultimately, the duration and amount of power held by the Big Man devolves largely on his personal initiative.

Some anthropologists have refined this gradualist view of social change further by offering a "punctuated equilibrium" model of transitions, from local-level groups, to chiefdoms, up to states.[11] According to Abrutyn and Lawrence, adding punctuated equilibrium tempos to the gradualism model helps explain certain empirical patterns.[12] There was a Mesopotamian jump, for example, from small villages prior to 4000 BCE (each containing a couple hundred people) to around 20,000 people in Uruk by as quickly as 3300 BCE.[13] Moreover, punctuated equilibrium pacing also helps explain the intensification of social hierarchy. Chiefdoms, for example, may arise more quickly than gradualist models suggest, because self-aggrandizing individuals and prestige mechanisms speed the process. And a particularly astute entrepreneurial warrior-shaman or a genius engineer of irrigation can be a game-changer in the pace of successful social growth. Nonetheless, our argument in this chapter—that emotions play a crucial role in both social stability and transformation—is unfazed by the resolution of the gradualism-versus-punctuated-equilibrium debate. We will hew more closely to the gradualism model (like that of

Johnson and Earle, for example) because it provides a clearer picture of the social taxonomy, but we concede that anthropologists may differ on details and pacing.

In a *regional polity,* chiefdoms (with communities of up to 500 people and polities of at least 1,000 people), develop in warring societies that require a leader to manage the risk of armed conquest and to incorporate defeated communities. For example, the ancient Incas of Peru were notorious for their sacrificial ceremonies and insatiable drive toward conquest.[14] Insofar as the act of conquest makes available confiscated goods that are above and beyond the subsistence needs of the residents of the polity, such a society needs financial technologies, such as elite investment, and thus wider-ranging control of propertied resources. Again, in chiefdoms we observe ceremonies that serve to legitimize the leadership of the elite set of families, which are directed by the chief. In this polity, rituals and ceremonies will occur in a particular space, such as a shrine; these spaces are controlled by elite figures, like priests or armed forces. To take one example, the Incan chief, by founding the state shrine and building a domicile for himself as a divine person, moved the regional capital to Cuzco in an act that united religious, political, and social groups.[15]

The *State* includes vast populations numbering in the millions. This population includes defeated communities as well as migrant workers and is thus ethnically and economically diverse. States function through the financial technology of finance economies that maximize surplus production, which is then translated into power and political survival for elite families. The issue of elite ownership of resources and technology becomes more acute, as a system (i.e., social technology) of legal property entrenches economic holdings. With such high stakes, the state requires great defense and conquest units like the army and the navy, or in ancient Rome, the praetorian guard.[16] Due to the amount of mouths to feed, technology must be developed to organize subsistence, and this brings about bureaucratic bodies, including regional law-enforcement units. In the State, ceremonies mark phases in the economic round, glorify fictive kinship structures, and legitimize unequal access to resources.[17] Consider the fervor of fictive kinship, qua nationalism, during the Olympic games. Or, if you are reading this from anywhere but the United States, the emotional bond that football clubs and national teams evoke during the Euro, Africa, and World Cup competitions. For an example of legitimizing the elite, one need look no further then Washington, D.C., in 2008, when a cabinet of economic experts was carted out to explain that multinational banks deserved a federal bailout for their complex usurious practices while individuals who signed subprime loans must face home foreclosures.[18]

Social Organization and Individuals

For the purposes of determining how the evolution of society relates to affective and conative forces, consider the following changes to an individual's daily life in the shift from small-scale (family-level) groups to industrial society (Local-, Regional-, and State-level groups). Ultimately, a nuclear family's duties become divided between a female domestic sphere on the one hand and a male productive commercial sphere on the other.[19] The time spent working on a daily basis, spread between men and women, increases, while time spent housekeeping grows, with women in the home for more time than men. In the domestic sphere, time spent manufacturing and repairing family possessions decreases by two-thirds, as the public market takes on a more substantial role.

In terms of warfare, in small-scale hamlets, aggression is personal and often leads to revenge-killing, while in local groups, warriors raid other villages as directed by Big Men. In the chiefdoms, systematic warfare is often waged against neighboring chiefdoms for land dominion and the acquisition of surplus goods. With increasing scale, technology, the social organization of production, and competition all come under the control of leadership, which generally acts to protect the surplus value-based economic inequality from which they benefit. In all forms of human society, institutions are sanctified by rituals and protected by taboos as a means of invoking awe to stabilize patterns of behavior.[20]

As economies develop, the basic elements of social interaction—reciprocity, redistribution, and exchange—are transformed. A subsistence economy, which satisfies root household needs for clothing, food, housing, and defense, is an attempt to maximize production but minimize the effort expended to meet needs.[21] Subsistence is a stable format, but group action and leadership become necessary as subsistence problems grow, and thus we have political economy (i.e., those institutions that, through political elaboration and rules, control free-riders in communities bigger than family-level groups). This type of economy includes finance and thus new forms of social complexity involving the exchange of goods and services in an integrated market society of interconnected families.

As a consequence of intensification (i.e., positive feedback between population growth and technology), problems arise; for example, holding reserve subsistence materials in community storage leads to opportunities for control. Also, raiding of and warfare over community storage and village land require an organized defense. Furthermore, technology needs require investment to maintain resources, and resource deficiencies lead to complex trade relations that require political economy in the form of bureaucratic institutions. All these factors bring about broad political-economic integration in an elite class.[22]

As political economy evolves, it becomes geared to mobilizing a surplus (or tax) from the subsistence economy that is used to pay for and justify elite ownership of productive resources and to finance social, political, and religious institutions.[23] Political economy then becomes growth-oriented, with power gained through investment that expands income in finance markets. Elites recognize limits to growth and try to overcome them by instituting major capital improvements funded by surplus. All the while, elites may use power, ceremony, and ritual to perpetuate their access to wealth; some in this position seek self-aggrandizement.[24] Meanwhile, individuals who do not participate (for any number of reasons, including being blocked out) in political economy come to be deemed outlaws. In this sordid segment of society, we see high rates of male death by violence and limited leadership of small groups in loose collectives, with frequent violence over territory and resources. Today, we call them gangs.[25]

Stages in the Evolution of Society

For ease of reference, we consider the evolution of society in the format of three stages of social institutions. These stages are all present in contemporary society, with some stages nested within others.

Stage one is the most basic economic unit: the level of nuclear families, households, and hamlets in a primarily subsistence economy. This is probably how early *Homo,* and indeed *Homo sapiens* lived, so it is no surprise that much of our emotional palette was built during the domestic-subsistence format of the nuclear family within hunter-gatherer communities. Some have taken this to mean that the human mind is a creation of the Pleistocene (i.e., the theory of "Environment of Evolutionary Adaptedness," or EEA) and gone on to reverse-engineer a raft of fixed cognitive modules that populate such a mind.[26] Our own view is that emotions are open-ended, domain-general, feeling/behavior matrices that adapt to the influence of culture and cognition to play a horizontal and vertical role in human and mammalian learning and social living.[27]

In stage two, we find local and regional groups, including the agrarian polity. In this stage there is population intensification, technological evolution, and more social interaction with strangers. The ecological pressures on our emotional systems during this stage are managed by the social technology of norms and ceremonialism, including myths and religious rituals. The shift from hunter-gatherer bands to agrarian states in the Holocene influenced a "release from proximity"—i.e., a loss of immediacy—which transformed enforced sharing and led to empathy taking on new forms in non-kin social groups and social norms.[28] These forms of social organization included (1) the creation of fictive kin (making "family" from non-blood conspecifics), which we argue is mediated by the CARE system; (2) awe/sanctity/reverence emotional relations to the chief/god/group, as mediated by the FEAR system; and (3) directed aggression in warfare, as mediated by the RAGE system.

Finally, stage three represents contemporary urban global society. Weber defines the State as a human society that has a monopoly on the use of legitimated force within a particular geographical area.[29]

We could consider this society as originating with industrial capitalism and continuing through the present digital revolution. This is a political economy, not a subsistence economy, and it includes intense urbanization, with daily interaction among a diverse set of strangers in cooperation and conflict. Further growth in the disparity between elites and a labor majority is observed, with major economic stress falling on the vulnerable family subsistence unit.[30] The emotional atmosphere here reflects a mix of retained emotional adaptations from our tribal period (stage one), emotions heavily modified by ecological and cultural factors (stage two), and even some new tertiary cognitive aesthetic emotions. Notice the highly flexible nature of contemporary tribalism during this stage (e.g., soccer team affiliations, music fan culture, etc.), suggesting emotional bonding is no longer dedicated exclusively to family. Nationalism is another stretched-out form of tribalism, marshaling all the emotional power for rather abstract social solidarity.

Running like a fault line through these stages is our ongoing question: "What motivates human behavior?" Is it reason? Emotions? Conditioning? When our ancestors chose to forage, or hunt, or mate, or make war, or burn a patch of crops, etc., was it a rational cost-benefit calculation or just imitation of respected members of their community? We have largely focused on the non-cognitive aspect of affect in this book, but cognition is clearly part of the equation at this point. We suspect that affective systems provide the dominant form of motivation in stage one, and maybe even in stage two, but cognition is online during stage two and, in the form of culture, probably dominates in stage three.

Organic social organizations that arise like gestalts from the interaction of individuals must be shaped not only by ecological factors like population density and geographic location but also by the underlying affective constraints and goads of individuals, and, subsequently, by the social norms that limn those behaviors. Arising from our seven basic primary and secondary affective drives is the development of tertiary processes, including honor, respect, pride, shame, trust, reciprocity, love, and courage. Below, we describe how humans use hierarchy, respect, reciprocity, identity, and ceremonies to constrain conflict in the same way that in Chapter 4 we described how bonobos use sex to neutralize conflict when food is available. That is to say, social norms are our most efficacious form of emotional management. In this chapter, we discuss the manifestations of these tertiary social emotions, but first we must address the notion of *Homo economicus*.

Homo Economicus?

In book II of Plato's *Republic,* Socrates and Glaucon argue about the origin of society. Glaucon expresses the somewhat pessimistic view that people only cooperate with each other to form social contracts, because they cannot fulfill their deeper selfish desires. The average person, according to the pessimist, is motivated to acquire the goods of others, take sexual advantage of whomever they like, and generally live like a despotic tyrant. That is the best of all possible worlds. Unfortunately, there is always some thug or bully who is bigger than you, and just when you're settling down to your spoils and ill-gotten gains, this bigger thug beats you until you relinquish your goods. That is the worst of all possible worlds. Glaucon summarizes the birth of cooperation: "This they affirm to be the origin and nature of justice;—it is a mean or compromise, between the best of all, which is to do injustice and not be punished, and the worst of all, which is to suffer injustice without the power of retaliation; and justice, being at a middle point between the two, is tolerated not as a good, but as the lesser evil, and honored by reason of the inability of men to do injustice."[31] This argument presages Freud's argument in *Civilization and Its Discontents,* which suggests an inevitable tension between the egoistic drives of the individual and the altruistic requirements of society.[32]

In response to this pessimistic social contract, Socrates offers a more optimistic origin story. Necessity is the mother of invention, Socrates says, suggesting that our need for food and shelter draws humans together into a small band of reciprocal providers. We naturally share and help each other.[33] For example, I am naturally skilled at finding tubers underground, and you are naturally good at building shelters. An alliance forms, and with a few other contributing members, we arrive at the original band of cooperators. A natural division of labor (based on innate skills mixed with training) produces a healthy micro-State, or hamlet, in which the members mutually benefit each other. Contrary to the Hobbesian pessimists, the community members of Socrates' imagined scenario do not merely endure each other as they repress their urges to rape, kill, and free-ride. Whether he manages to scale-up this tiny utopia is for readers of the whole *Republic* to judge—for example, it runs into trouble almost immediately when Socrates adds even humble luxury items.[34] But, for our purposes, it is interesting to notice that the pessimistic social-contract theory and Socrates' more optimistic version have different emotional-social ecologies but rather similar assumptions about rational choice.

Social interaction, in both these models, has an affective tone (e.g., in feelings of misery, frustration, joy, etc.), but the affective element is not taken as elemental. Theorists treat emotions as epiphenomena to the structural factors involved in social organization. Cost-benefit calculation, on the other hand, is taken as elemental or constitutive. The pessimistic social contract, proffered by Glaucon and Freud, is driven by a measure of costs (being pummeled by alpha members) against benefits (occasional success with stealing). And the Socratic version derives social organization from our individual judgments that greater benefits will accompany a rational barter system of skill-trading. Our own view is that affect binds and motivates the affiliative and antagonistic structures that make up society.

In 1965, economist Mancur Olson reopened the debate by questioning the optimistic assumption that individuals will take collective action whenever members are jointly benefited. As soon as the group size becomes larger than a tribe, some sort of coercion is needed, according to Olson, to make individuals act in their common interest. Olson claimed that "rational, self interested individuals will not act to achieve their common or group interests."[35] This "zero-contribution thesis" evolved into the "prisoner's dilemma" literature of the 1980s and still characterizes the collective-action problems discussed in contemporary social sciences.[36]

However, another economist, Elinor Ostrom, notes that the zero-contribution thesis is not borne out by observations of everyday life, wherein "many people vote, do not cheat on their taxes, and contribute effort to voluntary associations."[37] Whilst it is true that free-riders are always a potential problem, the participants in any long-term organized social group invest resources in monitoring and punishing violators, and thereby reduce the temptations and probability of widespread free-riding.[38] But Ostrom and other economists are puzzled by the gap between theoretically predicted selfishness (qua zero-contribution) and the empirical facts of widespread cooperative behavior. Her "central finding" to help resolve the puzzle is that "the world contains multiple types of individuals, some more willing than others to initiate reciprocity to achieve the benefits of collective action."[39]

We do not disagree with this central finding (indeed it seems almost trivially true), but we feel that an affective paradigm provides some of the mechanics behind such reciprocity motivation. Moreover, our theory that emotional systems generate human social behavior eliminates the mystery, or puzzle, of human collective action. Once the CARE system or the LUST system is triggered, it does not function with perfect fidelity on the optimal targets. It is not a rational choice that discriminates the best possible conspecific for the relevant activity. It is instead a serviceable and middling world of averages.[40] In the same way that we do not get to choose the best mate possible but one ready-at-hand, we also do not operate with perfect selfishness, perfect altruism, perfectly calibrated maternal care, well-focused aggression, or anything else. That's not how biology works.

There are several species of birds in which individual adults help take care of offspring that are not their own. Like human cooperation, this is equally mysterious to economists and game theorists who think that cooperation is incompatible with the selfishness of the "Darwinian" imperative. It is indeed costly for a bird, like the *Malurus,* to care for an offspring that is not his own, and he gets no direct reproductive benefit from doing so. One wonders how such cooperation evolved, since it does not maximize the fitness of the agent. But as Andrew Cockburn points out, the fierce competition over female birds results in many reproductively potent males being passed over.[41] Sexual selection produces some winners and some losers. But if every bird has evolved the behavioral potential to care for young, then the bachelors simply actualize that potential on any offspring they can. The system was built for one target (a genetic offspring) but can easily fasten on a different target (a conspecific's offspring). The process is not much different from that of human couples who adopt and love their non-biological offspring with full affection. The affective systems create action-patterns that are largely successful and adaptive but are nowhere near the idealized precision of rational-choice modeling.

The cooperative social contracts of the Holocene period seem to pose special challenges, because agriculture created the condition for accumulated wealth.[42] Prior to agriculture, humans could only accumulate the small resources of an individual hunt or season, and the lifestyle of the hunter-gatherer did not allow for multiyear control of fields—control that could be passed down through many generations. Economic inequality emerges in the Holocene and quickly contours the subsequent forms of social organization. For philosopher Kim Sterelny, the Holocene social contract is particularly mysterious because norms for property rights and coercive police powers did not yet exist, so how did the "have-nots" ever tolerate such an unlucky contract of servitude with the "haves"? How did the system get off the ground, so to speak?

Our affective approach may solve Sterelny's challenge. People already had a disposition to sacrifice their own interests for others in a group long before the rise of elites. Two sources of affective management are already in place before the rise of agriculture—namely, filial emotions and mammalian social submission. Chiefs, elites, and oligarchs can exploit the natural filial emotions that influence kin reciprocity and altruism (see our discussion of honor and respect below). Additionally, the dominant / submissive hierarchical tendency of social mammals (i.e., the class system of sexual and resource access) can be easily translated into the privations of the Holocene economy. Thousands of years before agriculture, there was an affective psychology that accompanied the social life of betas and omegas (just as there is an affective psychology of alphas).

We have no explicit evidence from the Holocene of how affection and fear scaled up to larger social norms and then laws, but we do have a compelling case study of a similar extrapolation during the Warring States period of ancient China (circa 453–247 BCE) and other Axial Age transformations. Confucianism, for example, sought to build the larger social ethic on the virtues and duties of the nuclear family. Being a good son, for example, instantiated high degrees of deference to the father and elders, as well as high degrees of altruistic alloparenting to younger siblings.[43] These privations were to be compensated at a later time, through virtue rewards, honor, and inheritance. No doubt, some of this self-sacrifice was the kind of emotionless cost-benefit calculation of the Thrasymachean pessimist, but most of it was premised on sincere affective filial devotion. Such a system might be easily exploited, but it reveals the inchoate accommodations of later forms of resource disparity. Large numbers of resource-poor citizens will submit to inequality if cultural-emotional mechanisms can convince them that they are all fictive kin, that their leader is alpha, and that some invading competitor group is at the gate. None of this cooperation and sacrifice may be an explicit cognitive conscious weighing of costs and benefits.

When we consider cooperation as an outgrowth of affective systems, our occasional irrationality and even self-defeating behavior is no great puzzle. A facile link between evolution theory and rational choice theory is tempting but should be resisted. Okasha and Binmore reveal interesting parallels but also important mistakes: they point out that in rational choice theory, agents make choices that maximize their utility, while natural selection also discriminates between alternative phenotypes according to the criterion of fitness maximization.[44] This assumption or association should not be treated as a solid foundation, and yet it has already led many evolutionary psychologists to mistake the model for the organism. Worse, it has led many economists to pin a model of rational man, *Homo economicus,* to evolution theory and to establish a mutually reinforcing circularity of game-theory hyper-rationalism.[45] If x does not maximize utility, then x disappears (as a human agent and as a phenotype), so any surviving agents or behaviors take their very existence from their success as utility maximizers. Hence, it is assumed, a proper evolutionary analysis is the same as a proper economic analysis. Every behavior becomes a rational response to a survival challenge.

Yes, rational choice theory assumes that the agent will be rational and therefore facilitate certain reliable predictions for the observer / theorizer, but that does not mean the biological creature will "live up to" this model-driven expectation. Rationality is embedded in a somatic affective substratum, and the substratum is further embedded in a contingent ecology. There are important contingencies that converge, creating states-of-affairs that are not optimal for decision-making. Also, phenotypic options and behavioral responses well-suited for a specific survival gauntlet may be unavailable for a variety of reasons. Instead, the affective systems, together with conditioned learning and relatively minimal cold cognition, provide us with better-than-random success responses to challenges. Lust and anger drive behavior more than rational utility maximization. And our view is not that scientific patterns are impossible to detect in emotional strategies—indeed, our adaptive view of the emotions requires that they be relatively reliable and subject to scientific scrutiny—but the rational choice approach already circumscribes the process too narrowly and over-idealizes the agent.

Our view is that evolution is contingent through and through, and that the organism is in a pluralistic causal matrix, where natural selection is dominant but interacts with other forces (i.e., coevolution, genetic drift, founder effect, etc.).[46] Moreover, natural selection itself should not be expected to produce either organisms that compute optimal solutions or organisms and behaviors that are perfectly adapted.[47] All adaptation is serviceable but need not be optimal. The fitness success of each phenotypic option cannot be measured against an idealized solution or abstract utility. Each success is only against its nearest competitor, and that competitor is a confluence of variables: the idiosyncratic behavior of the con-specific creature, its genetic mutation, its actualization of alternative potential, its dumb luck, and so on. You have to be only slightly better than your competitor—not a precision fit to the environment.

The Limits of Rationality

Our view of affective dominance throws doubt on the core assumptions of rational choice theory. We have doubts that rational choice theory mirrors the agent's true orientation to his environs (physical and social). The affectively oriented agent is not always maximizing utility. Moreover, the human being is not primarily a utilitarian egoist, but is (from well before the Holocene) deeply immersed in the filial bonds of kin relations.[48] Against the standard rational agent model, the affective systems complicate the causal story of social behaviors like cooperation and competition. For example, Robert H. Frank details the manner in which motivating emotions, like love and anger, can undo rational utility.[49] Many of us will be loyal to a friend even when it gives us no advantage and even when (less noted) we disagree with the friend that advantage will at least accrue to them. We simply enact fidelity (not merely signal it), no matter the cost-benefit calculus.

Similarly, revenge leads us, but leads us badly. We may burn every bridge in our social support system and compromise our own survival, all for the sake of vengeance. The rational choice theory will not help us understand such common human behavior. Gintis attempts to save the rational choice theory by adding the ingredients of strong reciprocity and revenge to human motivations, but this seems somewhat ad hoc (i.e., adding an extraneous hypothesis to the theory to save it from being falsified). It merely adds the tendencies without adding a convincing general theory about how those tendencies emerge from affective precursors.[50] Our view is that revenge is indeed an important addition to a successful behavioral model, but it is not a unique ingredient. Instead, it is one in a suite of affective drives that preexist rationality. It is the subcortical affect RAGE, refined and extended over time by the frontal, neocortical abilities of *sapiens* cognition.

While emotional action patterns can reduce and even obliterate rational calculations for rewards, they are not chaotic nonsequiturs. They are still mostly adaptive. Speaking metaphorically, emotions often have their own "sense of the good." Frank points out that emotions like love are well suited to solve commitment problems.[51] The affective experience of love, including illness symptoms (caused by cortisol-influenced contractions of the stomach's blood vessels), heart-racing (caused by increased adrenalin), feelings of empowerment (caused by endogenous oxytocin and opioids increasing pain tolerance), the desire for visual / physical contact (driven by dopamine), and so on, all converge to cement a pair-bonding relationship. These feelings don't just convince my lover that I am committed; they also convince me. Even when a rational calculation of a good partner might include how much money they have or how physically attractive they are, the feeling of love (once it locks in) will usually override such rational opportunism. Emotions like love may have been selected for in part because they trump simple egoistic cost-benefit agendas and provide affective mechanisms for prosocial allegiance. If you are too reasonable about love, you will not find it.

The now famous "ultimatum game" was first developed in the 1980s and has become an oft-repeated and well-known case of irrational economics. In the ultimatum game, two players are asked to divide \$10.[52] Player one proposes a specific division of the money—keeping amount x and offering amount y to player two. Player two then accepts or rejects the offer. If player two rejects the offer made by player one, both players lose all of the \$10. According to a purely rational agent theory, player two should always accept the offer from player one, because even a small payment is better than zero (the result of rejecting the offer). But player two frequently rejects a low offer, resulting in no money for both parties. Emotions explain this easily, whereas rationality (deliberative calculation based on utilitarian rules) does not. But anger or outrage in the face of some insultingly low offer is not just a maladaptive tantrum. Anger can be a precommitment mechanism that demands more fair distributions in the future. My anger is both a frustration of the endogenous SEEKING system (proximate cause) and a signal / threat to conspecifics that I require a substantial share of the resources (adaptive ultimate cause). Thinking about emotions like love, loyalty, and anger as commitment mechanisms makes sense of many human behaviors. It predicts many behaviors, like altruistic cooperation, indignant "bridge-burning," and so on, better than rational-actor theory.

Not only do emotional explanations solve anomalies of the rational agent model; they also explain many so-called maximal utility behaviors as well as or better than purely cognitive deliberation. Consider a rudimentary cooperative action of our ancestors, such as huddling behavior. We might make a rational decision that physical closeness and contiguity will keep a group of us warm through a cold night. But we might also feel, with our bodies directly, that huddling together increases warmth, and so the attraction of our bodies is an affordance process, not a deliberation. Your body is an affordance-generating, action-oriented representation for my body, and vice-versa. Our perception of each other comes loaded with imperative attraction in this case. Huddling together is driven by the homeostatic interests of our thermoregulation systems and the way such regulation feels to us. Both competing proximate causal scenarios—conscious deliberation and affective affordance—are compatible with and facilitate an ultimate causal explanation (huddling is adaptive evolved behavior). But just as there is no conscious, rational decision-making in natural selection—except the one we metaphorically project on it, so too there may be little or no conscious decision-making in the individual reciprocal warmth huddlers.

Of course, as social organizations scale up from huddled hunter-gatherers to tribes, chiefdoms, and even states, the ability to weigh costs and benefits in the virtual reality of representational mind takes on greater influence. Our advanced cognitive abilities can process increasingly complex social data from our tribes, and this improves our cooperative capacities.[53] But even our large-scale social ceremonies and ideologies are only as effective as their emotional component. Watch the stadium-sized synchronized movements, shouts, and displays of contemporary sports fans as fictive kin, and it seems that we are still, symbolically speaking, huddling together for warmth.[54]

Proximate and ultimate explanations of behavior are accurate if they pick out the real proximate and ultimate causal forces; otherwise they are intriguing redescriptions of phenomena.[55] As a redescription of the motivations behind our behavior, the rational actor theory is tolerable since it helps us make certain kinds of predictions by focusing on the intentionality of goal-directed agents, but rational agent theory is underdetermined by the facts of human cooperation. Those descriptive, observable facts will also admit a complex emotional / affordance theory, and with some advantages over the rational model (e.g., apes make similar social contracts and alliances with conspecifics, but without conscious, rational cost-benefit analysis). In fact, it is a telling point that the rational agent model of selfish behavior better predicts chimpanzee conduct than human conduct, and yet no one is suggesting that chimps are more rational than human beings.[56] This suggests that affective self-interest and affective other-interest (i.e., care) is a more accurate account of the true motivation for behavior and that rational calculation theory only captures the same data because the emotional mechanisms are also highly adaptive.

The Moral Emotions

A popular way to add a few outlier behaviors to the rational agent theory is to postulate dedicated modules. Some evolutionary psychologists argue that we have specific modules that detect whether another human is a genetic relation (thereby avoiding incest), and another module for detecting social cheaters.[57] This theoretical maneuver seems ad hoc to us, in the sense that it purports to explain a behavior by ascribing an innate black box to the mind that produces said behavior. According to this model, we may have a cooperation module and / or an altruism module. Hauser claims that we have a utilitarian moral module that computes the greatest good for the greatest number (or the least harm for the greatest number), and it automatically resolves many social behavior options.[58]

As we already established in Chapter 1, we are not sympathetic to the modular computational approach to mind. Just as we think domain-specific modularity mischaracterizes the flexibility and plasticity of nonsocial cognition, so, too, social intelligence seems an even less likely candidate for dedicated encapsulated circuitry.

If we are trying to explain cooperation or another universal aspect of human social intelligence, then modules are too mechanical and context-free, while moral judgments are too intellectual. The middle way is to focus on moral emotions (these are strong motivational psychological states, such as pride, pity, indignation, or guilt, that link perception of certain classes of social events to actions, resulting in judgments of right and wrong).

We think the investigation of moral emotions is productive and largely consilient with our view, but psychologists and philosophers who study moral emotions usually ignore their *development* and leap to create a taxonomy of such emotions or to discuss the interaction between emotion and rationality.[59] We want to emphasize the missing

developmental element of moral emotions—the way they are domes-
tications (i.e. secondary and tertiary levels) of primary level affects like
RAGE, LUST, CARE, and so on. The shaping influence here is not
just from the rational deliberation of instrumental reason, but, fundamentally, from the cultural context via mechanisms of *habituation*. We
see this as a continuation of the discussion we started in Chapter 5, in
regards to how the infant-caregiver relationship fixes our emotional
and inhibition networks through associative processes.

Tomasello is also skeptical of a modular approach or even a core-
knowledge nativism regarding altruism and cooperation, and he suggests a middle way.[60] Humans, he argues, have an early capacity for
shared intentionality and cooperation, which chimpanzees lack. But
while infants come into the world prepared for altruism (as evidenced
by various baby lab experiments), early experiential learning is the key
for its development and expression. More important than altruism for
the evolution of large-scale societies is coordination of behavior
through shared goals, or shared intentionality.

We are sympathetic to this "capacity approach" to social intelligence
and believe that the affective systems play a role in shared intentionality.
Curiosity, for example, has a cognitive closure impetus, but it also has
a high degree of positive affect (dopamine-based SEEKING and in-
trigue) and even tints of negative frustration at information incom-
pleteness, and so on.[61] Shared or cooperative projects feel good to
humans, creating an intrinsic attraction in addition to utilitarian effects. The affective approach adds a motivational causal element as the
source of shared intentionality.

As in every other scientific domain, there is more than one way to
proceed in order to understand social behavior: we can try to find the
exact causes that underlie an observable phenomenal pattern, or we
can find the model that best describes and predicts the phenomenon
and forget about whether it is also true, or whether it carves nature at
its joints, or whether it captures the exact causal mechanics. Most critical and indirect realists believe that the successful model takes its
success from our isolation of the real underlying causes (albeit with an
always deferred confirmation), but a purely instrumental scientific approach follows Newton's *hypotheses non fingo* and remains agnostic

about such causes.[62] The *Homo economicus* question is no different. Do humans really pursue their best interest in a rational manner, or do we just predict human behavior better if we treat them as if they were optimizing rational agents?

Kim Sterelny suggests that there are different models of human agency and that their success depends on properly correlating them with important changes in the evolution of social organization.[63] We agree with his distinction, already mentioned, between pre-Holocene and Holocene social demands and capacities, and his claim that the rational agent model is not a one-size-fits-all paradigm for understanding human behavior. Moreover, we agree with Sterelny that it is not just the model that changes; human agency itself transforms during the shift from being hunter-gatherers to stratified agriculturalists. But we have significant disagreements about how to characterize the mind of pre-Holocene humans.

In small bands, hunter-gatherers are living in "glass houses," in the sense that they can directly see the distribution of resources and skills. The intimate codependency of hunting and foraging, together with nomadic strategies, leave little room for individuals to deceive each other or sequester food surplus. Moreover, in this hostile world, what an agent *wants* closely limns what an agent *needs* (genetically speaking), because selective forces are unforgivingly prevalent. In this pre-Holocene world of small bands, the fitness and the utility of the agent are roughly the same. When the agent acts to maximize the usefulness of his niche (physical and social), he also maximizes his fitness (his chances of leaving progeny). Such behaviors have indirect influence on the population level and constitute a gene-culture coevolution with little or no mismatch between desires and adaptive behaviors. But as society grows bigger, more complex, and stratified, utility becomes decoupled from fitness. We cannot assume that the desires of the agent match the adaptive needs of the agent, because maladaptive desires can propagate in cultures where information flow is more horizontal (among peer relations) than vertical (from apprenticeship to parents) and where most survival needs are already met. In large state-level societies, however, the agent does not get as much immediate negative feedback from the environment or from the body (i.e., pain) so as to readjust behavior.

All of this seems correct to us, but Sterelny mischaracterizes the transitional mind of the Holocene when he states, "Despite the increased complexity of these new social worlds, our mechanisms of rational appraisal are intact."[64] He suggests that our rational cognitive power was a constant throughout this enormous transition (from hunter-gatherer culture to agricultural culture), whereas we think this transition actually engenders or accentuates the rise of the cold-cognition cost-benefit mind. In Sterelny's narrative, our rational deliberation skills are fully functional in the Pleistocene and are effectively tracking resources and means / end relations in the glass house of small-band reciprocity. Then as society stratifies, Holocene wealth concentrates, and family values diversify, humans lose sight of accurate fitness targets. Human appraisals of needful behaviors and resources become confused so that a kind of utility-maximization approach takes over from earlier fitness-maximization strategies. Sterelny imbues his descriptions of "utility" with connotations of egoistic hedonism, and this helps him claim that our previously healthy, rational agent capacities are corrupted into maladaptive pleasure-seeking during the transition to complex societies.

While we agree that large-scale civilizations can sustain maladaptive beliefs and behaviors, our narrative of these changes is quite different. In our affective model, much of the glass-house reciprocity of hunters and gatherers is based upon filial affective values, not economic rationality. Moreover, growing social complexity is not the confounder of an already well-tuned cost-benefit mind but rather the spark and the facilitator of a more utilitarian egocentric mind—one much more capable of Machiavellian deceptions and fitness decoupling.[65]

Our different narratives are partly a result of contrasting interpretations of empirical comparative ethology as well as different chronologies of filial values. Following Tomasello and Jensen, Sterelny points out that chimpanzees are more ruthless ultimatum-game players who maximize their personal interests, whereas humans actually enjoy helping each other and cooperating.[66] It may be the naked egoism of our ape cousins (i.e., chimpanzees) that lends weight to the assumption that our Pleistocene ancestors were primarily self-interested, efficient, rational agents. However, as we argued in Chapter 4, the anatomical changes in *Homo erectus* and the neotenous birth of offspring already suggest a radically different childhood and parenting structure for humans, well before the onset of large-scale societies. Following Hrdy, we find plausible the claim that *Homo erectus* had greater alloparent involvement, and this suggests affective decoupling of the CARE system.[67] If the emotional bonding of mother and offspring is loosened (decoupled) from the exclusive pattern of most primates and distributed across nuclear and extended kin, then filial devotion, affectively based reciprocity, and altruism are well in place during the Pleistocene. This renders the chimpanzee comparisons moot.

Just as Plato argues in book II of the *Republic,* Sterelny suggests that scaling up the complexity of social organization and introducing even minimal luxuries shifts the human agent's orientation away from *needs* and toward *wants* (which can be maladaptive). This destabilization occurs because multifaceted social sources reduce the tight vertical transmission of survival skills and values that previously existed between parents and offspring. In this model, the spread of maladaptive values and informational noise confounds the agent's well-being (decoupling fitness and utility) as well as the scientist's modeling of human behavior (i.e., as failure of rational agent theory). While we do not disagree with this view, one need not go so far to explain the unreasonable behavior of humans. Once we properly factor in the emotional systems (like LUST, or CARE, or RAGE), we already have agents acting in ways that defeat their own well-being (consider destructive emotional and erotic relationship patterns). Human agents inside nation-states, compared with those in local-level groups, have many more ways of behaving against their own health interests (e.g., violence, intoxicants, miscreant peer relations, gluttony, and other tempting vices). And the introduction of these temptations—furnished by finance economies that maximize surplus production—confuses the social scientist's theory of human behavior, when such a theory is founded on rational cost-benefit models of agency.

Finally, the *Homo economicus* model may not even be the productive heuristic ("as-if" model) that many assume. It appears to mask real affective causal mechanisms (emotions) underneath much of human social behavior. It obscures the real dynamic because it misreads the redundancy of motivational systems in the human mind.[68]

Social cognition and behavior rely on sensitivity to intention and context. This sensitivity seems to result from a rapid back-and-forth between our more immediate automatic responses and modulation or regulation. Zelazo and Cunningham suggest that social intelligence has an initial input sweep of information that is fast and unconscious, but that this strong bias phase is followed by additional modulation from top-down controls and conscious intent.[69] Feedback between the two systems is rapid and ongoing. Higher consciousness itself may have evolved originally as a policing modulator on lower impulse responses.[70]

Below the rational motives of the agent (cold cognition) lies the affective motives (hot cognition), and they are largely redundant or duplicated in the evolutionary kluge of the brain. Both hot and cold cognition are adaptive for the agent. These systems historically converge on similar targets, because they generally submit to fitness pressures. Arguably, the brain itself reflects some such division of labor, and the discrepancies between these systems are the places where the *Homo economicus* model fails; for example, in certain kinds of selfless cooperative behaviors as well as in negative affect trajectories—like revenge—that undermine agent flourishing.[71] The breakdowns of cold cognition are usually errors in identifying appropriate goals or errors in assessing the expediency of means to ends. The breakdowns of the affective systems are also cases of detrimental targeting or fixation (e.g. obsession or emotional imprinting on injurious objects / persons) as well as breakdowns of calibration. Often an adaptive emotion, like rage or lust, can become maladaptive by degree—by excess or deficiency. In a sense, it's this redundancy and discrepancy relationship that leads some theorists to see cold-cognition rationality as an adaptation to improve and error-correct the deleterious possibilities of a misdirected or miscalibrated hot-cognition system.

Ultimately, social organization is the result of several causal streams, and explanatory reductionism should be avoided. Kaplan, Hoover, and Gurven suggest that human organization is the result of several key ecological and economic variables, including a unique three-generational system of resource provisioning in families and long-term pair-bonding between men and women.[72] We see ourselves as adding to this causal pluralism in the sense that successful human social realities are motivated proximally by affective experiences and canalized into cultural constants by emotionally grounded normative rituals and traditions— from play to marriage, to gift-giving and shame and ostracism, affect becomes domesticated.

The Domestication of Affect

In comparing our bodies to those of other animals, it becomes apparent very quickly that we would not be the favorites in most one-on-one inter-species physical contests. But humans have essentially colonized the earth. How did it happen? Obviously, many factors are involved, but for our purposes, the most important is the efficacy and ubiquity of group action: our ultra-sociality. When cultural group selection produced tribal ingroup cooperation and outgroup hostility, cognitive capacities and emotional responses developed to adapt people to living in culturally defined cooperative groups.[73] It is our ability to work together in imaginative labor-sharing coalitions that brought about our ecological success; indeed, cooperation and reciprocity beyond kinship is crucial. Theories of the evolution of cooperation, if not based on kin selection or group selection (i.e., groups of cooperators are more successful than groups of defectors; cf. multi-level selection), focus on Robert Trivers's notion of reciprocal altruism (and the attendant developmental psychology of prosocial traits like friendship, gratitude, guilt, moralistic aggression, a sense of justice, and forgiveness).[74]

To understand reciprocal altruism, consider that membership in a non-kin group requires one be trustworthy so that any given individual in the group can be relied upon to share in not only the benefits but also the demanding work and suffering. How then does a group ensure that its members are committed and that individual members will not free-ride? Stabilized lifestyles made this possible, since living in stable social (local and regional) groups allows individuals to demonstrate, as well as evaluate one another's, trustworthiness over time. The format in which individuals may be judged in the public sphere is *reputation,* where what is on display as public knowledge is an individual's ability to cooperate, which demonstrates his eligibility as a member of the group. This is clearly demonstrated, for example, in the evolution of courage, as a long tail of altruism, wherein publicly risking life and limb for a conspecific demonstrates total commitment to the group.[75]

In human dyadic relationships, reputation may be mediated by a process of image scoring, in which every time an individual performs a cooperative action, his score goes up by one, and every time he does not donate aid, his score goes down by one.[76] Alternatively, it could be a matter of a cooperative action putting an individual in good standing and the opposite putting him in bad standing.[77] On the other hand, observing the punishment of defectors leads to moralistic strategies that maintain group cooperation.[78]

For our tertiary-level minds, reputation must be a fluid, complex bookkeeping affective tag, created by the sum of an individual actor's costly acts of altruism, commitments to the public good, conformation to local norms, and, of course, hours spent in gossip and hearsay. But how it works at the proximal level must be through the emotions, possibly a value mark or an affective nudge on immediate decision-making in appropriate social situations, which is brought about by the perceptual social affordance or indirect reciprocity of observing or speaking about a given individual.[79] This abstract role of affect within option outcomes is not the only functional locus for affective control; in fact, we believe that the public space of a social group is primarily maintained by domesticated social affects. They are the elastic bands—some pulled thin, some fraying, and some slack—that keep the dynamic structure together.

Even in experimental settings, emotions are the proximal cause of reaction in the ultimatum game, the public good game, and the prisoner's dilemma, not to mention in ecologically valid settings (i.e., the reason one gives loose change to an indigent person or engages in mild expletive-laden road rage).[80] In this context, emotions direct and deregulate an individual's attention to specific kinds of information, including possibilities for cooperative strategies.[81] The function of pride, for example, may be to provide subjective rewards for adhering to social norms, just as shame may provide a form of subjective punishment for nonadherence.[82] Accordingly, it has been shown that areas in the brain correlated with anger and disgust are activated by experimental situations in which fairness is involved.[83] What better than conscience to halt selfishness or what better than guilt to slow the appetites? And what better than empathy to stop us from giving pain, kindness to help people enjoy our company, and shame to pressure us into fixing our transgressions?[84]

We interpret the intense survival pressures on humans to act in groups as engendering the kinds of minds we have, and such engendering includes the domestication of primary and secondary affects into their tertiary forms as frameworks for social living. Successful social living (i.e., the ability to engage strategic, fluid coalitions within dominance hierarchies and over long periods of time) essentially depends upon activating the emotions of others appropriately and completing long-term flexible and imaginative analyses of one's behavior in the total social context. Reciprocal altruism is rare in the animal kingdom, suggesting that it is sociocultural arrangements in the tertiary cultural level that shape our affective responses and ultra-social behavior.[85] Social pressures on the mind are distinct; they seem to (1) create autonomous behavior that affords various social interactions involving cooperation and competition, (2) be most effectively manipulated by communicative signals, and (3) be used as sources of information through social learning.[86] Ethologist Frans de Waal observed that primate politics are a demonstration of the manipulation of emotions and social actors and serve a crucial role in mammalian social living.[87] Others describe the social mind as an arms race of Machiavellian intelligence technologies.[88] While the social mind in such a scenario can be considered cognitive, we have argued throughout this book that much of nonlinguistic social intelligence can be encapsulated in proximal, ecological, motor-perceptual affordances.

Our affective system seems to be at least as old as the mammalian clade, while our own highly social genus is about 2.5 million years old and our use of culture stems from around 160–200 KYA.[89] If, in fact, it was emotional social ecologies rather than rational choice modules that enabled fundamental social behaviors like cooperation and competition, we ought to enumerate the exact types of social behaviors built upon mammalian affective mechanisms that constitute social living. Below we describe the transformations our innate social affects (i.e., CARE, LUST, PLAY, GRIEF, SEEKING, RAGE, and FEAR) undertook during domestication. We begin with a brief consideration of the consolidation of self, or agency, and proceed by observing domesticated emotions as embedded in the ancient practices of ceremonialism, bodily adornment, and gift-giving.

The Evolution of the Self

While Cartesian theories of the self are dominant in contemporary psychology and philosophy, we contend that the self emerged in stages through the interaction between the mammalian social habitat and affective motivational processes. Panksepp has described an evolutionarily older bodily self primarily constituted by midbrain structures that mediate action tendencies and primitive affective sentience.[90] Along with Damasio, we posit three layers of the self: a primitive bodily self, a core self, and an autobiographical self.[91] The final item on this list is what we know as identity. For our purposes in this chapter on the evolution of society, the autobiographical self is most important since it enabled the adaptive function of the self; viz., navigating and being an actor in the social world of *Homo sapiens*.

We admire William James's, and subsequently G. H. Mead's, portrayal of the self as consisting of a knower (the "I") and a known (the "me") broken up into the material me—one's body and material possessions; the social me (i.e. one has as many selves as people he knows); and the spiritual me—or one's personality (i.e., the way one's mind works).[92] We find James's influential paradigm conceptually practical, yet it obscures how sentience (the "I") evolved from the body (material me).

We believe that in social mammals, the social me developed through its social niche into a practically functional and predictively useful unit (into the essentialist self, or personality) for the purposes of social bonding. At its most basic level, the self is accurately understood as a bodily image. But in social animals, this primitive action-tendency platform evolves a social identity, so that conspecifics can interact and so that memories and feelings have something to hang on. Crucial to this development is the ability to have memories of the past (i.e., episodic memory).

The evolutionarily older self we refer to as the Bodily Sentient Self; it accords with Panksepp's notion of SELF (Simple Ego Life Form).[93] It is present across nature, for any creature that moves must have an internal relation to its environment, which compares goal states, or innate programs, to its actions and surroundings.[94] This relationship is mediated by homeostatic valence in the feedback / feedforward loops we have discussed, which constitute the most primitive level of sentience. This system becomes elaborated in proportion to the range of movements, behaviors, and complexity of the organism. Some midbrain systems are geared toward basic regulatory functions like eating, sleeping, etc., while for social animals, coordinated relations with conspecifics are required for survival within their habitat. These social animals have homeostatic systems that include social bonding "apps." We argue that the social homeostasis needs for bonding and support are satisfied by another form of self, atop the Bodily Sentient Self, for the purpose of more effective social bonding.

Affective neuroscience reminds us of our phylogenetic homologies with other mammals, and so our biological identity should be found near the core of the brain—not the more recent neocortex. This archaic self would be a basic motor-mapping system—a template for action tendencies. Despite the inclination of philosophers to think about consciousness and subjectivity in terms of perceptions (like sense data qualia) or higher cognition, affective neuroscience reminds us that "a level of motor coherence had to exist before there would be utility for sensory guidance."[95] This archaic SELF would have to coordinate or integrate emotions from the periaqueductal gray (PAG) region of the brain and the perceptual somatosensory system. The centromedial zones of the brain (especially the deep layers of the colliculi and the PAG) answer to this requirement. Moreover, experimental work with mammals suggests that this area is much more relevant to biological intentional identity than higher neocortical areas. Experimentally induced lesions along the PAG are much more devastating to the intentionality or seeming *agency* of the animal than lesions in the higher areas of the brain. This archaic level of self is not cognitive per se, but resides in the intrinsic action-readiness of the biological system. Subjectivity resides first in the biological realm of action. It is not the disembodied Cartesian spectator.[96]

Nonetheless, the "spectator" aspects of self did eventually emerge (through social interaction and then imaginative power) and allow for meta-awareness of subjective states. There's no doubt that big-brained *Homo sapiens* can spin elaborate coherence out of disparate experience, using memory, discursive rationality, and intentional projections. But combinations of nonlinguistic perceptions, like visually based image schemas, together with engraved feeling dispositions, may be all that is necessary to begin some rudimentary *autobiography* of self. Animals with very impoverished symbolic and conceptual skills may nonetheless have the ability to sense (literally) their own personal history and then comport themselves into the near future (again, drawing on their affective entrenchments rather than on cognitive reflections).

This second form of the self is the Social Identity Self. It arose from interaction between social-habitat complexity and the affective motivational platform as an adaptation for the purposes of providing (1) an effective social unit (or platform) for bonding to conspecifics and (2) a personality locus for the prediction of behavior and psychology for the organism itself.[97] The core constituent of the Social Identity Self is episodic memory, a system evolved to provide an internal platform for prediction in the form of a set of reliable dispositions surrounded by specific exceptions.[98] This system provides the basis for personality, which can be characterized as a set of epigenetic behavioral and psychological dispositions. While fulfilling homeostatic social needs, the Social Identity Self also provides the basis for a personal narrative that energizes behaviors by giving meaning and values to our social actions through their constitutive relation to our identity within the social world.

The Social Identity Self is the platform for our reputation and stores identity relations within our position in the social hierarchy.[99] As such, representations and representational power are more constitutive of this social self, whereas the primordial self (Bodily Sentient Self) is more adequately understood as a form of consolidated mammalian agency. The Social Identity Self helps a person project himself into a virtual scenario of tomorrow's hunt or remember last year's ritual ceremony, but the Bodily Sentient Self helps the animal avoid a chasing predator, or helps the animal avoid eating parts of itself.

The following discussion of ceremony and ritual behavior helps us understand how the domestication of the emotions and the sculpting of the modern social self are intimately related developments.

Ceremonialism, Adornment, and Hierarchy

The earliest evidence of bodily adornment in our species is not entirely clear, as there is a wide gap between painted seashells and early jewelry around 75 KYA and suggestive red ochre pigment from 160 KYA, both in South Africa.[100] Meanwhile in Europe, features of culture marking apparent ethnic differences appear during the Upper Paleolithic transition, and around 35 KYA, greater population density created local groups and regional variation in subsistence strategies.[101] What is clear to us is that these artifacts had social purposes. While we do not have evidence per se of either linguistic capabilities or rational choice modules in early *Homo sapiens,* through these artifacts, we have inferential evidence that individuals were differentiated within social groups; that is to say, notions of identity and kinship must have been active for adornment to be meaningful.

For tens of thousands of years, humans lived in tribes, with cultural identities predicated on hard-to-fake rituals, technologies, and gestural communications. Concomitant emotions included status rivalry and hostility to outsiders, but also moralistic aggression against ingroup members who had broken social norms. But how is identity to be materialized and, like reputation, made publicly available? The answer is through the technology of adornment. The importance of bodily adornment (jewelry and scarification) is, first, that it identifies the worth and position of an individual. Symbolic cultural marks distinguish ingroups from outgroups, regulate empathy and altruism, and help direct cultural transmission of norms and models.[102] A walk through any historical museum will demonstrate that unique or aesthetically pleasing jewelry is a part of all cultures, from the Stone Age to the Bukhara Dynasty. A surface interpretation of the beauty of these artifacts, however, would miss out on their role as identity markers.

As Lionel Festinger pointed out, an individual cannot create a self-concept without social comparison; individuals, indeed, have a difficult time forming self-images in the absence of group identity and the technology necessary to reinforce these psychological passages.[103] In the Kalahari San groups (The !Kung, Nharo, G / wi, and !Xo) of eastern Botswana and northwest Namibia who share 90 percent of their material culture, we observe that beaded headbands display social identities within groups.[104] While sharing all the same pattern elements, the designs of these cherished objects (once made of rare ostrich eggshells) surprisingly varies most between adjacent tribes, thus acting to strengthen positive ties to kindred members in nearby and distant areas while carving out individual identities.[105]

Research on consumer culture has demonstrated that, even today, we turn to particular brands to maintain some class or political tribal affiliation. Consumer goods and the clothes and objects with which we adorn ourselves serve primarily as forms of social communication in our contemporary anonymous urban spaces.[106] The seashell necklace or bone earring from 75 KYA was similarly a sign of tribe identity; access to unique goods materialized one's ability to procure a certain extent of coalition membership and thus local power. Possession of unique objects was a fairly stable and hard-to-fake direct depiction of status and indicated the honor deserved by the owner of the goods.

One need not go much further to posit the connection between the perceptual cues of adornment as status symbols and our notion of social affordances. If reputation creates in an individual a subjective affective summary of the perceived person, which then modulates the social interactions afforded upon her, then identity markers may work in a similar fashion. The adornment tells us who someone is and thus gives us an idea of how to act toward her. As our social worlds grew larger and more complex in the shift from hamlet to local and regional group, adornment was a way to specify social action possibilities to an even finer degree. Even in the absence of local reputation, bodily adornment was a way to determine someone's importance, for example in the finery that regional leaders wore in various parts of Africa when on delegations to other regions (examples from Cameroon, Ghana, and Timbuktu readily come to mind). Such a tradition developed when Big Men and chiefdoms were in regular contact, and it served as a way to materialize wealth and regional coalitions. It also served as intimidation: luxury items mean that subsistence is well taken care of—remember, a successful potlatch ends with the burning of canoes and other valuables![107]

From the local-level group on, we see the social technology of the ceremony and can understand it as the depiction of the social landscape in which hierarchy is materially and behaviorally enacted through ancestral and social kinship identity, the display of productive forces, and the aggressive potential of the group. These features of leadership and fictive kin are demonstrated via adornment and coordinated performative displays of wealth as power, honor, and respect. Ceremonialism can be tracked in ever more elaborate forms, from the family setting, where the patriarch sits at the head of the dinner table, to the neighborhood, village, and region, where ritual displays of hierarchy describe the obligatory structure of respect and pride. Ceremonial rituals like the Pythia (Oracle of Delphi) in archaic Greece were able to induce honor for the soothsayer, sanctity for the modes of divination and relationships between man and the gods, and awe for founding myths that legitimated leadership.[108] Or, consider a given Hindu festival, the Thaipusam, celebrated in the Tamil community, where a class educated in the arts of recitation—the Brahmins—adorn their bodies and chant passages of the *RG Veda*. In these events, the Brahmins legitimate their superior position in the local social hierarchy through their honorable knowledge of the sacred texts, which they believe enables sanctified connection with ancient tradition and trance-like, awe-inducing contact with supernatural gods.[109]

When resource competition and population density become great enough that uncertainty arises about the use and control of property and resources, ceremonies can mediate power by reminding the community about historical tradition in a display that unites myth, religion, and political ideology. These public displays serve to solidify property relationships that could then be formalized into written laws in some judicial system. In Byzantium, every time the kingdom was imperiled by outside invaders, sacred objects, like pieces of the Holy Cross and bones of the Saints, were paraded down the streets.[110] A few hundred years later in those same streets, one could witness parades of holy memorabilia (like the shroud of Muhammed, which is still on display in the Topkapı Palace) at each occasion of a change in dynastic Sultans.[111]

If we zoom in on the domesticated emotions of *honor* and *respect,* we see that they uphold the function of reputation; in fact, it is these two affective bonds that make reputation matter. To have a good reputation is to deserve honor and respect; that is, to deserve to be treated in a wholly more beneficent manner. This nexus must have evolved first as a form of gratitude for care providers and for possibly wise, and therefore useful, elders. It also must be related to filial emotions and mammalian social submission, not to mention fear of the powerful members of society who can bestow, as well as withhold, benefits (cf. alpha males in primates). Accordingly, prestige has been correlated with wealth and reproductive success.[112] Honor and respect are, unfortunately, sometimes exploited as a cover of legitimation for baleful elite control, with examples ranging from lascivious cult leaders to violent, ideologically driven dictators.

The analogous primary-level affects that are domesticated by sociocultural living in honor and respect are CARE and FEAR. We could also consider how RAGE might be activated when dishonor occurs (e.g., in honor killings of unwed daughters) or GRIEF in the intense negative feelings of falling into dishonor (e.g., in the return of a defeated army or in the shame of being arraigned in a court of law for an unseemly act). It is also possible to consider how the "goad without a goal" of SEEKING could be directed toward securing

238 · THE EMOTIONAL MIND

prestige and honor through social-climbing or becoming hellbent on making up for a dishonorable action with recompensing behaviors.

In the *Descent of Man,* Darwin speculates on the prosocial influence of the honorable person and his ability to inspire altruistic activity in other tribe members.

> We may therefore conclude that primeval man, at a very remote period, would have been influenced by the praise and blame of his fellows. It is obvious, that the members of the same tribe would approve of conduct which appeared to them to be for the general good, and would reprobate that which appeared evil. To do good unto others—to do unto others as ye would they should do unto you—is the foundation-stone of morality. It is, therefore, hardly possible to exaggerate the importance during rude times of the love of praise and the dread of blame. A man who was not impelled by any deep, instinctive feeling, to sacrifice his life for the good of others, yet was roused to such actions by a sense of glory, would by his example excite the same wish for glory in other men, and would strengthen by exercise the noble feeling of admiration. He might thus do far more good to his tribe than by begetting offspring with a tendency to inherit his own high character.[113]

As Miss Mary Bennett intones in Austen's *Pride and Prejudice,* "[P]ride relates more to our opinion of ourselves, vanity to what we would have others think of us."[114] *Pride* is then holding oneself in honorable regard, which, in a benign form, leads to confidence and generosity of spirit, while in a malign form, it leads to self-serving behavior and narcissistic thought patterns. This domesticated emotion seems to share the sources of honor and respect, or, as Freud would have it, pride reflects self-love, where feelings that were evolved to track conspecifics are cathected onanistically. Love itself can be considered a domesticated mixture of LUST and CARE for chosen kin. Of course, our cultural understanding of this has been greatly transformed by cultural narratives of romantic love and subsequent changes in the social function of pair-bonding.

Hierarchy

Honor, respect, and *pride* all exist within the context of a hierarchical social structure. Across species, there are patterns in how pecking order is established. Specifically, when several individuals who do not know each other are placed together in a group, they tend to engage in contests for dominance (usually one-on-one). Sometimes the contests are violent and sometimes there is a passive recognition between dominant and subordinate. Initially, dominance contests are very frequent, but they tend to happen less and less over time, being gradually replaced by a stable, implicitly agreed-upon, linear hierarchy among the members.[115] The linear hierarchy, which is the classic pecking order, appears in nature and in experimental scenarios with many insect species, crustaceans, fish, birds, mammals, and human children and adolescents.[116] These social structures consist of dominance relations between all pairs of individuals in the tribe, where the individual who won one dominance event is likely to keep winning: those who win, win (sometimes called "the loser and winner effect").[117]

For analogues in nonhuman primates, we observe that captive chimpanzees are more likely to share food with former groomers than with others and are less likely to behave aggressively toward attempts to share by former grooming partners than others.[118] Also, the grooming time that females engage in to gain access to infants seems to depend on the relative rank of the mother and the handler; hierarchy will play a role when mothers are groomed longer by lower-ranking than by higher-ranking females who want to handle their infants.[119] Based on a review of extant literature, Tomasello describes how primates seemingly understand not just their own relation to other groupmates but also the relationships that groupmates have with one another. This reveals awareness of both the vertical dominance dimension of social structure as well as the horizontal dimension of kinship and friendship.[120]

But how are dominance hierarchies formed in the first place? What may seem to be the most obvious answer—that individual characteristics such as size, hormone levels, ranking in previous hierarchies, weight, seniority, relative maturity, and aggressiveness structure the relationships—

is not confirmed by evidence from chickens, primates, or human adolescents.[121] Several innovative studies of the development of social
relationships in various nonhuman primate species suggest interactions
between one dyad have important implications for the interactions between all other dyads and, thus, for the overall formation of group social
relationships.[122] Hierarchy formation, therefore, can be best viewed as a
developmental process, where preceding dominance interactions influence succeeding ones.[123] While being the winner or loser of pairwise
dominance contests is important, it is also crucial that the other members
of the group are present as bystanders for the dominance hierarchy to
become implemented.[124] As bystanders, it turns out, we are experts at
tracking hierarchies. Consider the complexity of the vertical and horizontal dimensions of the social world, where most animals cannot explicitly remember the episodes that shaped the hierarchy but still maintain
appropriate relations with others. For example, if your child is bullied by
the young thug of his cohort in an empty bathroom between classes, his
place in the dominance hierarchy would be less likely to diminish than if
the bullying occurred in a public space like the dining hall or the worldwide web, where other members of the community observe and can
then perpetuate the hierarchy demonstrated in the bullying incident.
It is our contention that some form of affective control mediated by developed social affects like respect, honor, pride, shame, guilt, trust, and
reciprocity motivate our social behaviors in cases like this and many
others. It may be that we accumulate and tag dominance information to
individuals that can be congealed into a type of a perceived character trait
or virtue; these are then subjectively experienced as somatic markers or
social affordances in future interactions.

Reciprocity and Gift-Giving

Gift-giving provides an opportunity to study conscious and unconscious affect tags that track reciprocity within the context of feelings
of trust, pride, shame, honor, and even love. From Marcel Mauss, we
learn that gift-giving is always a visible public drama made up of both
giving and the obligations to give, to receive, and to repay.[125] Similar
to the way body adornment and ceremonies are material artifacts of

social status, gifts materialize relationships between individuals and between groups. There could be an emotional engine that translates this rational process of balanced reciprocity, tracking long-term punishments and rewards over transient, short-term gains.[126] But to understand the impact of a gift requires understanding the social norms of the given cultural niche.

. . .

Consider a schematic scenario: one family harvests a surplus of corn and they decide (and this could be for any number of reasons) to give some of the surplus to their neighbor. This neighbor, upon receiving the corn, may decide to give them some of his surplus wheat in return, maybe with a bit extra thrown in. What this illustrates is that a few domesticated emotional bonds (i.e., feelings of indebtedness and generosity) reflect an ethical dimension of mankind, namely the norm of reciprocity (or what the Chinese call *guanxi*). It also illuminates the practice of gift-giving—the sense of honor and shame that binds giver and receiver, sometimes engendering respect or pride and sometimes resentment. The role of trust in a relationship, or the possibility of shame were the receiver unable to reciprocate, are also a part of the scenario. While we can all acknowledge that these emotional bonds exist within the giver-receiver dyad, it will behoove our argument to provide an exploration of the exact nature of trust and reciprocity.

The norm of reciprocity (i.e., that resources given or taken will be recouped) is evident in every relationship of premodern society.[127] Humans reciprocate even when the initial giver will not know about their reciprocation, suggesting that it is an internalized norm.[128] But resources traded in a social exchange are not the same as tender; rather, their value is the social meaning of the action itself, and that may be one reason why reciprocity functions not only to maintain prosocial actions but also to quickly suppress antisocial actions via escalating retaliation.[129]

Why do we reciprocate? It could be that a lack of reciprocation is connected with the fear of being considered a free-loader and, thereafter, getting a bad reputation or low standing.[130] The "shadow of indebtedness" that is an element of reciprocity is a fundamental principle in shaping social interactions and furthering social integration.[131] It could be that reciprocity developed within a group-selection context and simply stabilizes the social structure, or it could be that it is mediated by an individual's personal conscience.[132]

If reciprocity is rare in the animal kingdom, then we have to imagine that human domestic living (or hearth-based living) played a role in creating the pressure for such fundamental altruism.[133] Quite possibly, the norm and / or internal pressure toward reciprocity is but a piece of the development of reputation-based identity markers within cooperative group action, whether in the hamlet or in local- and regional-level groups. Of course, it could also have been a regular element of kin selection that, as we came to live in larger groups, we became gradually decoupled onto non-kin tribesmen.

Consider the following possible trajectory of phases in the norm of reciprocity. First, mother and father (or primary caregivers) share resources with their child, as mediated by CARE and filial emotions. The child reciprocates with positive emotions and resource-sharing with parents as he develops into a productive member of the household. Next, imagine a public square, where families living around the square set up informal economic coalitions with friends or just sit around and gossip. In this scenario, we observe dyadic friendship in reciprocal social grooming behaviors. In the public space, any individual who benefits must also do something to maintain the shared resource, whether that be cleaning up, mediating conflict, or just giving time and audience to other members of the community. Consider now the nation-state, where there is a somewhat removed reciprocal trade between paying taxes and receiving public services (including military protection). At each level, the same norm of reciprocity is being abstracted from its kin-selection origins in response to various ecological factors that take the form of sociocultural arrangements. The norm of reciprocity is well studied, but how about other norms and the ways in which norms come to form a moral universe for our actions? We turn now to this topic.

Social Norms, Morality, and Adaptive Emotion

Human morality has clear adaptive benefits, even while it sometimes thwarts the fitness of individuals. Like speech or pictorial communication skills, ethical norms can be understood as biocultural phenomena that help *Homo sapiens* survive and flourish. There is still controversy surrounding the naturalizing of ethics, but not as much as when E. O. Wilson first suggested that "the time has come for ethics to be removed temporarily from the hands of the philosophers and biologicized."[134]

Like many evolutionary theorists, we are interested in the antecedents to morality—the component parts that first emerged and made morality possible. Some have argued that strong reciprocity or altruism is the key ingredient that gave rise to morality, while others have emphasized the evolution of reason, and yet others have stressed the emergence of empathy in our ape lineage.[135] We agree with aspects of these theories (particularly that of de Waal) but wish to add the uniquely affective ingredient to the mix. It makes little theoretical progress to ascribe morality to altruism but to leave altruism as a vaguely analyzed occult force—describing its ubiquitous spread throughout diverse populations, but not its mechanics or its felt motivational origins. Reason, as we've already seen, is too late on the scene to be the driving force behind ethical norms, unless we steer our notion of "reasoning" away from declarative propositional deliberation and toward sensorimotor reckoning. Reason seems better equipped to *guide* motivated action than to originate it.[136]

There is no compelling reason to think that ethics is a special kind of normative reality, distinct and unrelated to other kinds of behavioral customs, like hygiene, food preparation, and mating conventions. The evolution of morality is a species of the larger genus; namely, the evolution of social norms. Most social science research on norms has centered on their functions and their efficiency, stressing the exogenous aspect of norms as external behavioral constraints.[137] The flip side of this approach—the endogenous aspect—is often discussed in philosophical literature (e.g., utilitarianism, deontology, virtue ethics, etc.); it stresses the agent's mental relation to group expectations or disembodied standards (including beliefs about those norms, beliefs about the self and harmony with those norms, second-person self-monitoring about motivations, etc.).

Recently, anthropologists have begun to consider the genealogical, phylogenetic aspects of human social norms. Using better scientific tools, researchers like Michael Tomasello try to provide a more fine-grained explanation of the human cooperative tendency that first intrigued Plato in the *Republic* and Kongzi in the *Analects*. According to Tomasello, there are two evolutionary steps that bring us to the uniquely human level of cooperative social success. The unique mind of human beings is capable of *objective-reflective-normative thinking*. And the two steps that produce such normative mind are (1) the evolution of *joint intentionality* and (2) the evolution of *collective intentionality*.[138]

Beyond conditioned learning, primates (and many mammals) run simulations of possible behaviors in something like an imaginative faculty, anticipating (sensing as much as inferring) the effects of their behaviors. Some internal self-monitoring allows an animal to pre-correct his errors and avoid pain and possible death. As we have argued throughout, this kind of thinking can be quite rich with nothing more than affectively coded somatic markers and associational schemas (together with proto-negation and proto-conditional "logic").[139] We regularly observe animals like squirrels preparing to leap from a branch on one tree to a branch on a neighboring tree. The muscles tense, the squirrel does a get-ready-set motion, and after several preparatory almost-jumps, the squirrel abandons the plan and goes down the trunk to ascend the other tree.[140] Presumably, the animal has done some kind of evaluation of an internal jump simulation, using sensorimotor affordance information, and then stopped himself from a possibly deleterious behavior. Behavioral and cognitive self-monitoring, according to Tomasello, entail impulse control, attentional control, and emotional regulation and can be observed regularly in ape behavior.[141]

Something new emerges, however, in the transition from a two-year-old human child to a three- or four-year-old. If researchers try to coax a two-year-old away from a collaborative game with another child, the first child will be easily drawn away to a more exciting or interesting game. But after age three, the child is resistant to abandon his little colleague and makes apologies if he departs the joint project.[142] After age three, the two children feel some obligation to each other, understanding their common ground and sensing that their respective subplans add up to a larger plan. Our great ape cousins do not possess this shared intentionality. Moreover, children of this age (three and above) begin to understand that their collaborative tasks have a division of labor and that most roles are interchangeable. One child can collect the blocks and another can stack them, but these roles can be swapped. This suggests, according to Tomasello, the emergence of a "bird's eye view," in which the children are able to include their varying task roles in the same representation. In order to do this, the child sees himself—if only momentarily—in the second person, or he sees himself as being interchangeable with the other. This relatively simple detachment of the agent from the functional role scales up to a sense of *self-other equivalence,* because it provides humans with a view from nowhere (from no biased agency). The tyranny of animal egocentrism is loosened, and nemocentrism (nobody at the center) is born. For Tomasello, this kind of perspective-taking is a precondition for normativity, and he sees the spark of egalitarian ethics in this developmental step.

The second step, collective intentionality, reinforces the first. The evolution of language enables greater representational decoupling, creating the increased reification of symbols, abstract behaviors, standards, and cultural norms. All this facilitates the development of institutions—first, simple skill guilds, and then educational systems, and so on. All this gives humans the unique cognitive ability that we call "objectivity"—a shared, agent-neutral world.

While we agree with much of Tomasello's account, we think that he oversells the importance of attention as an enriched cognitive capacity but shortchanges the fact that attention needs desire and social-affective motivation to give it more than an epistemic function. It is because I CARE (in Panksepp's sense of CARE) about my kin or ally that *his* project becomes *my* project. It is because I LUST after this conspecific that her intentions become my intentions. It is because I share in the FEAR of this enemy that we find joint and collective intentionality against them.

Our view is consilient with Tomasello's, and we suggest that emotion may be a necessary ingredient. Tomasello admits that we don't know what forms the underlying structure of joint "we-ness" and holds out hope for some deeper causal picture.[143] He suggests, for example, that some kind of recursive mind-reading (TOM) is somewhere implicit in shared intentionality. But the affective components, imagination, and social identity self must be added as well.

In order for our ancestors to share goals in the unique human way, they needed an extrapolated bonding system (the opioid and oxytocin care system) that turned the altruistic helping behaviors of mothers into a flexible alliance system across wider kin circles and eventually non-kin circles. It actually feels good to help other people or work alongside other individuals to achieve something. That feeling is a large part of the proximate motivational causation that can be selected for indirectly through the successful behaviors it produces. But auxiliary emotions seem to have evolved in tandem with the accomplishing of goals (both as individuals and groups). We have already detailed the positive affect of the anticipation or SEEKING system. And in the same way that SEEKING has a satisfying satiation or completion phase, which provides the agent with positive affect as it returns the organism to homeostasis, so, too, humans have higher-level forms of enjoyment that pay out during the pursuit and completion of social, technological, and even aesthetic goals. As Paul Ekman's and the Dalai Lama's joint "Atlas of Emotions" suggests, there are fine-grained forms of enjoyment like *naches* (a Yiddish word for feelings of pride at the accomplishments of one's offspring) or *feiro* (an Italian word that describes the pleasure of accomplishing a goal that challenges one's capabilities).[144] Furthermore, evidence shows that human collaborative groups are more dysfunctional if emotions between participants are negative, and in such cases, improvement in positive emotion between members significantly increases successful collaboration.[145]

Our view is that conscious awareness of *objectivity* (having a concept of a view from nowhere) plays a much more recent cultural role in the development of human norms (within the last 5,000 years), and an older affective structure predates it. An alternative to Tomasello's overly cognitive account of norm evolution can be found in Darwin himself. And following Darwin's lead does not obviously tilt toward the optimistic assumption that *Homo sapiens* is uniquely egalitarian in his cognitive orientation. Tomasello offers a narrative arc from egocentric orientation (shared with other primates) to nemocentric orientation (uniquely human), but this may be too tinted with liberal optimism.[146]

To be fair, Tomasello makes no clear commitment on the incremental steps from self-centered primate to objective-minded collective human, but we suspect the evolution of social mind neither aimed for nor achieved egalitarianism. After all, filial affect (e.g., Chinese *xiao*), which is highly preferential toward kin and tribe, is a different path to normativity—bridging the distance between great ape egocentrism and egalitarian altruism.

Norms are conventional in the sense that each society creates and conforms to different customs and standards. Hindus do not eat cows, but they do eat pigs, while Muslims, Jews, and Seventh Day Adventists do not eat pork. But while norms can have an idiosyncratic constructivist dimension, they are often bounded or constrained severely from two directions. One direction is the top-down constraint of specific ecological potential. As Marvin Harris showed, the prohibition of pork correlates with specific ecological challenges in which nomadic people must resist the temptation to farm pigs in suboptimal climates, when pigs are in fact dietary competitors with human beings.[147] Raising pigs for food is tempting because they deliver high-quality fats, but they also severely compromise underground calorie-storage organs like tubers, from which humans benefit more-than-occasional fat blasts. This ecological pressure—not the oft-cited disease and hygiene issue—created the need for a cultural norm against eating pork. We have already seen other ecological influences on social customs, like the way plentiful fruit reduces hunting in bonobos in Chapter 4, and many scholars have detailed the influence of factor endowments (labor, land, and capital) on sociocultural customs.[148]

The other direction of constraint on sociocultural norms is bottom-up endogenous feeling. In particular, Darwin argues that sentiments like sympathy and shame are the emotional engines that drive social norms like ethics. Feeling motivates us, and habit secures or stabilizes behavioral patterns. In *Descent of Man,* Darwin points out that "however great weight we may attribute to public opinion, our regard for the approbation and disapprobation of our fellows depends on sympathy, which . . . forms an essential part of the social instinct, and is indeed its foundation stone."[149] The specific norms of a social group may be related to ecology and eventually to education, charismatic leadership, and even caprice; but emotional forces like love, shame and fear generate the lion's share of normative social life.

We agree with Darwin (and subsequently Panksepp, Ekman, Davidson, de Waal, etc.) that emotional connection with the other is instinctual and genetically underwritten. It may not be an elemental natural kind, as Darwin seems to suggest; it may, instead, be a mirroring system that allows for contagion of said emotions, or it may be a downstream instance of CARE decoupling.[150] More recently there has been a somewhat useful refinement of terminology, suggesting that sympathy is feeling pity or compassion for another creature's suffering (or joy, etc.), while empathy contains some further ability to share directly in the emotions of the other.[151] In addition to sharing the relevant feelings, the empathic person may be experiencing a cognitive move that allows her to adopt the perspective of the other. Darwin's use of "sympathy" is clearly mixing sympathy and empathy, and he makes no attempt to distinguish them. Indeed, it's possible that the distinction between sympathy and empathy does not reveal some underlying faculty difference of kind, but rather a difference of degree. The important point is that social animals have a simulation system (possibly rooted in the mirror neuron system) that enables them to catch the emotions of others (originally as "contagion" and later as "empathy"), and this system is enough to generate normativity.[152]

Following the sentimentalist tradition of David Hume and Adam Smith, Darwin argued that ethics starts in feeling rather than in rational calculation. But he reconfigures the psychological sentiments of the Enlightenment thinkers to be innate instincts bequeathed by evolution. This changes the orientation of my sympathy. It is more than just a psychological state that arises whenever I see miscellaneous suffering and that can be conditioned via learning. Evolution has sculpted empathy (CARE), according to Darwin, to have intentional, dedicated targets. Those targets will vary by individual experience, and the system is flexible, but sympathy is highly preferential, not egalitarian. According to Darwin, this explains why "sympathy is excited, in an immeasurably stronger degree, by a beloved, than by an indifferent person."[153] There is no reason to think that the innate affect comes into the world with a fixed target, but there is every reason to think that early bonding experience with caregivers, as we described in Chapter 5, dials in a heavily preferential system of special feelings, loyalties, and duties.

Many social animals, beside ourselves, set aside their selfish urges to engage in altruistic behaviors (like social grooming, food sharing, protection, and task cooperation). Kin selection is a way to understand how seemingly "disadvantaging behavior" could have invisible but powerful genetic advantages. Social insects, like ants or bees, are good examples of groups that flourish even when some individuals are disadvantaged (e.g., workers are sterile and have no way to pass on their genes directly but their larger gene pool—the hive or colony—benefits from their many sacrifices). Ground squirrels, for example, that call out alert chirps to warn their companions of an impending threat can save a clan of squirrels, but the chirp itself brings the predator's attention to the chirper.[154] How could altruistic behavior be selected for if natural selection only operated at the level of each individual? Chirping would be wiped out quickly, unless the family of related squirrels benefitted substantially and lived on to reproduce the virtually same gene pool (including the altruistic chirping trait). As Stephen Jay Gould put it, "By benefitting relatives, altruistic acts preserve an altruist's genes even if the altruist himself will not be the one to perpetuate them."[155] These cases are just the hard cases (extreme altruism) that prove the rule: kin will frequently compromise their own advantage in favor of their relatives.[156]

When biologists studied ground squirrel and prairie dog calling behaviors, they independently confirmed that warning calls become much more frequent and intense when kin are nearby.[157] Animals will also engage in "rescue missions" against predators if their own relatives have been captured or cornered. In short, preferential behaviors—stemming from biological favoritism—can be life-saving behaviors. Attachments must be uniquely intense—etched in the mammalian family experience—if they are to compete with an animal's egocentric tendencies. Presumably, blood nepotism evolved first, and this chemically-based reciprocal altruism developed into wider (non-blood related) forms of networked social cohesion (see our discussion of reciprocity above).[158]

The correlations of kin selection have been known for decades. An old joke suggests this idea: "I'll throw myself on a hand grenade for two brothers or eight cousins." But now, through affective neuroscience, we are beginning to understand the emotional or affective springs that trigger the behaviors associated with kin selection (at least for mammals). A human does not perform a rational calculation when a hand grenade falls in the room; still less does a prairie dog do math to figure out if it should chirp more warnings when its siblings are nearby. The engines of adaptive nepotistic action are not rational but emotional.

We suggest that affective systems are originally engraved in the brain and body of the animal and that ontogenetic development, environmental conditioning, and perhaps epigenetics then sculpt those feelings into complex behavioral tendencies or dispositions. In other words, *inner feelings* (always treated skeptically by behaviorists) guide the preferential behaviors of altruistic animals. The chirping squirrel *feels* more panic when its own family is in danger and so increases its alert calls. As we discussed in Chapter 5, intentionality fixes on individuals in the infant socialization period.

The emotional brain, the limbic system, is a natural nepotist. The rational neocortex, however, is much more *principled*. The idea that everyone deserves equal treatment, or the idea that everyone has equal claim upon resources, or the idea that everyone has equal value as my kin, are all foreign to the intrinsically hierarchical emotional brain.[159] This is because our experience-based *values* (whether of an object, person, or idea) are originally encoded in the course of psychological development by *feelings* (e.g., chemically grounded oxytocin, or opioid, or dopamine patterns).

The biological process of filial favoritism is a *pattern* or a *happening,* or a mammalian *tendency,* not a principle. Indeed, many high principles take their start from lowly patterns or tendencies, but our smart neocortical rationalizations abstract these principles out of the more humble experiential patterns.[160] By this process, an affective experience of bonding with my favorites (my allies) can be inductively extrapolated into a rough-and-ready rule of action—a principle we might call *loyalty*.[161]

How Norms Stem from Emotions: *Shame* and *Guilt*

How are early feelings or emotions scaled up to norms? In part, the answer is that they are always normative, so the question is misguided. Social relations do seem to be regulated by norms embedded in belief systems.[162] As we have already established, our PPRs and AORs are simultaneously indicative and imperative. In fact, even when the indicative cognitive recognition system is compromised, the imperative aspect, or evaluative appraisal, continues to mark our perception. In other words, perceiving or recognizing my mother (indicative) comes automatically with duty / loyalty feelings (imperative norms). A certain deference, protectiveness, comportment, and regard accompany my recognition of my mother, in part because the recognition / perceptual platforms are often simultaneously visual, tactile, olfactory, etc., with hot and cold cognition neural pathways. Once I have a rudimentary mnemonic system, the memory representations of my mother will also carry somatic markers and behavioral affordance tendencies. So, in a sense, my family provides the earliest training of not just the associational system but also the impulse control system. And, crudely put, ethics (and many other kinds of normative behavior) are patterns of impulse control.

In another sense, of course, feelings scale up to norms through explicit training. Emotions can be socially adaptive because associations can be managed by habituation, thereby becoming second-order dispositions. Darwin describes the moral sense as a sequence of feelings and feedback responses:

> At the moment of action, man will no doubt be apt to follow the strongest impulse; and though this may occasionally prompt him to the noblest deeds, it will more commonly lead him to gratify his own desires at the expense of other men. But after their gratification when past and weaker impressions are judged by the ever-enduring social instinct, and by his deep regard for the good opinion of his fellows, retribution will surely come. He will then feel remorse, repentance, regret, or shame; this latter feeling, however, relates almost exclusively to the judgment of others. He will consequently resolve more or less firmly to act differently for the future; and this

is conscience; for conscience looks backwards, and serves as a guide for the future.[163]

Shame and *guilt* are crucial ingredients in the development of normativity, because they arise whenever the social contract has been violated. Essentially, shame is the addition and association of a negative, affective, somatic marker to an object; viz., a represented or remembered behavior. Shame or guilt may be derived from more fundamental affects like disgust, in this case leveled at the self. For example, shame has all the earmarks of hygienic pollution (grounded in disgust), but now it is applied to one's self or one's behaviors. This mechanism has the ability to mark past behaviors, but it also influences future behaviors. As populated by memory and cultural narratives, our representational narrative mind expanded, and we were able to do pre-guilt forecasting and pre-error behavioral shaping. Guilt creates an internal and relatively automatic way for social cooperators like us to avoid the many daily temptations to exploit the group for our own advantage. As Pascal Boyer puts it, "Guilt is a punishment we incur for cheating or generally not living up to our advertised standards of honest cooperation with others. But then a feeling of guilt is also useful if it balances the benefits of cheating, making it less tempting. Prospective guilt provides negative rewards that help us brush aside opportunities to cheat, a capacity that is crucial in organisms that constantly plan future behavior. . . ."[164] In short, conscience in the form of *shame* and *guilt* makes exogenous constraints into endogenous constraints.

Shame has been studied in over a thousand cultures and shows the same basic physiological processes.[165] Some researchers have suggested that the prototype of the affect of shame is when the mother fails to mirror the infant's facial expression.[166] There is a strong correlation of shame and activation of the parasympathetic nervous system, which suggests that its phylogenetic history may put shame in the adaptive category of demobilization, or freezing behaviors.[167] Cultures depict shame for very different things, but shame itself is a relatively universal emotion (as is the desire for vindictive retribution). Shame and the desire for revenge are affective forces that keep most people from vio-

lating established customs or norms. The ecological necessities motivate the norms from the top, and the affects motivate them from below,
as seen in the primate behavior we discussed in Chapter 4.

The most obvious mechanisms that force exogenous constraints into
endogenous conscience are socially engineered rewards and punishments.[168] As anthropologists have long noticed, the social and the economic are deeply entwined in forager social groups.[169] Gifting is
structured across individuals and groups so that reciprocating returns
can be collected in times of need.[170] Gifting and resource sharing are
important ingredients in strengthening centripetal forces, but punishment is equally important (certainly more common) in staving off the
centrifugal forces of antisocial behavior.

Norms that support cooperation include mutual kin obligations, a
willingness to share, and respect for another's possession of material
goods and mates.[171] In a wide-ranging study of Ju / hoansi (!Kung)
Bushmen of the Kalahari region, Wiessner found that punishment was
eight times more frequent than praise / rewards when enforcing norms
of cooperation. Most punishment consisted of social-reputation levers
like shaming, ridiculing, and ostracizing. There were four major techniques: put-downs through pantomime and mocking, mild criticism
and complaint, harsh criticism, and finally criticism plus violence.
These forms of punishment were designed to bring offenders in line,
curbing free-riders and other violators. Verbal jousting is a way of
managing behaviors and emotions without the more costly forms of
physical violence. As such, we can consider language and gestural confrontation as important forms of adaptive affect-management in the
service of gossip and reputation modification.

Social transgressions (e.g., free-riding, big-shot behavior, drunken-
ness, etc.) create anger in the group. The anger is channeled into loud
verbal complaints in the quiet evening, when all members can hear.
A coalition of critics usually forms. This is also accompanied by refusal
to share food with the target of the anger. "If anger persisted, there
was the risk that a segment of the camp would break off and depart for
weeks or months."[172] The costs of unmanaged affect can be very high,
so every attempt is made to handle centrifugal forces carefully. The
common response by the target of punishment is silence. The person

being criticized is usually quiet and seems to ignore the punishment. Behavioral ameliorations by the target in the days following usually reduces or ends further criticism. If the criticism is intense and unrelenting, the target may slaughter a cow (on some pretense; not in an explicit acknowledgment of wrongdoing) and distribute meat to the group. These reconciliative behaviors are not very different from primate food-sharing, hand-tapping reassurance, grooming, and embracing. Social cohesion is restored after punishment.

Some theorists have suggested that punishment behavior has evolved as a form of virtue signaling—an honest (though somewhat costly) way of demonstrating the punisher's quality as a potential mate or coalition partner.[173] In this view, the punisher demonstrates his fitness as a norm follower and enforcer. The "law and order" zeal of the punisher will be attractive to potential mates, according to this theory. This is a compelling possibility, but Wiessner argues that it does not fit with her observations of the !Kung Bushmen, because the most active punishers are not those in the mating age range, and most mates are drawn from neighboring tribes where the potential mates would not be witness to the supposed virtue-signaling. Mating aside, it might still be the case that punishers make attractive coalition partners, although the !Kung seemed as irritated as anyone by repeated busy-body critics (i.e., the "Ned Flanders effect").

A more obvious explanation of punishment prevalence is affective motivation. It is not principle or mating strategy that guides criticism of the free-rider, but plain old anger. If you are trying to muscle-in on my wife or my food, or take credit for someone else's work, or get drunk when we need your vigilance, then I become enraged. I am not enraged because it looks good or recommends me to others, or even because it violates some deep social contract. Your free-riding might indeed so violate the social contract, but I need no such theoretical grasp or subterfuge rationale to be angry and punish you. It might also be added here that the affective dimensions of punishing or taking revenge are complex and somewhat embarrassing, as compared with the rational model. Brain studies have revealed, for example, that punishers take great pleasure in the anticipation and delivery of revenge justice.[174]

Frans de Waal's and Sarah Brosnan's famous experiment with capuchin monkeys is often held to "prove" the innate primate instinct of fairness.[175] Two capuchins, in adjacent cages, were trained to take a token from a trainer and then trade the token back for a piece of food. Each monkey could easily witness the barter of the other. The food reward for this barter was usually a slice of cucumber, which capuchins like to eat. But grapes are loved by capuchins as a delicacy. If one monkey bartered her token and received only a cucumber slice, but then watched as the other monkey received a grape for the same kind of token, the first monkey would become incensed—refusing to continue playing, throwing cucumbers back at the experimenter, protesting, and even punishing the lucky grape recipient.

This experiment has been overinterpreted by journalists and even scientists themselves to illustrate a supposed fairness module in primates. Brosnan and de Waal have a nuanced view of their experiment, but some well-intentioned liberal academics regularly reference the experiment as an ancient ethical position that extends from grape payment to Occupy Wall Street and beyond.[176] We submit that the experiment does not illustrate an ancestral fairness module or even a fairness instinct. The actions of the capuchin monkeys are simply an expression of mammalian emotional systems and do not appear to be moral per se—although they represent the homological seeds of more generic normativity. Envy and resentment are powerful in social animals, and while they may eventually scale up to social contracts, they are not moral per se.

As we have already discussed, neuroscientists like Jaak Panksepp and Kent Berridge have shown that primates like us, as well as other mammals, have a very strong seeking or wanting system (driven by dopamine).[177] Once this desire system is triggered, it ratchets up expectation—motivating the mammal in powerful ways (toward food, sex, etc.). This goad or generic drive can be tethered to specific reward pursuits. The dopamine flood is at the high watermark just before attaining the goal, not during the reward consumption. Thwarting or frustrating the culmination of that seeking drive immediately activates the RAGE system and results in behaviors as trivial as tantrums or profound as punishing criticism and even murder. It is the same emotional neurochemistry, albeit ramped up, that propels more extreme forms of revenge behaviors. When we add this neuroscience of frustration to primate research, we observe that exasperation amplifies exponentially when agents perceive that they are not satisfying their desires while others are. That perception transforms frustration into resentment and rage.

It may be difficult to admit that our high moral principles may be descended from temper tantrums, but just as a grown man is no longer the child he was, so too our tantrums grow up and get converted into healthier social norms. Our tantrums evolve from personal feelings to become reasoned principles with cognitive justifications for egalitarian notions of the good. The oak is not just the acorn, and cognition must build on what precedes it.

Long before higher cognition turns our emotions into principled philosophies, it is the work of simpler social and cultural "institutions" to shape and sculpt our feelings into adaptive resources. Anthropologist Polly Wiessner studies how informal institutions, like speech patterns, behavioral traditions, rituals, and so on, shape emotion and cognition.[178]

In her studies, songs, for example, are important mechanisms in cultivating and directing adaptive emotions. The Enge peoples of New Guinea, for instance, solve ecological and political challenges, in part, by musical group manipulation. When a group of friends split into two hostile factions, as sometimes happens during competition for resources, the newly opposing groups will rile up violence by singing songs that demonize their new enemies—songs that describe the opponent families practicing incest and the opponent women having thorns in their vaginas, along with other dehumanizing narratives. But after six months of warfare and three dead, the enemies grow weary and begin to sing peacemaking songs and songs of consolation. The new songs down-tune the anger and shift the emotional state to one of reconciliation. This leads to expressions of care, and then meals are shared together between the groups. The songs and the meals pacify the rage and foster prosocial emotions and behaviors.[179]

Likewise, the !Kung peoples' evening tradition of "firelight talk" is an important informal institution that structures emotion and fosters cognitive complexity, as social reputation and hierarchy transform and grow more complicated. As Wiessner points out, night activities around a common fire steer the whole group away from daytime tensions and toward prosocial bonding activities, such as songs, grooming, dancing, and storytelling. These traditions, still observable today, may have developed during the rise of Pleistocene hearth culture. "Night talk plays an important role in evoking higher orders of TOM via the imagination, conveying attributes of people in broad networks (virtual communities), and transmitting the 'big picture' of cultural institutions that generate regularity of behavior, cooperation, and trust at the regional level."[180] So in addition to the natural emotional bonding that occurs during firelight social traditions, a space is created for slower-paced reflections that seek to better understand the thoughts and emotions of agents who are not immediately present. These public assessments of reputation, in the context of honor and respect relative to hierarchy, can build steam—to the benefit or the detriment of the individual being discussed. This is a kind of cultural expansion and enrichment of empathy, or TOM. And such informal institutions are helpful in the expansion and enrichment of cooperation.

Other Factors

These informal, high-repetition institutions, like night-talk or gossip, do the majority of the norming work that humans need in order to cooperate. Such interactive institutions bond the group, define the transgressions, enact the punishments, and, epistemically speaking, clarify the important ecological demands (i.e., resource acquisition, predator protection, etc.) from the perceptual, cognitive, and emotional noise that is always present.

Analysis of social organization has often suffered from the same biases as rational agent theory; namely, purely functional analysis, hyper-optimization assumptions, and simple cost-benefit modeling. Some anthropologists have offered a more pluralistic and nuanced approach, suggesting that multiple smaller innovations contributed to the social success of forager groups.[181] Forager success is characterized by a three-generational system of resource provisioning within families, long-term pair-bonding between men and women, and high levels of cooperation between kin and non-kin. These innovations rest upon ecological and economic factors, including the role of skill in resource production (the scale of apprentice learning / teaching) and the degree of complementary division of labor in males and females. For example, it takes around two decades of daily hunting for male apprentices to become master hunters, and this requires a context of relatively significant social stability. Division of labor patterns in ten foraging societies reveal that males secure an average of 68 percent of the group's calorie intake (88 percent of the necessary protein), while women acquire the remaining 32 percent (the remaining 12 percent protein).[182]

We agree with the pluralism approach to causes of social evolution, but once again, we need to add the affective / emotional component, which is more foundational than previous research has recognized. It is the emotional niche (as well as the cognitive niche) that allows for the apprentice learning and pair-bonding successes mentioned here. An overly hostile social environment, for example, compromises certain kinds of cooperative endeavors. In informal, high-repetition institutions, the emotional niche is also crucial in the formation and enforcement of norms through *shame* and *guilt*.

One of the features of an emotional niche is *impulse control*. Members of groups must practice impulse control upon their personal pleasure-seeking behaviors (their hedonic agendas). They must defer gratification and surplus consummation in order to provision offspring, partners, parents, allies, and so on. They must impulse-control anger or rage in order to sustain coalitions and stabilize apprentice learning. They must impulse-control lust so as to avoid disruptive intermale aggression competition at every turn. They must impulse-control fear, and so on, with every affective system.

Moreover, the emotional niche requires input from higher cognition and social habituation for regular adjustments. Healthy emotional management entails regular *target correction*—in the sense that anger, for example, can spill in all directions without cognitive focus on relevant targets and without informal institutions (like songs, gossip, etc.) to concentrate adaptive emotion. Additionally a successful emotional niche requires appropriate *intensity calibration* (e.g., minor infractions cannot be met with frenetic or hysterical responses). Part of the role of informal institutions, like songs and firelight talks, is to help strengthen these aspects of the malleable emotional niche.

Nowhere is the relationship between norms and adaptive impulse control more obvious than in the case of the Maasai peoples (in northern Kenya and Tanzania). In an age-old rite of passage, we find the interdependency of emotional control and the economic / social future. *Emuratare* is the circumcision ritual performed on every boy sometime around the age of sixteen. An elaborate pre-circumcision ritual *(enkipaata),* which includes living apart from the village and engaging in endurance tests, leads up to the circumcision.[183] The *emuratare* itself is considered the most important ritual for Maasai society, and it is how boys transition to men. All such coming-of-age rituals are complex stages in the emotional reconfiguration of boys to men and girls to women. They are not just symbolic adjustments in how the tribe sees the individual, but actual transformations of the agent's emotional disposition (and cognitive self-conception). Boys, for example, often re-create their emotional disposition from the highly expressive style of youth to the Stoic style of householders and warriors. Their very survival in the group requires such emotional management.

In the Maasai *emuratare,* the young man must endure the excruciating pain of having his foreskin removed without anesthesia. But more astonishing, he must do the ceremony before an audience of tribal villagers who watch his face carefully for any signs of flinching or grimace.[184] If the boy winces and shows pain on his face during this procedure, he will fail his greatest test of manhood. There is no second chance. If the boy sustains a Stoic visage, then his father awards him with a goat or some other livestock. This is economically and socially crucial.[185] The goat means that the young man now has wealth, and wealth means marriage and membership in the community. If the boy flinches and fails this circumcision ritual, his father gives him nothing. Without the goat, he has no economic or social footing in the community, and he is ostracized and banished from the group. He must leave the protection and opportunity of the social group and live alone in the bush.

It is hard to imagine a more high-stakes example of socially adaptive impulse control. Failure to retrain one's emotional expressions can mean social death. Most humans evolved in similarly high-stakes ritual cultures, designed to harden householders to the many privations of adult life. Life in a hostile environment is tough, and the agents need to be emotionally calibrated for it. The emotional niche constructed for juveniles is not the emotional niche for adults, and social groups devise ways to transition members across the life-history bridges that span from childhood to adulthood to elderhood.

The conceit of modern ethical philosophy ignores the naturalistic underpinnings of human normative life, preferring to derive ethics from categorical imperatives of logic (deontology) or from hedonic calculations (consequentialism). But the emergence of this kind of principled ethical reflection is so recent (emerging in very recent history, rather than in deep prehistory) that it has almost no relevance to the evolution question. Kant, Mill, and Rawls might be the way *forward,* but they are not very relevant to understanding how we got here.

Our view is that the broad genus of normative cognition (of which morality is but a species) relies upon four major domains of human problem-solving. These domains would have come online piecemeal throughout the Pleistocene and Holocene periods: (1) prosocial emotions (sometimes called moral sentiments); (2) ritualized behavior patterns or informal institutions (involving prosocial rewards and antisocial punishments); (3) TOM, and / or imagination; and finally (4) abstract reasoning.

As we have already established, the normative emotions are crucial in the sense that they motivate the agent, tag the social environment with values, and strengthen social bonds. Brain research, including fMRI, EEG, and lesion studies have converged on the idea that moral emotions are processed in the prefrontal cortex (PFC), and in the orbitofrontal cortex (OFC) in particular.[186] This makes functional and evolutionary sense because the PFC receives crucial inputs from the older limbic and sensory areas but also assists contiguous executive impulse-control areas (i.e., anterior cingulate gyrus). Note also the role OFC plays in shaping emotional control and decoupling of emotional tags, as discussed in an earlier chapter concerning the somatic-marker thesis.[187]

With the onset of better imaging technology, neuroscience has drawn a more detailed map of the prefrontal cortex and its division of labor. Using fMRI, we can see that the ventromedial prefrontal cortex (vmPFC) is implicated in automatic, immediate evaluations.[188] And the dorsolateral prefrontal cortex (dlPFC) is implicated in slower, controlled deliberation and self-restraint.[189] For some theorists, these cortical areas have been correlated with the "dual processing" model of mind, suggesting that "thinking fast" (hot cognition) happens in the vmPFC and limbic areas, while "thinking slow" (cold cognition) occurs in the dlPFC.

This work has been extended into questions of prosociality—specifically cooperation, reciprocity, and ethics. The results are intriguing, if somewhat preliminary, and contentious. The perennial philosophical disagreement, as to whether ethics comes from natural cooperative sentiments or deliberative restraint of our selfish impulses, has been projected onto the brain. Some data, derived from various ultimatum games, seem to support the *restraint model* of prosociality, suggesting that cooperative behavior results when the dlPFC is actively "deliberating" and preferencing the group over the individual.[190] When I deliberate (employing the dlPFC), I restrain my default egotism. This model predicts that compromising the deliberative cognitive processor (dlPFC) should reduce cooperation. Evidence in favor of this model comes from several sources, including the use of transcranial magnetic stimulation (TMS) and, more recently, lesion studies.[191]

Alternately, the *intuition model* of prosociality suggests that humans are cooperative by default. In this model, the hot cognition path of the dual-processing mind, activated in the vmPFC and amygdala, already predisposes us toward altruistic behaviors, and it is the deliberative cold-cognition system that turns us away from cooperation and toward selfishness.[192] Using many different kinds of economic games, David Rand finds that subjects who are forced to solve cooperation problems quickly, through hot cognition (vmPFC and amygdala) are more cooperative and charitable than those who deliberate more slowly (dlPFC). He interprets this to mean that humans are intuitively cooperative, and increased reflection and reasoning only diminish our natural ethical orientation to the group.

This tension between models of ethical processing will presumably be resolved with more empirical work, but on the face of it, our affect-based model of prosociality is more congruent with the *intuition model* of cooperation. Still, the issue is extremely complex, and the meaning of the recent data is unclear. Even after neuroscience clarifies whether the dlPFC or the vmPFC is more active during ethical activity such as sharing, we will need to understand the relation between these cortical areas and the more hidden deep-brain affective systems (like FEAR and CARE, for example). These older emotional networks were crucial in mammalian affiliative behaviors and norms long before frontal executive forces came online.

Is the dlPFC actually involved in creating values or evaluations, or is it processing / associating values that were generated elsewhere (e.g., limbic areas)? Some provisional work suggests that the dlPFC may indeed play a more constitutive role in affective appraisal.[193] But our argument throughout this book has been that subcortical systems, like CARE, provide the original roots of mammalian values—a view borne out by the copious data on mammalian kin bonding. These conserved subcortical systems provide the affective psychology (i.e., rewards and homeostatic reset) that motivates resource sharing and cooperative behavior. The pathway for this more ancient prosocial system connects perception quickly to the nucleus accumbens, the amygdala, the hippocampus, the hypothalamus, and so on. In rats, the process of bonding and values-formation is rooted in olfactory learning, but in humans the visual system and higher cognitive capacities are enlisted. It would not surprise us, then, if executive brain areas like the dlPFC were found to be crucial in values "processing" (e.g., acting as amplifier, or salience lens, or limiter, or associational manager, or even as logical sketchpad via working memory). Indeed, it may be that cortical regions like the dlPFC play a crucial role in the way *Homo sapiens* stretch the tribal values of kin cooperation out to wider circles of egalitarian prosociality. In the evolution of the brain, we find no clear dlPFC homologies with non-primate mammals, and this suggests that primate prosociality (and eventually our own style of cooperation) may owe some of its character to this executive cortical innovation, together with cultural enrichments.[194]

Brain-based social emotions emerged together with behavioral innovations (shaped by ecological pressures) and even technical innovations (e.g., fire and hearth culture), providing *Homo erectus* with enough normative sophistication to bootstrap increasing cooperation.[195] Civilization is, in part, emotional work. It is the cause and effect of affective patterns, emerging during the Pleistocene and converging in the Holocene. In this chapter, we shared a vision of how social organization and the social emotions may have evolved, and in the next chapter we will do the same for some of the highest aspects of the mind; namely, religion, mythology, and art.

9

Religion, Mythology, and Art

IN 1899, anthropologist Edward Burnett Tylor defined culture as "That complex whole which includes knowledge, belief, art, morals, custom and any other capabilities and habits acquired by man as a member of society."[1] While we explored custom, norms, and morals in Chapter 8, here we focus on the role of the *spiritual emotions* of awe, wonder, and transcendence in art and religion. We fold our consideration of these unique emotions into a broader discussion of cooperation but do not find them easily reducible to purely adaptationist logic. Moreover, we are comfortable with that irreducibility.

In this slightly more speculative chapter, we explore the spiritual emotions through prehistoric art, shamanism, aesthetics, and religion. We recognize that we're sometimes moving beyond data here, but there is always good speculation (staying close to the facts) and bad speculation (just-so stories). Extrapolation that builds upon reliable epistemology, metaphysics, and evolutionary and historical facts paves the way for new research and is not intended as corroborated fact or entrenched theory. We engage with empirical work throughout but expect the reader to track this chapter with a healthy mix of charity and skepticism.

Taking their energy and drive from basic affective sources like the SEEKING and PLAY systems, spiritual emotions function to temper intense feelings of FEAR and GRIEF in the format of culture. The emotionally saturated state of spiritual emotions is immanent in our neuropsychology, but the ways in which we communicate our transformative experiences in the neocortical imaginative elaborations of art are unique. Taking place in the association-rich cultural spaces of affective modernity, shamans, artists, and mystical individuals sought to aesthetically materialize their unique experiences in ways that produced faith, belief, and further transcendent states for people in their communities. While scientists seek the cognitive processes behind wonder and curiosity, we urge a consideration of the emotional landscape that gives this sphere of culture its indelible importance to groups and individuals. Art and religion coalesce formats of emotional manipulation that are prosocial, but the mystical experience, as William James wrote in 1902, produces an "added dimension of emotion . . . and a new reach of freedom for us . . . performing a function which no other portion of our nature can so successfully fulfill."[2]

Transcendence and the Spiritual Emotions

We define awe, wonder, and transcendence as *spiritual emotions* because they induce heightened sensations of meaning and call out for further interpretation. As if they provide access to a truer reality, mystical states add a layer of meaning to everyday consciousness and thus tend to become powerful factors in determining an individual's belief system. Awe, that state of reverential respect, can be provoked by natural beauty and art or by overwhelming experiences of love and fear. Wonder similarly makes us childlike in a mysterious world where almost anything is possible. Transcendence is a mystical experience that some people spend their whole lives seeking out, hoping its raw power will overtake the meager mind and grant peace to one's whole being. Even if, as Kant argued, the sublime turns out to be a subjective experience caused by a failure or confounding of cognitive categories, we still recognize that such an experience is highly motivating and transformative for many individuals and propagates through populations as a cultural ideal.[3]

It is possible that not every reader has experienced a mystical state of transcendence, and thus it is worthwhile to describe some basic lineaments through firsthand accounts and a summary of its major components.[4] Writer Arthur Koestler, echoing the experiences of a varied set of individuals who have provided written descriptions of this state, wrote of his own mystical experience: "The 'I' ceases to exist because it has, by a kind of mental osmosis, established communication with, and been dissolved in, the universal pool. It is the process of dissolution and limitless expansion which is sensed as the oceanic feeling, as the draining of all tension, the absolute catharsis, the peace that passeth all understanding."[5] This instance may have been generated by any number of causes, including meditation, sensory deprivation, bodily mortification, synesthesia, sexual ecstasy, or a drug-induced altered state of consciousness.[6] While spiritual emotions can be described as pleasurable, ecstatic, manic, blissful, peaceful, tranquil, sublime, or vital, the depersonalization of ego-dissolution can also induce negative affects, such as the feeling of drowning, a panic attack, an overwhelming sense of dread, or a feeling of horror of annihilation.[7]

The neural causes of this oceanic feeling may reflect changes in brain metabolism in fronto-parietal networks responsible for the formation of a coherent sense of self in time and space.[8] The pleasure associated with oceanic boundlessness, in turn, may be due to a decrease in amygdala activity.[9] We see commonalities in these states of spiritual emotions in terms of their particular relation between diminished conscious thought and a high level of emotional saturation (more on this below).

What is pertinent to us here is that similar feeling states can be interpreted differently, depending on the neocortical imagination space of a given cultural milieu. For example, Koestler's portrayal barely differs from diary entries describing transcendent states by Saint Teresa in sixteenth-century Avila and by Sufi scholar Al-Ghazzali in eleventh- to twelfth-century Persia. In each case, the cultural milieu suffuses the interpretation, with Koestler emphasizing mysticism, Saint Teresa the Christian God, and Al-Ghazzali Allah of the Qur'an. Similarly, the trance states described by prehistoric shamans, as we describe below, reflected and coalesced with cultural narratives about the vital powers immanent in nature. This is another demonstration of how primary- and secondary-level affective states are inextricably tied to tertiary-level neocortical elaboration. Observe, for example, how landscape and culture shape mythology and rock art in the Dinwoody images of the Shoshone in Wyoming, where tribes that live below 1,650 m. have differing images (like frogs and turtles) from those up to 1,800 m. (bison and horned figures) and those up to 2,100 m. (owls and eagles).[10] Or consider a currently popular interpretation of the positive elements of spiritual emotions, dubbed "flow state," which emphasizes the feeling of immersion in an activity.[11] It is no surprise to us that in this secular age of productivity and efficiency, such a state of spiritual emotions is interpreted vis-à-vis the completion of an activity and cognitive fluidity.

While art and creativity are intimately connected to spiritual emotions, there are other prosocial elements of ritualized affective sculpting, which are evident in prehistoric visual art, the narrative format of literature, religion, and the ecstatic bonds of music.[12]

Art and Evolution

We will focus on prehistoric art for several reasons: one being that we wanted to avoid the context of contemporary art appreciation and creation, as it is heavily influenced by the very recent advent of the museum, which distorts our understanding of the situatedness and cultural specificity of any work of material representation.[13] Prehistoric art allows us to focus on the visceral relation of spiritual emotions to material representation and its social function.

While the author of the most widely used textbook of art, Ernst Gombrich, begins his study claiming that "there is no such thing as art, only artists," philosopher Denis Dutton set out to encapsulate an exhaustive criteria to define art in its artifacts—sculpture, painting, decoration (including tools or body-painting)—and performances—dance, music, recitation, and stories.[14] His list is as follows: For something to be art, it must (1) give direct pleasure, (2) display skill and virtuosity, (3) be created in a style, (4) display novelty and creativity, (5) exist alongside criticism and appreciation, (6) utilize representation, (7) be separate from ordinary life, (8) display some expression of individuality, (9) lead to emotional transcendence, (10) provide an intellectual challenge, (11) be part of a tradition or institution, and (12) provide an imaginative experience for both creator and audience.

Meanwhile, anthropologists have emphasized how both production and appreciation of art is a matter of the mind's receptivity to form. In *Primitive Art,* Franz Boas discusses how aesthetic pleasure is not necessarily emotionally expressive; rather, in ornamental art, the viewer is responding to the regularity of form and the evenness of surface, insofar as they display the virtuosity and mastery of the creator of the piece.[15] Leder and Nadal and their colleagues at the Center for Empirical Aesthetics at University of Vienna investigated the psychological mechanisms of the aesthetic appreciation of art.[16] They found that "(1) An aesthetic experience has an evaluative dimension, in the sense that it involves the valuation of an object; (2) it has a phenomenological or affective dimension, in that it is subjectively felt and savoured as it draws our attention; (3) it has a semantic dimension, in that an aesthetic experience is a meaningful experience, it is not mere sensation."[17] Some claim that aesthetic awe is "the most pronounced, the ultimate, aesthetic response, in all ways similar to the fundamental emotions."[18] This resonates with the depiction we provide below, albeit somewhat speculative, of the creation and appreciation of prehistoric trance art as arising from evaluative emotions that are elaborated within neocortical cultural matrices. We see this focus on the spiritual emotions of awe and transcendence as crucial to the affective roots and evolutionary function of material representation.[19]

As far as *why* art evolved, Dutton suggests it has a role in sexual selection, in that, through their squandering of resources, works of art demonstrate an artist's or patron's fitness.[20] His evidence is that art is made of rare or expensive material, is time-consuming to create, requires fine skills, may have no possible use, demonstrates a sense of waste, and requires special creative effort to create. Another standard analysis of the evolutionary purpose of imagination is that it arose primarily as an ability to engage in thought experiments and make suppositions about the future, with art as a sort of byproduct or elaboration for not-necessarily adaptive reasons.[21] Imagination as the source of artistic enterprise is thus a type of pretend play or counterfactual-generating pastime.[22] Some claim aesthetics is part of habitat selection or related to the skill of deciding whether or not a face is attractive; others claim it is an exaptation.[23]

One post hoc explanation for the evolution of art practices is that representational art in nonliterate societies (including all prehistoric societies) serves to record, in a lasting and public form, information that would otherwise be lost, creating history and precedents for future generations across various structures of meaning.[24] Art, in this scenario, is a form of cultural materialization of lifeways and ways of being in the world that can be shared by communities both synchronically and diachronically.[25] Also, we note that the greater the evocative, emotional aspects of a given work, the more effective it will be in remaining a part of the system of cultural meaning of a given community. In contrast to the limited use of visual media in nonhuman primates, the abstraction of form in prehistoric art curtails nicely with our discussion of the evolution of the symbolic and associationist mind in imagination and representation.[26] Indeed, use of symbols in prehistoric art may be a reflection of the advent of spiritual and religious practices emerging from dream and daydream states.[27] According to Marvin Harris, art, religion, and magic express out-of-the-ordinary emotions that impart a sense of mastery or communion with unpredictable events and unseen powers.[28] By imposing human meaning on an indifferent world, we seek to penetrate to a true level of cosmic significance. Material representation, then, allows us to memorialize these structures of meaning, perpetuating "truths" and providing models for future use.

Neurologically, we can discuss the relation between the reward system, primary and secondary affective systems, and the distinct contributions of tertiary cortical regions to the production and appreciation of art objects.[29] Our analysis focuses on how art gives rise to affective responses by employing symbols within a culturally established set of emotional associations and, through aesthetics, depicts emotion-arousing events, entities, and persons in ways that enlist participation in the viewer.[30] This aesthetic sense of art objects and paintings reveals a dimension of mind that is appreciative and cognizant of visual meaning in a cultural context and toward some social end, in addition to being a form of gratification in itself.[31]

Since we have no reason to believe that neurological changes are what caused the cultural use of material representations in the Upper Paleolithic, we must assume it was powered by the changes in social organization that we discussed in Chapter 8 and by encounters with other human groups, wherein cultural innovations proffered selection advantages. According to anthropologist Clifford Geertz, "certain activities everywhere seem specifically designed to demonstrate that ideas are visible, audible, and . . . tactile, that they can be cast in forms where the senses, and through the senses' emotions, can reflectively address them."[32] It may be that humans need shared representations of their history, future, beliefs, and heroes to experience social solidarity as a form of fictive kinship through mythical ontology.[33] Consider the Australian Aboriginal songs about Dreamtime that link together spirit ancestors, chart a road that links places where food and water can be found, and outline the rituals that must be performed in those spaces.[34] Or the group-trance dances that are continuous with cave art depictions of the mythological origins of the great Mantis / Kaggen and the explanations of rain and dreaming depicted by Kalahari San tribes in the last few thousand years.[35] These aesthetically depicted material representations create a shared social and cultural space.

Prehistoric Art

The low-density family-level *Homo sapiens* of the prehistoric age satisfied the necessities of life with simple technology based on hunting, gathering, and plowless agriculture. Their social, economic, and political specialization was, likewise, relatively simple and depended on the task at hand rather than on a status hierarchy.[36] In terms of their material culture, over 160 KYA, we observe the construction of blades, grindstones, and pigments, while around 70 KYA, we observe remnants of ornamental beads, presumably for social purposes. But it isn't till the Aurignacian culture (35–25 KYA) of the Upper Paleolithic (40–11 KYA) in Western Europe that the first instances of images appear.[37] The historical record (about 300 sites of Paleolithic parietal art that we know of) goes on to reveal representation in painting, engraving, sculpture, and jewelry, as well as fragments of bone, antlers, and ivory that have patterned markings, including, notably, flutes made of vulture bones.[38] While prehistoric art served many purposes, including mythic and aesthetic functions, the relationship between art and experience in the evolution of social technologies seems to indicate the engendering and memorializing of spiritual emotions.

From the identity-marker ornamental beads to etched spear-throwers and parietal art (i.e., rock or cave art), the varied objects of the Upper Paleolithic suggest new functions in the creation and appreciation of material representation. While language can maintain complex social relations, material representations endure over many generations, linking people together in shared knowledge, beliefs, and metaphorical associations; for example. the interlocking pattern used in ivory-bead manufacture at Sungir (circa 32–28 KYA) lasted over 1,000 years.[39] In Gravettian culture (28–20 KYA), we observe not only representations of animals like ibex and horses, but also human handprints and figurines of pregnant women (as in the Dordogne sites in France, or in the later Natufian culture in the Levant about 13 KYA), which may have been fertility charms. We observe painted hands in the Aurignacian, possibly commemorating the touching of a subterranean tier of the cosmos, and vulvas, as fertility symbols, become popular in the Solutrean (22–17 KYA) and Magdalenian eras (starting around 18 KYA).[40] Magdalenian culture included an expansion of projectile weaponry, stone tools, and open-air campsites with portable representational objects such as thousands of engraved and painted blocks of geometric signs. These signs have been interpreted as structural social divisions or conceptual, affective, and cosmological states.[41] Over the 20,000 years of parietal art in Spain and the Ariège Pyrenees, we observe many examples of painted signs, namely claviforms and tectiforms that most probably signified ethnic or territorial markings of tribal identity.[42]

Also notable is that rarely hunted animals like horses, bison, and ibex are represented far more often than reindeer, which were the main source of sustenance. This suggests that the sympathetic depictions of horses, bison, and ibex in Western Europe had special status, possibly providing eschatological meaning as dream-helpers during shamanic ritual. Similarly, the Eland and depictions of leonine shamans, which served as spirit animals for the Kalahari San tribes, may have been believed to come under the tribe's control by being depicted in cave art.[43] The Australian Aboriginals (30 KYA to present) represent mythical figures in their cave art; for example, the humanlike Wandjina, who control fertility, the regeneration of life, and cyclone storms, or the dangerous Nabulwinjbulwinj, who kills women by hitting them with a yam.[44] Such cultural elaborations, which often include therianthropic depictions (i.e., creatures with both human and animal characteristics, as in Chauvet and Les Trois-freres caves), are sometimes ways to conceptualize our place in nature in relation to the most important determinants of survival; namely, the weather, fertility, and other animals. Indeed, in San thought, which we can piece together through rock paintings, nineteenth-century phonetic texts, and transliterations of these texts into English, religious experience, and ritual form a closely related network.[45] The abundance of imagery emphasizing transformation in San and other prehistoric art—all the way up to modern human examples like Ovid's *Metamorphoses* and the *Grimm's Fairy Tales*—reveal the importance of alterity, identity, and self-image. The transcendent experience of transformation seems hardwired into the human nervous system, yet it is up to the cultural traditions to inform the meaning of the experience of the spiritual emotions of transformation.[46] Possibly, the fronto-parietal networks that deliver a notion of identity are crucial, but we may also consider how the bodily integrity of the midbrain SELF complex may be attenuated during trance and leave room for cultural interpretations of the altered state after the fact. Additionally, as we continue to speculate below, transformations often result from cognitive mismatches of basic domain categories, and these supply art and religion with a veritable menagerie of imaginatively elaborated supernatural hybrids that materialize ways of thinking about our place in nature.[47]

Shamanism and Cave Art

Art in the form of song, dance, painting and sculpture is present in all cultures, as are magic and taboo.[48] In the early twentieth century, the influential European anthropologists Abbé Henri Breuil, Count Henri Begouen, Bando, Maringer, Obermaier, Lantier, and Salomon Reinach saw prehistoric cave painting as a form of sympathetic magic and totemism that facilitated hunting and fertility.[49] This interpretation claims that representing an animal means that one can master the form of the double of the animal and thus control it.[50]

Practices by which desired ends may be achieved or harm and evil avoided through a prescribed formula for contacting a spirit or supernatural power are forms of magic. Of course, magic as such is not held in high regard in the urbanized West, but the motivations for such practices remain recognizable and altogether human. Magic— whether it be through divination or ritual—provides some form of security and protection; while maintaining equilibrium and balance in the environmental niche, it also satisfies a desire for continuity and stability in relation to the ancestors. Tribes in Indonesia and Melanesia believe that a portion of the power and energy of the spirit released from the body at the time of death can be ritually persuaded to take up abode in a sanctified wooden carving.[51] Fetish objects, like the Melanesian carved totem, are said to have an indwelling power to cure, destroy, or protect. Also, judging by the quality of fetish objects and rituals, it appears they may satisfy an additional or intertwined need for the creation of forms and aesthetic expression.[52] While the hunting/sympathetic magic interpretations of cave art have come under scrutiny in the last fifty years of scholarship, there is plenty of evidence that prehistoric cave art had magical purposes within shamanic rituals.[53] Contemporary animistic practices (e.g., spirit houses in Southeast Asia) also lend credence to the theory that aesthetic objects can have magical functions. To be precise, it appears that during the Upper Paleolithic and early Holocene, no distinction was made between art and religion.[54]

Each cave is unique, and thus consideration must be made for the mix of subject matter, technique, and color, as well as for the form of the cave in surfaces, textures, light conditions, volumes, and acoustic qualities.[55] Yet there seem to be great similarities in the function of cave art in Western Europe during the Upper Paleolithic, San Bushmen cave art in South Africa in the last several thousand years, and the 6,000 open-air rock carvings in Coa, Portugal (dating uncertain). Many of the West European cave images of animals or therianthropic images tend to be located in difficult-to-access parts of caves that were not living sites. Also, the major cave systems at La Volpe, Lascaux, and Chauvet seem to each follow singular grand plans, with the conglomeration of images relating to each other vis-à-vis the order of visiting the cave, with an anteroom, a main chamber, and smaller side rooms.[56] The barb-like branching signs (at Grotte de Niaux and other locations) and depictions of animals and of the use of tectiforms seem systematic and seem to form part of larger cultural systems of meaning. For example, in the Magadalenian cave of Les Trois-Frères, zigzag signs are associated to bison, while curvilinear signs are on animal tails and angular signs grace animal flanks.[57] There is a cohesive sense of meaning in each of these sites, and much of it is probably generated by the spiritual leader of prehistoric communities: the shaman.

Shamans are individuals who, on behalf of the community, enter altered states of consciousness to heal sickness, see into the future, control weather, and visit supernatural realms.[58] The shamanistic interpretation we favor suggests that caves were considered underground portals to parallel spiritual universes, most likely materially representing revelatory hallucinations experienced during states of trance.[59] Shamanism may have been the earliest animistic religion and, as a practice, it may have motivated the first art, because "gods talk to shamans" in wisdom and descriptions of the future or in the shape of the cosmos during their trances.[60] This early religion reflects the role of the shaman as both healer and sorcerer in a cosmos that includes darkness and light, death and life, where a vital force in all things must be balanced through the torturous journey of the uncertain mystic figure. These journeys usually involve experiencing death, bodily transformation, drowning, combat, sexual arousal, and a vivid sense of movement and flight.[61] They explore the range of spiritual emotions available to the human mind. In this way, shamans—like monks and holy men in our more recent history—are truly genii of the mind.

One group that has a deep historical record of trance-derived visions, geometric figures in engravings, and isolated painted rock shelters in the Kalahari Desert is the San peoples, who have a group shamanic tradition via trance dance. According to Lewis-Williams:

> The main focus of San art was the building up, through generations of painters, of a cumulative manifestation of the spirit world, its fear, pain, beings, creatures and power, and . . . its pervasive infiltration of the material world . . . San painters played not only a religious but also a social role, establishing and contesting a variety of human relationships through manipulation of the elements, symbols and experiences of a shamanistic cosmos.[62]

It is possible that symbols on cave art materially represented and imprinted experiences and knowledge gained from spiritual journeys and transcendent experiences.[63] Shamanic actors could use the painted panels for their images of the past that could then, in an altered state, be revivified with new insight and an enhanced shamanic cosmos as forms of cultural elaboration.[64] These parietal panels are similar to stained-glass windows in medieval cathedrals, wherein the light of spiritual realms shines through the images, animating them as manifestations of symbols.[65] Each Western European rock-art cave is itself an ensemble, a space for the performance of ritual.[66] The rock art and the myths are tied together by the spiritual emotions of the trance state and straddle the divide between the material world and the spirit realm, clarifying elusive relationships and the spiritual emotions they elicit in the people.[67] One could also think of the caves as the cultural informing of the trance participants insofar as the embellished chambers were vestibules to prepare the vision-questers for the solitude of the smaller spaces, stocking the mind with cultural imagery that upheld foundational myths of the tribe.[68]

Spiritual emotions seem to play a role in the creation of prehistoric art, whether it be in a trance or during an altered state of consciousness induced by extended periods of dancing, drumming, clapping, drugs (e.g., tobacco, jimsonweed, swallowed harvester ants, peyote, yage, psilocybin mushrooms, Morning Glory seeds, San Pedro cactus), dehydration, pain (self-flagellation), or extended periods of light deprivation. These experiences are forms of emotional saturation that connect the affective roots of spiritual emotions like awe and transcendence with art and religion. Mystical visions themselves may be generated by entoptic (i.e., in the eye) trance visions and their subsequent cultural elaboration.[69] Interestingly, no matter the cultural background, humans report that during altered states, they experience either flying or entering the ground through tunnels.[70] These altered states are interpreted as religious in particular cultures, whether it be as spirit-possession or as soul-loss; or, for the San bushman, whether it be in mystical flight or down through the vortex of underground travel between the levels of the shamanic cosmos.[71] The cosmos of the San bushmen was conceived as being split into three levels—the level of daily life (where they hunt and gather plants) sits beneath the level of spiritual things (where gods and spirits and the gods' herds live), and all of this sits atop the subterranean spirit realm, which is accessible by holes in the ground, where "rain animals" and other spirits dwell.[72] The purpose of the San shamans' journeys between the three levels was to plead for the sick, fight off malevolent shamans, obtain information about the location of animals for hunting purposes, and find ways to make the rains fall.[73] In all, the goal is to aid the community in satisfying the basic necessities of life through material representation in ritual cave spaces that served to affectively sculpt cultural and social norms.

In the Great Basin rock art of Arizona and California (during the last 2,000 years), shamans produced images after depicting vision quests in altered states of consciousness, in which they learned how to cure an individual's malady, how to call forth rain, and how to enact sorcery. In this tradition, the rock art sites were sacred places that served as portals to supernatural realms; sometimes the journey was undertaken to learn new knowledge and sometimes it was part of a puberty-initiation ritual where moral, religious, and cultural codes were instilled.[74] While the preeminent book on shamans emphasizes their mastery of the ecstasy of trance, other scholars emphasize their mastery of the spirits during the emotionally violent effects of trance.[75] Such experiences are subsequently expressed in fearful awe-inducing material representations that evoke counterintuitive concepts and beliefs in the viewer.[76] This capturing of the spirit world gave meaning to impossible worlds that solved existential problems; this organization of beliefs, paired with our innate spiritual emotions, enabled shamans to create tales, songs, and dances that permanently materialized their spirit contacts. Maybe the shaman was the master of spirits and trance states, a role in which he provided and appropriately modified the cultural background by which individuals would be able to structure (i.e., culturally interpret) their own transcendent emotional states.

Finally, a word on the affective nature of the individual shaman figure: the emotional distress of the calling to be a shaman, which is usually passed down via bloodline, originates in a period of illness or madness that almost disintegrates the personality and can only be cured by accepting the calling.[77] While Robert Sapolsky focuses on the hallucinatory dimension of the vocation, arguing that shamans suffer from schizotypal personality disorder, other salient diagnoses are manic depression (i.e., bipolar disorder) or cyclothymic conditions.[78] The latter affective disorders are more common in artists than in the normal population and they reflect tremendous creative output and fluency, not to mention flexibility of associational thought during manic sleepless periods.[79] Mild depression, on the other hand, produces concentration and dedication to a single problem, which may lead to creative insight.[80] Whitley claims that shamans were crucial in our shift toward modernity, not because of their rationality, but because the full emotional range of the human mind, including mental sickness, was manifested in material representations of their experiences. The spiritual emotions—putatively experienced as a journey to the world of spirits—are on display in prehistoric art as being evoked by a ritual that demonstrates and sculpts for the viewer the range of affective processes available in the modern human mind.

Fiction, Mythology, and Music

Cave art is site-specific, but nomadic humans developed portable-format material representations of literature, song, and story, which they fleshed out during leisure time between hunting expeditions. Boas highlights fixed forms, such as conversation, poetry, and chants, as well as the repetitive rhythmic formats of war songs and religious songs (like those of the Kwakiutl in British Columbia, Canada) that resemble the proverb, the riddle, the animal tale (cf. Aesop, Jataka Tales, Ovid, La Fontaine), and even the epic poem (e.g., *Son-Jara, Gilgamesh, The Illiad*).[81] These fictions—whether they be in the form of novels, plays, movies, or videogames—may serve as a form of pretend play that provides a low-risk simulation; a life lesson that can prepare one for possible life experiences.[82]

The spiritual emotions, like wonder, are important for creativity, providing a goad and lifting the fetters of inhibition on imagination. Emotions of awe, or what psychologist Jonathan Haidt calls "elevation," are feelings of inspiration and expansion, which come from awe-inspiring experiences with nature, art, or spiritual beauty.[83] Art frequently engages the emotions and thus helps individuals navigate life by teaching control over rather than enslavement to these powerful forces.[84]

Spiritual emotions also can serve to provide a bond between the reader and the text; for example, the famous revelation of Krishna in the *Bhagavad Gita* brings the reader face-to-face with the awesome mystery and grandeur of reality. In the tenth teaching, Krishna incites awe by describing his extent:

> I am the thunderbolt among weapons,
> among cattle, the magical wish-granting cow;
> I am the procreative god of love,
> the king of snakes.
>
> I am the endless cosmic serpent,
> the lord of all sea creatures;
> I am the chief of the ancestral fathers;
> of restraints, I am death.

I am the pious son of demons;
of measures, I am time;
I am the lion among wild animals,
the eagle among birds.[85]

In the eleventh teaching, Arjuna replies, trembling in awe:

I see no beginning
or middle or end to you;
only boundless strength
in your endless arms,
the moon and sun in your eyes,
your mouths of consuming flames,
your own brilliance
scorching this universe.

You alone
fill the space
between heaven and earth
and all the directions;
seeing this awesome,
terrible form of yours,
Great Soul,
the three worlds
tremble.[86]

Here we experience a mediation of fear through an aesthetic rendering of the confrontation between a creator of terrible majesty and a mere mortal. We can all empathize with the puny man's humble position in the face of nature. This same kind of confrontation was a recurrent theme for the European Romantics. For example, Percy Bysshe Shelley (1792–1822), in his poem on Mont Blanc (1817), exclaims in awe:

A loud, lone sound no other sound can tame;
Thou art pervaded with that ceaseless motion,

Thou art the path of that unresting sound—
Dizzy Ravine! and when I gaze on thee
I seem as in a trance sublime and strange
To muse on my own separate fantasy,
My own, my human mind, which passively
Now renders and receives fast influencings,
Holding an unremitting interchange
With the clear universe of things around;
One legion of wild thoughts, whose wandering wings
Now float above thy darkness, and now rest
Where that or thou art no unbidden guest,
In the still cave of the witch Poesy,
Seeking among the shadows that pass by
Ghosts of all things that are, some shade of thee,
Some phantom, some faint image; till the breast
From which they fled recalls them, thou art there.

The British and French landscape paintings of Arcadia (see Thomas Gainsborough, J. M. W. Turner, Claude Lorrain, or François Boucher) may have had a similar purpose of aesthetically depicting the exhaustive extent of creation relative to man. And the Daoist and Zen landscape paintings *(shān shuǐ)* of the Song Dynasty period similarly infuse the natural world with transcendental import.

Poetry—that most subtle drama of words and spirit—can also dramatize overpowering emotions like grief,[87] as shown by Dylan Thomas (1914–1953) in this famous poem in response to the death of his father: "Do not go gentle into that good night." Or consider the ancient tradition of epitaphs collected in the Greek anthology, where we read aesthetic reflections on mortality, like that of Callimachus: "This stony sepulchre you pass is / now the tomb of Callimachus; / he could sing and he could laugh / while the red wine filled the glasses." Or the pithy Julian of Egypt, who wrote his own epitaph: "I often said, and still I say 'Drink up!' Like me, you'll soon be clay."[88]

Also, the ecstatic Sufi poetry of figures like Hafez, Sa'adi, Rumi, Kabir, and Ghalib were able to give us a glimpse of the ecstatic state through the beauty of their lyric:

Hafiz, from whom did you learn these words, that Fate
has wrapped your verse in gold to wear around its neck?[89]

Fiction can be instructive; it teaches the ability to explore points
of view and the motivations of other minds. In this way, stories can
regulate social behavior by providing positive templates (cf. the Bil-
dungsroman, like *Wilhelm Meister's Apprenticeship* [1795] by Johann
Wolfgang von Goethe or *A Portrait of the Artist as a Young Man* [1916] by
James Joyce), and negative templates (e.g., Yukio Mishima's *Temple of
the Golden Pavilion* [1956] or *A Hero of Our Time* [1840] by Mikhail Ler-
montov). Stories function as a form of social communication, where
author, audience, and fictive characters interact. Sometimes fiction can
also be a delicious form of reputation gossip. Consider Jun'icihrō Tani-
zaki's *The Makioka Sisters* (1949) or Jane Austen's *Pride and Prejudice*
(1813), where the reader becomes embroiled in the drama that sur-
rounds the marrying of daughters. Of course, as we discussed in the last
chapter, gossip is integral to reputation and thus an adaptive element in
how we navigate and create our social niche, suggesting that fiction can
be a continuation of our everyday social lives, not only in the sharing
and producing of spiritual emotions, but also in simulative play.

The field of narrative studies has attempted to describe how narratives construct reality and give meaning to our lives.[90] Psychologist Jerome Bruner claims we organize our past experiences in our memory in
the form of narratives.[91] Narratives occur across time, linking particular
events with characters who act with intention. These life narratives are
always contextually sensitive to a particular cultural background, but
some scholars suggest that all stories boil down to seven plots: over-
coming the monster, rags to riches, quest stories, voyage and return,
comedy, tragedy, rebirth, and rebellion.[92]

As we know from the foundational text of Western tragedy, Aristotle's *Poetics* (335 BC), every story needs a plot with motivation, in
which there must be desire against obstruction and in which human
intention plays a major role. If we take a closer look at the extant Greek
tragedies (seven each by Aeschylus and Sophocles, and eighteen by
Euripides) we find the following themes: love (erotic and familial),
honor, civic loyalty, the experience of facing death, the relationship

between genders, satire, fate, the relationship between gods and man, and a few others. These themes line up with the domesticated emotions of stabilized communities, which we discussed in Chapter 8; namely, honor, respect, pride, shame, trust, fairness / reciprocity, love, and courage. Stories, from folk tales to literature, act as normative forces to map behaviors for community members. And from the distant perspective of the historian, cultural stories provide a window to the valued emotions of a given people and era.

There is a special form of stories called myth that has excited anthropologists, because the content of these narratives reveals deep cultural values; for example, the dreamtime we described earlier. A metalinguistic approach of myth would seek out where such stories came from and how they dramatize the moral codes of the community. Another approach is to investigate their internal structure. The latter structural approach has revealed that the mythology of any given community is made up of fundamental building blocks, such as symbols, material objects, mental experiences, repeated movements, and binary oppositions.[93] The essence of any given tale is the way that the basic building blocks are put together in the context of a greater cultural narrative.[94] Narratives that involve origin stories and transformations are the sine qua non of myths and are especially important for us to understand, as they evoke images in nonliterate cultures. We emphasize here that a successful story, or myth, needs both appropriate cognitive mismatch, like the notion of a supernatural agent, and, more importantly, an affective manipulation that makes the mismatch effective.[95] Mythology plays a similar role to cosmology in prehistoric art, sometimes in smaller ethical situations (cf. Aesop's fables) and sometimes in the larger context of human endeavor (cf. the *Mahabharata*). Myths aesthetically depict the shared cultural heritage of the community.

Music and Dance

Music is one of the most ancient of human practices; indeed, we observe multi-holed wind instruments in Gravettian era sites in Europe, with the first instruments dating to between 30–53 KYA.[96] Dance and the physical coordination necessary to create music is wholly social, engaging our empathetic faculties, cooperation, and social solidarity. Case in point: a seasoned jazz musician told me in between sets, "For us to sound good, imagine the rhythm section is in a three-legged race. The point is not being right; the point is to be together." Part of the thrill of dance and music in fast-paced interactions between individuals and groups is that they so obviously engage the emotion of PLAY. Consider the intricate shuffle between soloist and accompanist in jazz, where the melody can shift the accompanying chord extensions and bump across the rhythm with accents and bursts of polyphonic motion. There is an ecstatic bond that musicians and listeners experience during a concert. Regardless of musical training, thousands of people of all ages attend concerts, and many find in this a rare space for transcendent emotions. Sometimes these feelings are created by the communal nature of the gathering; at other times, some deep nostalgia generated by sound and song brings about warm sentiments. It is these intrinsically emotive elements of sound that compel the spirit.

The forms of sound have an enormous range; from trance-like rhythms and repeating motifs to short, catchy popular tunes to expansive meditations on the variations of complex phrases, and much more. While the different forms have differing social functions—dance music for trance and the expression of joy, often within the courting ritual; church music to soften the mind into contemplation of greater things; war music to intimidate with cacophony—music as an art form achieves its greatest effects by exciting the affective system. Indeed, sometimes rational contemplation of the technical achievement of a composition or performance can itself be moving, but we argue that it is the feeling of awe and wonder at the ingenuity and achievement of the work that is the real locus of the pleasure in this experience. Music engages the imagistic (voluntary and involuntary) imagination as a form of thinking and feeling without words.[97] This meditation allows space for the transcendent emotions of awe and wonder.[98]

The lyric can also express an elaboration of PLAY systems that cut through to our deepest emotions. They can make us laugh or cry—or both, as in the case of Hank Williams's "Long Gone Lonesome Blues" (1950):

I went down to the river to watch the fish swim by
But I got to the river so lonesome I wanted to die, oh Lord
And then I jumped in the river, but the doggone river was dry
She's long gone, and now I'm lonesome blue

Upon a bed of melodious (or aggressive) musical sounds, song lyrics—
like the lyric poetry from which they are derived—are potently effective in delivering us into a mood and sphere of dispositions. Consider the vainglorious moments of a stadium singing Queen's "We Are
the Champions" (1977) or of the chorus taking up the "Ode to Joy"
in Beethoven's Ninth Symphony (1824). These are moments of great
emotional catharsis, of an overflowing and saturation of our affective
faculties. It is in this state that humans describe feeling like they have,
to take one descriptive phrase, "gone beyond themselves." In this both
ultimately vulnerable and experientially invincible transcendent state,
the human mind is open to imprinting. This may take the form of a
key moment that gives meaning to their lives—whether in conceptualizing the experience as an anchor in their emotional life ("That was
when I felt most in the moment") or as a vehicle for making space for
the fixing of their beliefs. In the latter case, we know that while faith
can be learned or decided upon rationally, or through rote repetition
(for example in the case of propaganda), it generally requires an emotional event to fix the belief. In some cases, this emotional event comes
to symbolize the moment when a divinity spoke to you, as it was with
St. Paul on his way to Damascus; in other cases, the emotional event
is when one feels receptive to make or accept conclusions concerning
metaphysical questions. This receptivity could lead to a conversion or
it could be something less dramatic, like a confirmation, or a state of
resolution, or a perfect stasis. Whether induced by trance, music, drugs,
visual art, or religious incantations, the transcendent emotion makes
way for an acute experience of background field-consciousness, which
feels apodictic due to both its rarity and its intensity.

Of course, because they engage deep spiritual emotions of awe,
fervor, sanctity, and joy, the power of the arts has been successfully
patronized (to great success) by religious institutions; take, for example,
the sublime passions of Johann Sebastian Bach or the Arezzo murals

by Piero Della Francesca. And with that thought, let us turn now to the affective sources of religion.

Religion

Just as it is difficult for linguistically shaped minds like ours to think our way into the prelinguistic mind, it is difficult for monotheistic Westerners to think their way into animism, pantheism, and polytheism. But any understanding of the evolution of religion, especially as it relates to affect, will need to go beyond the familiar Axial Age religions, to consider some of the more spontaneous, preliterate forms of spiritualism. What most religions have in common, whether monotheism or animism, is a network of supernatural commitments and related devotional practices. Recently, anthropologists and psychologists have begun to explain the adaptive (and maladaptive) aspects of religion as part of a larger Darwinian view of culture.[99]

The current anthropology and psychology of religion take their lead from Emile Durkheim's famous argument in 1912, which says that religion is best understood as a form of social life and not as theology per se. To the theory that religion creates community, more recent scholars have sought to supply confirmatory empirical data.[100] Much of this work has acknowledged a role for affects / emotions, but we feel the importance of emotions and the feedback mechanics of their influence have been woefully underappreciated. Religious culture structures or organizes affective systems (e.g., controlling lust), and affect also motivates features of any specific religious niche (e.g., rage motivates many forms of orthodoxy and blasphemy-policing). Indeed, it is probably safe to say that religion forms one of the oldest and longest-lasting forms of culture in human history and is therefore not just a product of evolution but a highly efficient system of selective pressures.

In this section on religion, we will examine the familiar "Big God" systems but also the animism and polytheism traditions that preexisted monotheism and continue their robust influence around the globe— often sublated within monotheisms. We will examine the so-called puzzle of Holocene cooperation and the role that religion may have played in solving the puzzle. In this analysis, we will suggest that affective systems need to be added to the fitness package of religion. We will explain several proximal ways that religion facilitates emotional management such that the Holocene successes make sense. These include (1) anxiety assuagement, (2) domestication of antisocial emotions, and (3) fear-based norm-policing via transcendent force (i.e., a deity or karma).

Belief and Affect

As we saw in Chapter 8, the Holocene marks a significant shift in human social organization. Our ancestors went from living in small-scale tribal bands of known members to inhabiting large-scale, increasingly anonymous towns and cities.[101] Holocene era cooperation is puzzling because traditional culpability mechanisms of close-quarter proximal living disappear in the informational noise of large-scale communities.[102] Humans in small groups can keep track of each other's reputations, trustworthiness, and reciprocity debts, but large-scale societies magnify free-rider problems. So, how were we able to sustain and even increase cooperation in mega-societies of strangers, without the ability to track distant consequences (temporal and geographic) and without the blood ties of kin cooperation?

As a response to this cooperation problem, religion served to minimize aggression among non-kin communities.[103] Norenzayan et al argue that religion increased and promoted fitness (fertility) and group solidarity, effectively creating the cultural scaffolding for cooperation among strangers. In particular, they argue that a "package" of related beliefs and behaviors, loosely called "religion," provided that very framework.[104] The package includes "potent moralizing," "supernatural agents," "credible displays of faith," and so on. Groups that moralized heavily, for example, succeeded over their competitor groups and thereby spread the package more widely.[105] Eventually, the adaptive success of religion transformed it from a mere byproduct of mind to a target for increased cultural selection. According to Soltis et al., such cultural triumph over competitors via intergroup contest is slow but effective and can be modeled readily.[106] Moreover, other mechanisms of cultural evolution, like prestige-biased group selection, can speed up the spread of adaptive ideologies and behaviors.

In what sense is religion a byproduct of mind? Several recent evolutionary anthropologists, and now the Cultural Evolution of Religion Research Consortium (at the University of British Columbia), have argued that religion emerges out of cognitive precursors (or preadaptations) like folk physics, TOM, categorical schemas, and innate essentialism.[107] If our folk-taxonomic mental categories carve the world into basic domains, like animate and inanimate, or into more complex schema like animal, artifact, and human, then occasional category mismatches can spark unique cognitive arousal, producing supernatural counter-intuition. For example, artifacts that are thought to speak, or dead creatures that are thought to live again, consist in relatively simple category transpositions or violations. A talisman, for example, is a physical object that can "hear" your wishes and influence the world toward those ends. It is an inanimate object blended with agency properties that are usually reserved for creatures. Religions around the world contain similar category violations, and this suggests that innate cognitive architecture can, with slight distortions, produce the paradoxical or counter-intuitive entities of religion (i.e., animistic river spirits, virgin mothers, resurrected messiahs, invisible men).[108] Once the phylogenetic Theory of Mind Mechanism or mentation evolved or came online, it enabled our ancestor to project mind onto nonpersons as well—onto any natural object or even artifact. These false-positive attributions of mind generate the thick world of invisible agents that religious devotees believe to be immanent in nature. In these ways, it is thought, religion emerges as an exaptation or byproduct of other cognitive adaptations for navigating the world. This analysis gives us the mechanical prerequisites of religious thinking but not the meaning or function of religious thinking.

The most recent scientific study of religion has stressed the centripetal social function of religion, with only modest reference to the existential or consoling aspects of religion.[109] Indeed, some scholars suggest that the terrifying aspects of some religious imagery and narratives (e.g., hellfire, Satan, Shiva, etc.) stand as strong counterevidence to the consolation argument for religion.[110] But we will endeavor to show that the "religion as consolation argument" is alive and well, despite the occasional fear-inducing aspects of some religious iconography.

Indeed, one of the most important emotional aspects of religion is anxiety and stress reduction.

Diamond typifies the common view that the main functions of religion include providing an explanation of nature, establishing political obedience (especially in the Holocene), teaching moral precepts (which is important in large-scale stranger societies), and finally, justifying war (i.e., murder is wrong inside our community, but not against heathens / pagans).[111] Our view is that religion has many additional functions, loosely housed under the aegis of emotional management. And even the functions on Diamond's list are largely facilitated by religion's ability to harness and exploit our affective systems.[112]

In the same way that cognitive pre-adaptations (i.e., categorical schemas, etc.) preceded religion proper and enabled the many functions of religion, even more so did the affective precursors feed into imaginative supernatural systems. This part of our evolutionary story has not been properly explored. As we've articulated in previous chapters, concepts have affective or emotional content. Whether it's action-oriented-representations (AORs), vertical-association PPRs, or somatic-marker tags on memories and impressions, thinking itself is entangled in affective tone and in the intentionality of affective systems.[113]

William James noticed that a concept has essential or defining properties only in light of the goals or purposes we are pursuing, and the same holds true for supernatural concepts. "What now is a *conception?* It is a *teleological instrument.* It is a partial aspect of a thing which *for our purpose* we regard as its essential aspect, as the representative of the entire thing. . . . But," he continues, "the essence, the ground of conception, varies with the end we have in view. A substance like oil has as many different essences as it has uses to different individuals. One man conceives it as a combustible, another as a lubricator, another as a food; the chemist thinks of it as a hydro-carbon; the furniture maker as a darkener of wood. . . ."[114]

It is not enough, as the cognitivists suggest, that a conceptual domain like "inanimate object" gets a binary switch from "unconscious" to "conscious" and suddenly produces a religious supernatural entity.

We must remember that the content of these cognitive categories—which often dictate metaphysical commitments—contain affective tone from the low-level approach / avoid affect to high-level layers of ennui, angst, and wonder. When we conceptualize God or spirits or other features of religious cognition, those concepts are infused with FEAR, or CARE, or RAGE, and with complex mixtures of these systems. Thinking about God is not primarily an amodal inferential processing of symbols. If the cognitive science approach to religious concepts focused more on developmental learning, it would appreciate the sensorimotor-affective dimensions of our supernatural commitments. By analogy, experiential associations inform whether a child sees his father as a frightening threat or a benevolent force (or some specific mix of the two). Experiential associations with religious representations are less direct, but no less somatic. Contemplating images of the passion of the Christ, for example, is not a disembodied or disinterested meditation but frequently brings tears to the eyes of believers. And, in another example, Jonathan Edwards' stories of a vengeful God struck fear into the hearts of New England Puritans. Put differently, it is only because we are afraid of the spirits or love the spirits that they mean anything at all to us. On that emotional foundation, most of the adaptive aspects of religion are built.

Additionally, starting with Freud, many scholars take religion to be a set of obsessive-compulsive behaviors rather than a set of ideas in the head. The body of the believer is frequently engaged in repetitive perseverations that reflect the kind of embodied PPRs that we first discussed in Chapter 3—invested with meaning by shared cultural representations at the tertiary level. Consider how similar bodily supplication gestures appear across cultures. Acts of supplication recreate or simulate the bodily submissions that entreat some favor from a dominant force. And as we shall explore shortly, when the repetitive behavior programs are not successful manipulations of the external world, then they often continue as forms of adjunctive consolation for the devotee.

Holocene Cooperation and Religion

We agree that a functional approach is helpful in understanding the evolution of religion, but the Holocene cooperation problem may not be the best frame for appreciating the depth of religious culture. As we intimated in Chapter 8, we are not entirely convinced that the Holocene represents the radical cooperation challenge so commonly accepted by contemporary anthropologists.[115] Remember that the Holocene cooperation puzzle asks how increasingly large and anonymous societies maintain cohesion and did not succumb to either massive free-riding or constant revolution. There are two major reasons for thinking the Holocene "cooperation problem" is more contrived than real. First, the expansion of the CARE system from maternal oxytocin bonding to alloparents and beyond was already well underway during the Pleistocene.[116] This decoupling and expansion of the CARE circle provided a means by which the cooperative motivations of kin were extended well beyond the immediate family. These emotional affiliations would have been amplified dramatically by pre-Holocene forms of body-ornamentation, scarification, and other cultural signs of in-group identity. The decoupled CARE system already predisposed humans to altruism—independently of cost/benefit reciprocity tracking. These changes in the affective system provide default motivations for human cooperation that can be witnessed in contemporary baby lab experiments.[117] And this made cooperation robust in the transition to a signal-noisy State-level social ecology. No doubt "big gods" religion further elaborated cooperation, but it did so by better harnessing affective systems like CARE and FEAR.

Another reason for some skepticism about the "Holocene problem" frame is that *Homo sapiens* don't really live in an anonymous State. That is to say, the State level is an abstraction for most humans, who still live their day-to-day lives in smaller tribal groups that are sublated within anonymous cities and states. It is somewhat misleading to think that our altruism has transformed from tribal dedication to stranger dedication, just because we live in mega-societies. Yes, we have to tolerate strangers on the bus and pay taxes, but our cooperative altruism is still tribally oriented rather than citizen-oriented. Even in cases where soldiers appear to fight and die for huge abstract tribes like "America," further examination reveals the very *local* heart of their real motivation—the parents back home, the sweetheart, the newborn daughter, the little bungalow neighborhood, and so on.[118] The point here is that a widespread, sudden appearance of stranger or citizen ethics may never have emerged, and therefore needs no special explanation in the form of Statist religions.

Additionally, the new science of religion is fond of the following two paradoxes or puzzles that we feel are red herrings.[119] Religion is paradoxical because it is expensive (consuming vast individual and community resources) without providing clear benefits. So why should it persist and spread through populations? Second, religious beliefs are concerned with unverifiable entities (e.g., gods, souls, spirits, etc.), and it seems puzzling that humans would spend so much time, energy, and resources on unverifiable entities.

Here again we find this framing wrong or highly artificial—at the very least, it looks like religion observed and studied from a great distance. First, believers obviously receive enormous benefits from religion. Not only do theists receive the many consolations of future transcendent life, but they frequently receive mundane help with the needs and challenges of this life. In the 1996 Center for Advanced Study's report "Buddhism in Cambodia," anthropologist Peter Gyallay-Pap describes the comprehensive quotidian benefits of the Theravadan Buddhist *wat* (temple-complex).

> Apart from serving as moral-religious and educational centers, the *wats* were also the foci for villagers' social and cultural activities. *Wats* were the symbolic centers for all community festivities and ceremonies. They were places for learning and performing such applied arts as dance, music, shadow puppetry, carving, pottery, theatre, and poetry. . . . The *wats* provided social services by housing male orphans and children from disadvantaged families as temple boys who received basic education in literacy and numeracy. They provided for the elderly, particularly women who chose to become lay devotee nuns *(doun chee)* in their declining years while serving the *wat* according to their capacities. The *wats* served as important health and counseling centers as well as courts

of justice where disputes were conciliated and mediated. *Wats* were regarded as sacred spaces and gardens of peace *(wataram)* for concentration and meditation.[120]

Historically, churches, synagogues, and mosques also provided this kind of temporal or mundane support and provision. In some communities, they still do this, but in the contemporary developed world, it is hard to remember the necessary and extensive nature of benefits provided by religious institutions. Spending time in the developing world gives one all the evidence one needs for asserting that religion has direct benefits on individuals and on community. This fact is harder to see in developed, prosperous countries, where surplus capital ensures survival and religion is relegated to occasional observance. However, Richard Sosis's comparative research on utopian societies also offers compelling evidence that religious communities are more likely to survive than secular communities.[121] And according to the Philanthropy Roundtable group, religious faith has an enormous influence on giving. Religious people are much more likely than the non-religious people to donate to charitable causes—including, surprisingly, secular causes—and they give much more.[122]

Besides tool industries like Acheulean (1.7–0.1 MYA), one of the earliest cultural systems to evolve was religion. Prehistoric religion, animism, and even Axial Age religion is not like our anemic contemporary Western version. It was—and is—a totalizing cultural nexus of social norms, ritual practices, ingroup and spiritual individual identity, metaphysical commitments, and economic relations. Religion was not a rational choice of beliefs amidst other competing beliefs in a pluralistic marketplace of options. Rather, religion was part of the lived environment—the social, cognitive, and emotional niche of *Homo sapiens*.

Moreover, we are also not convinced that unverifiable entities are a special problem or paradox for religion, since they have always occupied every aspect of folk life.[123] Following Durkheim, we recognize that the notion of "supernatural" is a relatively modern invention, arising largely in contradistinction to the "mechanization of the world picture" (to quote Dijksterhuis's useful phrase).[124] As we discussed earlier, regarding shamanism, before naturalism emerged as a corollary metaphysics to the so-called scientific revolution, everything was supernatural. In many parts of the world, it is still so. Most people reading this book live comfortably in the de-sacralized world of impersonal laws of nature, but this is a very recent and provincial worldview, as most of human history has experienced nature as a capricious tumult of invisible agencies.

In animistic traditions, spirit agencies are not centralized but are rather immanent forces distributed throughout local ecologies. They do not provide an overarching coherent purpose to life (as in eschatological monotheism) but rather mix and mingle in the quotidian challenges of survival. In animism, we are trying to cooperate with other agents (albeit invisible), whereas in monotheism, God is the always-watching policeman that inspires mundane cooperation. The former is more interactive on a daily basis, while the latter is more remote.[125]

Anxiety Management

Animistic beliefs loom large in the everyday lives of Southeast Asians. Animism can be defined as the belief that there are many kinds of persons in this world, only some of whom are human. One's task, as an animist, is to placate and honor these other spirit-persons. There are local spirits, called *neak ta* in Cambodia, inhabiting almost every farm, home, river, road, and large tree. The Thais usually refer to these local spirits as *phii,* and the Burmese refer to them as *nats.* Even the shortest visit to this part of the world will make one familiar with the ever-present "spirit houses" that serve these tutelary spirits. When people build a home or open a business, for example, they must make offerings to the local spirits, otherwise these beings may cause misfortunes for the humans.

Everywhere in Cambodia, Laos, and Thailand you'll find these miniature wooden houses with offerings dutifully placed inside them. The spirit houses are usually colored according to which day the owner was born, and they often contain miniature carved people who will act as servants to the spirits who take up residence there. The offerings can be incense, flowers, or fruit, or anything valuable and precious, but the spirits are particularly pleased by shot glasses of whiskey or other liquors. Quality alcohol is an expensive sacrifice and demonstrates the seriousness of the devotee.

The offerings are designed to please *neak ta* and *phii* but also to distract and pull mischievous spirits into the mini-homes, thereby sparing the real homes from malady and misfortune. Most businesses and homes in Cambodia and Thailand, for example, have these shrines. The mix with Buddhism is so complete that monks frequently make offerings to these spirits and Buddhist pagodas actually have little spirit shrines built into one corner of the temple. The more recent Buddhist religion is built on top of this much older animistic system—animism was never supplanted by modern beliefs.[126]

The belief that nature is loaded with invisible spirits that live in local flora, fauna, and environmental landmarks is generally characterized by Westerners as "primitive" and highly irrational. And even religious devotees of monotheism in the developed West look down their noses at animism. The developed world usually tells a familiar Comtean narrative: early religion was animistic and polytheistic, but then we progressed to monotheism as we became better educated and advanced; meanwhile, the less developed pagans kept on with their anachronistic animism. In this progress myth, animism is way down on the ladder, but a more nuanced appreciation of religion around the globe reveals both the ubiquity and the intelligence of animism.

Most people in the world are best characterized as animists. In actual numbers and geographic spread, belief in nature spirits outweighs the familiar Western Axial Age contenders. Almost all of Africa, Southeast Asia, rural China, Tibet, Japan, rural South and Central America, and the indigenous Pacific Islands—pretty much everywhere except Western Europe, the Middle East, and North America—are all dominated by animistic beliefs. Even places where later religions like Buddhism and Catholicism enjoy *formal* recognition as national faiths, much older forms of animism comprise the daily concerns and rituals of the people.

The well-traveled Darwin himself noted the universality of animism in the *Descent of Man*, when he said, "I am aware that the assumed instinctive belief in God has been used by many persons as an argument for His existence. But this is a rash argument, as we should thus be compelled to believe in the existence of many cruel and malignant spirits, only a little more powerful than man; for the belief in them is far more general than belief in a beneficent Deity."[127]

Some Western theists assume that animists are highly uneducated and unscientific, and that, eventually, they will "evolve" into a more scientific view of God—a rational God of natural laws (who is also omniscient, omnibenevolent, and omnipotent). For their part, many secularists contend that all theists will eventually join the evolution beyond even monotheism. Humans all previously believed that storms, floods, bad crops, and diseases were caused by irritated local spirits (invisible persons who were angry with us for one reason or another), but now we know that weather and microbes behave according to predictable laws and that there are no "intentions" behind them. This view of nature as "lawful" and "predictable" and "value neutral" has given those of us in the developed world power, freedom, choice, and self-determination. This power is real, and we can be sincerely thankful to benefit from dentistry, cell theory, antibiotics, birth control, and anesthesia.

But two things are worth noting. Contrary to its own self-image, monotheism has not really been "rational" since the brief reign of Deism during the Enlightenment; in fact, actual monotheism of the last few hundred years has been staunchly fideist, or faith-based. Martin Luther's suggestion, that a good Christian should "tear the eyes out of his reason," is still a default position for many believers. And the second point worth making here is that animistic explanations of one's daily experiences may be every bit as *empirical* and *rational* as Western science, *if* we take a closer look at life in the developing world.

We are not making the usual relativist argument that indigenous people have an equally good epistemology and logic—though entirely different from our own.[128] On the contrary, we want to suggest that a universally recognizable rationality (in animists) can be read as an understandable (albeit unfamiliar) response to radically different data (i.e., daily experience with a capricious Nature). Living in a hostile world produces the unique affective and cognitive responses of animism.

It is important to remember, however, that the daily lives of people in the developing world (or even in the world before the industrial revolution and urbanization a couple hundred years ago) are not filled with the kinds of independence, predictability, and freedom that those in the developed world enjoy. Frequently, one does not choose one's spouse, one's work, one's number of children—in fact, one does not choose much of anything when one is very poor and tied to the survival of one's family. In *that* world, where life is truly capricious and out of one's control, animism seems quite reasonable.

It makes more sense to say that a spiteful spirit is temporarily bringing misery or a benevolent ghost is granting favor than to say that seamless neutral and predictable laws of nature are unfolding according to some invisible logic. Unless one could demonstrate the real advantages of an impersonal, lawful view of nature (e.g., by having a long-term well-funded medical facility in the village), the requisite experiential data to overcome the animistic view will never manifest. The First World model of neutral, predictable laws will be an inferior causal theory for explaining the chaos of everyday Third World life. In the developing world, animism literally makes more sense.

Is this worldwide superstitious animism a mere "opiate," as secularists from Marx to Dawkins have argued? Well, yes, but not a *mere* opiate. Opiates can be highly adaptive and consoling—helpful in pain management. The endogenous opioid system is an innate pain-relieving system. Cultural forms of analgesia, like religion, can be as helpful as endogenous systems.

Religion provides some order, coherence, respite, peace, and traction against the fates; and perhaps most importantly, it quells the emotional distress of human *vulnerability*. Many superstitious beliefs and ritualistic practices provide some genuine relief for anxiety and agony. As Roger Scruton says, "[T]he consolation of imaginary things is not an imaginary consolation."[129]

The main function of religion—especially animism—is not to serve as a pathway to morality; nor can it substitute for a scientific understanding of nature. Its chief virtue is as a coping mechanism for human suffering. Consolation has a significant physiological aspect and should not be overintellectualized. Yes, contemplative reflection on death and immortality in the material representations of art is a uniquely human activity, and religion alleviates such existential anxiety via philosophical ideations. But most daily religious activity calms and / or distracts the devotee through adjunctive behaviors, which act as positive self-stimulation, or "stimming." All humans engage in stimming behaviors (e.g., nail biting, hair twirling, constantly checking our smartphones, etc.), but more intense forms can be seen in those on the autism spectrum. Temple Grandin points out, "When I did stims such as dribbling sand through my fingers, it calmed me down. When I stimmed, sounds that hurt my ears stopped. Most kids with autism do these repetitive behaviors because it feels good in some way. It may counteract an overwhelming sensory environment, or alleviate the high levels of internal anxiety these kids typically feel every day. Individuals with autism exhibit a variety of stims; they may rock, flap, spin themselves or items such as coins, pace, hit themselves, or repeat words over and over (verbal stims)."[130]

Calming or distracting forms of self-stimulation are part of the larger phenomenon described in animal ethology as adjunctive behavior.[131] In the course of natural and artificial conditioning, animals key into behaviors that are regularly paired with rewards but that have no necessary connection or causal role in delivering the said reward. Lab pigeons, for example, will peck a small light that turns on just before a food pellet is released. Eventually, this irrelevant light-pecking behavior becomes a time-consuming, intensive behavior in between feeding events.[132]

Pavlovian adjunctive behavior takes on a life of its own, persisting even after experimenters omit the food reward whenever light-pecking occurs.[133] Similarly, rodents, cows, and other animals will engage in adjunctive water consumption when hungry. Ordinarily, water consumption is naturally paired with eating, but when experimenters reduce food or extend the temporal intervals between food rewards, animals will drink double and triple the amounts that thirst would dictate. Adjunctive drinking is a kind of stimming behavior that may reduce stress and regulate dopamine while the animal is in a frustrated SEEKING phase of anticipation. The stimming behavior is palliative when more productive or inducing behaviors do not work or cannot be executed.

Religion is filled with ritualized behaviors, including ceremonial body movements, routinized manipulations of prayer beads, talismans, totems, candle lighting, supplications and prostrations, prayer recitations, collective singing, holy water rites, pilgrimages, sacrifices, and so on. When our SEEKING systems are aroused (e.g., with the promise of food, etc.), but there is no way to satisfy it, then we engage in adjunctive behaviors.[134] But this is also true of long-range human seeking or the teleological projects that we might generally call "hope." Our prayers are hopes for family benefit and other successes, both mundane and spiritual.

Adjunctive repetitive behaviors may be byproducts of adaptive behaviors, since certain kinds of repetitive actions ordinarily produce helpful results—as in the case when repeated pecking on shiny trash releases food crumbs for city pigeons. Adjunctive light-pecking behavior in laboratory pigeons may be a byproduct of otherwise adaptive routines. In the case of humans, our limbic-based SEEKING system does not reflect on or think about experience, but our neocortex does, and it tries to detect cause and effect patterns from all of the possible correlating associations.[135] Panksepp suggests that prayer and other rituals like rain dances are forms of adjunctive behavior that make people feel better in situations where they have no better action possibilities. They make us feel better because our affective system (SEEKING, RAGE, etc.) has been ramped up to accomplish something, but there is nothing we can do in this circumstance—the pigeon cannot make the food come faster, the farmer cannot make the rain come, the mother cannot bring her baby back to life. "Could praying be an adjunctive behavior," Panksepp asks, "that gives human beings the illusion that they are somehow able to magically change their fates?"[136] These are the neural roots of religious beliefs that need further study.

We are suggesting that religious rituals are partly adjunctive behaviors (culturally sanctioned and transmitted) that help devotees manage their emotional lives (e.g., hopes and vulnerabilities); this view makes sense out of many of the seemingly paradoxical behaviors of religion when considered from the inadequate rational-agent model. Among other things, religion is a culturally structured set of psycho-behavioral perseverations, often providing some return to homeostasis when other resources and consummatory acts cannot do the work. When a loved one dies, we feel an overwhelming need to "do something." But, really, there is nothing to be done. Religion is helpful in those moments, not because it solves problems, or enlightens, or anesthetizes, but because it gives us something—usually very precise and elaborate—to do.

Oftentimes, the explicitly noetic aspects of religion reinforce and further articulate the physiological forms of anxiety reduction. The grooming touch of a warm hand on the grieving or anxious person is recreated in linguistic format to bring relief to the troubled heart. Even a cursory survey of gospel lyrics will make the case obvious. The consoling balm of a friend's or parent's care is transferred to God and made totalizing and absolute in the cultural poetry of religious song.

> The Lord's our Rock, in Him we hide,
> A Shelter in the time of storm;
> Secure whatever ill betide,
> A Shelter in the time of storm.
> Oh, Jesus is a Rock in a weary land,
> A weary land, a weary land;
> Oh, Jesus is a Rock in a weary land,
> A Shelter in the time of storm.[137]

Domesticated Emotions

We have been arguing that the emotional modernity of cooperative human beings may have arisen much earlier than the Holocene, dating back to the rise of alloparenting and the decoupling of the affective CARE system. Genetically based affective tendencies may be indefinitely channeled and structured by cultural mechanisms, once those cultural mechanisms are strong enough and transmission is robust. But anatomical changes, particularly feminization of the skull in the Late Stone Age (starting circa 50 KYA) suggest dramatic testosterone reduction in modern *Homo sapiens*. The emotional life of Late Stone Age humans was probably less violent and more tolerant. That new emotional niche is reinforced and expanded by religion and local-level social life.

The brow ridge of Late Stone Age male humans became less prominent and sexual dimorphism of the skull reduced generally.[138] This suggests that testosterone lowered during this period, which probably reduced male-male violence, facilitated safer childhoods, and furnished a more secure social life. A more pacific social life, whether through lowered testosterone and / or better, more steady forms of subsistence, would increase learning opportunities for young apprentices, strengthen pair bonds, and create new value systems. When physical retaliatory violence is reduced, tolerance increases, and alternative forms of conflict resolution become operative. Aspects of intelligence and creativity may become more adaptive and more prized by a less pugilistic social group.

A similar story of feminization or the reduction of sexual dimorphism is clear in the case of bonobos.[139] The "self-domestication hypothesis" suggests that bonobos experienced selection against aggression and that this reduced the size of male bodies. These changes (i.e., reduction in testosterone, smaller male bodies, reduced inter-male violence, etc.) are knock-on effects of relaxed feeding competition. The result of these changes is a more cooperative and seemingly more charitable bonobo culture. They are, for example, much more likely to share food with strangers than chimps.[140]

Like bonobos, humans may have been domesticated, and may have domesticated themselves, through sexual and cultural selection of less pugilistic traits. Religion has a role to play in dialing down RAGE and offering alternative mechanisms of redress and catharsis. It is unclear that animism—in its theological structure—prefers diplomacy over violence, but its social cooperative assembly may have inadvertently deferred disputes and anxieties to shaman or Big Man mediation and sacrifice rather than to combat resolution. In any case, it is clear from the domesticating norms of Axial Age religions that we should forgive our transgressors, help and love strangers, and otherwise turn the other cheek, repressing our hedonism and tribalism in favor of the brotherhood of man. Presumably pre-Axial Age religions, now largely lost to us, would have offered a similar cultural pressure to "get along." In places where food was more plentiful or surplus storage more reliable, virtues like creativity and even compassion could be selected for. Aggressive males, who could pummel competitors, may have been slowly replaced by gnostic virtuosi, who comfortably navigate the unseen imaginary world of invisible spirits. And "blessed are the meek" rather than the strong became a way of thinking in the increasingly large-scale cooperatives that continue to grow and organize in the Holocene. Emotions like shame, guilt, and forgiveness are fostered in religion because they are highly adaptive in shaping behaviors within large-scale cooperative social ecologies.

Our tertiary layer of emotion contains cognitive ingredients, but the feelings of *shame* and *guilt* are, at bottom, the secondary-level limbic feelings of privation and social alienation. These feelings might be explicit in the overt ostracism of the community, or they might be implicit in the virtual ostracism of the penitent's mind.

Religion is a cultural organization of this tertiary layer of emotion (which also includes the lower layers). Religion, for example, expands the reach of superego scrutiny. It expands the realm of things for which we are sorry and need forgiveness. Now it is not just murder, but lying too. And then it is not just lying, but thinking bad thoughts, and so on. Then we have religious ways to supplicate and roll over with submission: asking for forgiveness, lighting candle offerings, saying penance, giving of money, time, and resources. All these might mask the deep function of guilt and penance—and no doubt guilt takes on a life of its own and can become neurotically maladaptive—but, essentially, the goal is to knit the group back together. Guilt and forgiveness become increasingly necessary in the evolution of human social groups and the resulting multiplication of possibilities for conflict over resources.

Many primates, not just humans, engage in reconciliation or forgiveness behaviors. When two primates have a fight, they often follow up in the next hour with some kind of affiliative or friendly grooming or touch behaviors, maybe even food sharing.[141] Neuroscientist Robert Sapolsky points out that male baboons never reconcile: "[T]here is nothing that looks like forgiveness in them."[142] But some kind of reconciliation behavior has been observed in over twenty-five different species. And forgiveness behavior is increased when the two animals share longer history together. "There is a pattern that looks perfectly human, which is that not everybody reconciles with each other, but some pairings are more likely to reconcile. What we are seeing is that the more valuable your relationship with the other animal, the more likely you are to make up afterwards."[143]

The larger tribal society of primates, and especially human primates, functions better when its inner squabbles are minimized and negative emotions are directed outside the ingroup. Forgiveness helps this happen, because it reduces ingroup anger, the thirst for vengeance, and the ingroup grudge; all of which reduce cooperative efficiency.

Religious principles, like forgiveness, join with ritualized expressions of CARE to overcome egoistic drives and free-rider tendencies. It is common in religious ceremonies, for example, for strangers or distant acquaintances to touch each other in a formalized greeting or peacemaking gesture. In Catholic church services, for example, there is a moment when the priest invites people standing near each other to shake hands, or embrace, or otherwise demonstrate physical affection. It is a simple gesture but fairly powerful in activating real care. It would be interesting to test oxytocin levels before and after such ritualized grooming, which reduces antisocial emotions and increases prosocial emotions.[144] Similarly, consider the peace greeting after Muslim prayers. At the end of most prayer sessions, Muslims turn to the right and the left, sending God's peace on those who are most proximate. "Peace be upon you and the mercy of Allah." This salutation brings the obligatory prayer to an end. Devotees of these and other similar rituals receive tiny immunity boosters of anti-aggression sentiment.

Empirical research by David Desteno reveals how tertiary-level emotions like *gratitude,* which religion often inculcates, dramatically improve certain adaptive behaviors.[145] Humans have a well-documented bias for immediate rewards over delayed gratifications, but the emotion of gratitude actually reverses or inhibits this natural bias. Desteno ran a series of reward tests on human adults using small cash payments. These were similar to the famous Stanford marshmallow tests, which were originally performed in the 1960s and have been frequently replicated. In that classic experiment, a marshmallow is placed before a child (4–6 years old), but she is told that if she can delay eating the treat, she will get a much greater reward (many treats) when the experimenter returns to the room in fifteen minutes. Delayed gratification is one of the important features of executive control, and the kids who disciplined themselves well did better on later measures of adult success (e.g., SAT scores, educational attainment, income level, etc.). Forty years after the initial tests, a team of psychologists submitted fifty-nine of the original test subjects to fMRI tests.[146] The brain scans on the subjects, now in their fifties, revealed that strongly disciplined people had more active prefrontal cortex areas (i.e., executive control areas), whereas the weak-willed subjects had more active limbic regions.

Desteno's research, however, suggests that the emotion of gratitude may be a more powerful form of impulse control than both habit and rationality. Moreover, while habit can be very effective in motivating specific behaviors like going to the gym to exercise, giving donations to homeless people, eating healthfully, and so on, habits do not produce much cross-over to other domains. In other words, my habit of making my bed does not—contrary to many self-help bestsellers—give me greater impulse control in all the other areas of my life. But gratitude motivates impulse control across many different domains.

In Desteno's reward test, subjects were told that they could have a small cash award now, or they could wait and get a larger reward later. Subjects overwhelmingly took the short-term payoff. When the subjects were primed with feelings of gratitude, however, they switched and preferred the long-term reward. In many cases, they displayed this switch across domains (other forms of delayed gratification). Gratitude makes people more willing and able to discipline their desires. And Desteno suggests that social emotions like gratitude evolved to tie humans into longer term cooperative alliances. In other words, if I feel that I owe you, then I am bonded to you and I am less likely to free-ride. We interpret Desteno's work as yet another confirmation that emotions (in this case, religiously cultivated emotions) like gratitude are affective roots of group reciprocity and individual hedonic management.

Fear and Transcendent Force

As we mentioned at the top of this chapter, art and religion cultivate unique spiritual emotions like awe, wonder, and transcendence. We have shown that some of these experiences are well grounded in affective sources like PLAY, CARE, FEAR, and GRIEF. FEAR, for example, colors many sublime experiences of the divine, whether it be Krishna's terrible revelation or Hieronymus Bosch's *The Last Judgment*. But is there another emotional system—or possibly the momentary cessation of affect—that correlates and motivates the cosmic sense of unity of transcendental emotions?

Many have noticed that sexual orgasm—what the French call *la petite mort*—releases one temporarily from the tyranny of the ego.[147] Is it appropriate to think of the LUST system as terminating in a separate emotional awareness of transcendent release, or is such an empty moment simply the resetting to homeostasis? In this view, the sense of egoless equanimity is just an epiphenomenon or residue of a more persistent set of drives. It is easy to see how pleasure motivates the reward-seeking organism and figures into the adaptive struggle, but it is harder to see how a feeling of transcendence could be advantageous and targeted by natural selection. Cultural selection, however, has clearly favored it in certain contexts—developing high degrees of inward contemplation and meditation.

These mystical aspects of religion do not attempt to track nature for accurate predictions; nor do they obviously improve the thriving of the person. Rather, some mystical forms of religion endeavor to acknowledge the unique field-consciousness that seems to be independent of background for particular contents of consciousness.

Robert Bellah argues, following Jerome Bruner's work in developmental psychology, that humans can engage in different modes of representational thought. In religious thinking, we employ these same modes of enactive thinking, symbolic thinking, and conceptual thinking.[148] As young children, we develop enactive representations that pertain to our bodily actions. And in rituals, we act out religious meaning with our bodies; we bow, supplicate, kneel, dance, and so on. Then we learn to think symbolically using iconography and story, to represent virtue, vice, heaven, hell, spirits, etc. And finally, we think conceptually. In the religious domain, conceptual thinking drives theology and makes possible a more systematic representation of universals.

Before these modes of representation, however, lies what Piaget and Inhelder called the "adualism" of the child—the state of consciousness before an integrated self.[149] The distinction or boundary between self and world is not an early datum for the infant; it emerges slowly through experience. Bellah calls this the "unitive" experience and suggests that it precedes the more content-rich representational modes, serving as a crucial touchstone for later religious thought. "The unitive event, then, is a kind of ground zero with respect to religious representations. It transcends them yet it requires them if it is to be communicable at all. Christian negative theology and the Buddhist teaching of emptiness *(sunyata)* attempt to express this paradoxically by speaking of nothingness, the void, silence, or emptiness. Yet the very negative terms themselves are symbolic forms, are representations, and therefore introduce an element of dualism into the unitive event even when they are trying to overcome the dualism of representation."[150]

It's safe to say that mysticism, supernaturalism, and religion generally have had a great interest in celebrating this presymbolic, unmediated experience. Whether it is the *sensus divinitatis,* the oceanic experience, or the Atman / Brahman consummation, religion and art affirm the transcendental experience. Such an experience need not have a true referent in order for it to be transformative. But it must have some emotional content, if only in the reentry phase of consciousness. William James suggests four characteristics of mystical experiences.[151] First, they are ineffable and cannot be named. Second, they have noetic content—that is to say, the person learns something. Third, the mystic is passive and the event happens to the person. And fourth, the experience is transient, passing quickly. A more intimate familiarity with global mystical traditions suggests that both passivity and transience are not necessary, but the first two criteria seem essential.

We would like to see a new line of inquiry develop, such that the emotive aspects of the unitive experience—culled from the phenomenology of mystical involvement—be correlated with affective

neuroscience and joined together by an investigation of the epistemology of curiosity and wonder. There is nothing in the seven mammalian affective systems that generates a cosmic unitive feeling, so the
question is whether it is a late arrival of big-brain consciousness (i.e., a
rarified frisson of ego detachment) or a psychological return to a very
ancient pre-symbolic, unconscious mind (a psychological case of ontogeny recapitulating phylogeny). Neuroscientist Richard Davidson
and the Dalai Lama have been working together to isolate neural aspects of Buddhist dharma, but these efforts have been focused on psychological aspects of compassion and seem less concerned with the
oceanic experience per se.[152]

Self-awareness seems crucial for the unitive experience because one
leaves the ego during the apophatic moment and then returns to it. It
is possible that we are disengaging the neocortical "autobiographical
self"[153] or "narrative self"[154] but retaining the ancient mammalian
agency of the periaqueductal gray and other subcortical regions we
discussed as the bodily self in the Chapter 8.

Another interesting possibility also suggests itself. Setting aside the
momentary obliteration of the self, for a moment, we recognize that
the perceived world also looks strange to the mystic. The mystical experience of cosmic unity that is so celebrated by ecstatic traditions
may be the rare glimpse of the world without its usual somatic markers,
vertical associations, action-oriented representation, or imperative motivations. This unique experience may be the purely indicative (rather
than imperative) perception of the world, with none of the usual affective tilting or preferences. It is the world without emotional coloration or values. In this rare modality, the world even lacks teleological
structure.

Robert Wright suggests that Buddhist meditation, which instructs
the mind to empty our perception of essences, is really stripping affect
from our normal perception of the world.[155] The resulting experience is
emptiness—empty of craving, but also empty of even low-level approach / avoid orientation. The garbage and the flower are equally valuable (Or is it valueless?) for the accomplished Buddhist meditator.

Our entire book is devoted to the idea that emotional penetration
of perception is much deeper than most scholars have noticed, and yet

we concede the possibility that, in humans, a kind of difficult and cultivated detachment is occasionally possible. However, contrary to the spiritualist's claim that such experience is more real—that is, more metaphysically objective (due to its independence from our evaluations and appraisals)—we are inclined to think of it as less so. Detaching emotional evaluation from the world is helpful in varying degrees in the sciences as well as in religion, but it is produced by a willful distortion of the mind's engagement with the world. The objects divested of affective coding and entanglement are more abstract (like geometric models of physical objects), rather than more real. They are rarified simulations of simulations. We are not Platonists about form, nor are we realists about cosmic commune per se, but we do believe that spiritual emotions are real, that they are not easily reducible to simpler states, and that they deserve better attention in the future of evolutionary studies.

Departing from these abstract, mystical territories, we find that most religious experiences of the transcendent realm come tinged with dread, fear, and trepidation. This is because the transcendent God or karmic power is a force—with teleological interests of its own. It is not just the abstract unity of all things. In contrast to the purely unitive experience, this anxious modality still has anthropomorphic residue, and its adaptive value can be unpacked.

"Thou shalt not commit adultery," is a well-known edict from the Pentateuch's Ten Commandments (Exodus 20:14). Along with many of the other commandments, it strictly manages social behavior and can be seen as yet another form of cultural domestication (consistent with our earlier discussion of reduced aggression in the late Paleolithic). Divine commandments from a transcendent realm help humans dial down intragroup aggression, or RAGE. Violent competition for mates becomes culturally managed by religious rules, because bonds are strenuously protected from interlopers. This, together with elaborate creeds of purity, chastity, asceticism, and guilt, stabilizes the group significantly. But this regulation of LUST and RAGE is often only possible by the counterbalancing weight of neocortically extended FEAR.

Religion, particularly monotheism, up-regulates fear in adaptive ways. Freud recognized the strict surrogate father in monotheism, and we now consider it almost obvious that belief in a strict patriarchal God restrains human violence and fosters civilization.[156] Religion helps dial fear up to paranoid levels, because the tribe comes to believe that an invisible all-seeing God is watching our every move. As we discussed earlier, there is great difficulty in violating your neighbors in the "glass houses" context of small-band moral reputation and hunter-gatherer lifeways; so, too, monotheism creates the ultimate glass house of supernatural omniscience.

This transcendence tradition is not just an oceanic experience, but is also a distinctly threatening metaphysics. God is so much more powerful than any earthly chief or Strong Man that he acts as a constant check on power. Before modern law, God was the ultimate insurance against the unchecked tyrant—and the Chinese notion of the Mandate of Heaven *(tiānmìng)* did the same normative work in the Warring States period.

Specific virtues of humility, rather than pride and hubris, are consistent and inevitable when even the biggest Big Man realizes that he is not an alpha after all, but a mere beta. The ultimate anti-bullying mechanism is to have an invisible giant bully in a more dominant position than the local alphas. God is that giant, and awe is his affective signal.

As Paul Seabright points out, even in hunter-gatherer contexts, we have "coalitions of losers" who check the power of aggrandizers.[157] Countervailing power is the way that many fragile cooperatives (including the over-romanticized egalitarianism of the pre-Holocene) are sustained. Power unseats entrenched power. In small bands, coalitions of losers can keep aggrandizers from ruining the collective, but much bigger societies can benefit from an invisible and totalizing apparatus.[158] We have already pointed out that a simple "Big Gods" theory has many lacunae, not least of which is the need to incorporate the evolution of "virtue modification."

Virtue modification—what Nietzsche called a genealogy of morals—can be a path of increased domestication or belligerence, depending on the culture's adaptive strategy. Art and religion combine to do this virtue-modification work via stories, images, and rituals. But such modification is largely emotion- or affect-sculpting, and it is much more subtle than just stimulating fear of Big Gods. Consider, for example, the reversal of virtues that Christianity performed on paganism.

Virtues of heroic strength are very adaptive in eras and regions of tribal skirmish and war. Add to this the intragroup violence of males competing for females and you have a machismo virtue culture. Machismo virtues have an upside too, as they inspire adaptive fighting, protection, resource acquisition, and defense. In some regions, families and tribes evolved in dangerous environments that often required brutes *inside* the clan to fight off the various brutes *outside* the clan.

Accordingly, hero pride was a favored impulse in the pre-Christian era, even if it came with flaws of excess and immoderation. But the Biblical tradition brought a new ethic: "Blessed are the meek." One could argue, in fact, that the main theme of the Old Testament is "submission" to Yahweh, while the New Testament resounds with the call to humbleness. The "hero" of Christianity "ends" in the ignoble position of suffering on a cross.[159]

A new kind of hero was invented in Christianity—one who suffers, like heroes of the ancient world. But unlike them, the suffering *is* the heroism. Victory no longer comes when the hero is standing over the slain monster. It comes in the next life, after one has lived humbly and proven oneself by accommodating large amounts of unjust suffering. Traditional heroes, like Beowulf, or Hercules, or Odysseus, were acknowledged for their strength and achievements, but their prideful humanism—their attempts to personally bring justice into the world—were devalued during the rise of Christianity.

One of the most impressive aspects of the character of Beowulf is his embodiment of what J. R. R. Tolkien calls "Northern courage." Beowulf embodies a "theory of courage" that puts the "unyielding will" at the center of heroic narrative. The Norse imagination, filled with the philosophy of absolute resistance, was properly tamed in England, according to Tolkien, by contact with Christianity. The Viking commitment to "martial heroism as its own end" was unmasked by Christianity as just hopeless nihilism—something to be overcome and remedied. Tolkien argues that the poet [or the scrivener] of *Beowulf* saw clearly that "the wages of heroism is death." The Christian looks back over the course of pagan history and finds that all the "glory" won by heroes and kings and warriors is for naught, because it is only of this earthly temporal world.

Without Christian "virtue modification," monster-killers and warriors of old are either hopeless existential heroes, trying by pathetic human effort to rid the world of evil, or they are themselves monstrous aggrandizers amidst a flock of righteous and meek devotees. Hercules, for example, is judged by medieval Christians as an abomination to be dethroned from his traditional place of adulation. Medievalist Andy Orchard, for example, quotes Aelfric's tenth-century *Lives of the Saints,* which asks, "What holiness was in that hateful Hercules, the huge giant, who slaughtered all his neighbors, and burned himself alive in the fire, after he had killed men, and the lion, and the great serpent?"[160]

Alexander the Great's ancient heroism was also reconfigured in the Medieval era. He courageously kills monsters in the Alexander romances but is also regularly humbled by wise sages who point out his prideful ambition. In the medieval story of Alexander's "Journey to Paradise," for example, he is given a Judeo-Christian lesson in humility. After being surprised by a small mystical jewel that could outweigh hundreds of gold coins, he was told by "a very aged Jew named Papas" that the jewel was a supernatural gift. The disembodied spirits who are waiting for Judgement Day (when they will get their bodies back) have offered this jewel to Alexander. Papas tells Alexander, "These spirits, who are enthusiastic for human salvation, sent you this stone as a memento of your blessed fortune, both to protect you and to constrain the inordinate and inappropriate urgings of your ambition." He continues, "You are oppressed with want, nothing is enough for you. . . ." And after more critique of his vaulting ambition, Alexander is converted to meekness and charity. "At once he put an end to his own desires and ambitions and made room for the exercise of generosity and noble behavior."[161] The new "nobility" is quite different from the old pagan nobility, and it serves as but one example of religious virtue-modification through emotional manipulation at the tertiary level.

Conclusion

Cultural forces, like art and religion, work effectively but imperfectly. They function, along with other social forces, to strengthen the social organism. We have argued that they do this primarily by managing the emotions of individuals toward therapeutic health, transcendence, cooperation, and domestication. But it is important to remember that, despite many protests to the contrary, the State is not for us. That is to say, the State is not necessarily designed for the happiness of the individuals within it.

The State is an organism in competition with others, so the happiness of its internal citizens is sometimes denied, delayed, or otherwise sublimated to the larger survival struggle. Yes, the telos of many cultural traditions is cooperation among citizenry, but the telos of good cooperation is advantageous fitness over competitor groups. When friends complained in *The Republic* that Socrates had described a very harsh and Spartan lifestyle for the warrior class in his ideal state, he replied by pointing out the holistic benefits for the entire state. If you make the warriors too comfortable, then they will not be good warriors. He likens it to an artist who must resist the temptation to paint every part of a human figure with the most beautiful colors, else the whole figure should turn out chaotic and incoherent. "Really, now. You don't think we should paint the eyes so beautifully that they no longer appear to be eyes, and the same with the other parts, do you? Instead, look and see whether we're providing what's appropriate to each part and so making the whole beautiful . . ." (420c1–d5).

Evolutionary analysis of social and cultural mechanisms cannot mistake the goals of liberalism with the goals of nature. In the same way that Social Darwinism misinterpreted nature as an engine of Hobbesian hierarchy, so too the pendulum has swung in the opposite direction, where we find Rousseauian anthropology asserting a natural human preference for egalitarianism. Careful study of many cultures reveals, however, that human groups have a toolkit of cultural options, allowing them to employ the most adaptive structures for the unique ecological conditions. Sometimes we live in top-down authoritarian cultures (perpetuated by the propaganda of religion and art, as they modulate fear, care, lust, etc.), and sometimes we live in liberal egalitarian collectives. Such alternating is heavily constrained by social histories, however, and while human cultures are faster to change than genes, they are still slow-moving creatures. But from the evolutionary point of view, one of these sociocultural systems is not absolutely better than another; they are only relatively better given local ecological resources.[162]

In short, there are different resource ecologies that elicit different behavioral strategies, which, in turn, tap into different emotional motivators. Yes, CARE is a powerful glue for kin and fictive-kin bonding and cooperation, but RAGE is also a highly useful energy (directed by art and religion, as well as politics) when responding to outgroup threats. As Plato (1992, 373e–377b) recognized in *The Republic,* the job of cultural education (paideia) is, in part, to balance the emotional contradictions of love and hate in the same person or people. The successful state, Socrates says, requires an army of some kind to guard the city, and these guardians must be "like dogs" in the sense that they love their fellow citizens but hate their outside enemies. In order to achieve this tricky balance, the guardians must be educated or cultured very carefully, because they need to protect not only the people, but also the laws and customs as well. This means that only certain kinds of stories, myths, religion and art should be used to inculcate the adaptive emotional / behavioral tendencies.

The cultural project is, in part, therapeutic. We've been suggesting that, from an evolutionary perspective, one of the central functions of culture itself is to guide somatic affective associations into prosocial pathways (using ritual, art, and religion). In this sense, the somatic affective approach to the evolution of the mind in the context of culture and society modernizes Freud's repression thesis (in *Civilization and Its Discontents*) that prosocial harmony comes at the cost of a difficult retraining of individual affects. Retraining our "pleasure principle" to accommodate other egos and conform to the "reality principle" is accomplished by affective associational reeducation—not merely by cognitive reassessment. One must, according to Freud, learn to loathe one's selfish desires, and not merely calculate their social liability.

Assigning the experience of horror, for example, to antisocial feelings and behaviors (e.g., aggression, incest, murder) is part of culture's job. Somatic markers are socially engineered (mostly unconsciously, but sometimes consciously, as in propaganda) by dedicating the flexible affective systems of FEAR and RAGE to specific "enemies" (of the tribe or state). The epistemology of flexible horror is not merely an academic question. It may seem trivial to track the way ancient biological fears become reassigned to creatures in horror myths, novels, and movies (default snake phobias, for example, are heightened by Hollywood horror films, etc.), but the triviality fades when we realize that racism and xenophobia are subspecies of the same epistemic processing. Art and religion can help turn populations toward fear or toward alliance, and—as is often forgotten in a purely functional account—toward wonder and transcendence.

Yet culture is more than a mere survival strategy and cannot be reduced to adaptationism. Alva Noë makes an interesting argument about cultural technology and custom when he says "Our lives are structured by organization. Art is a practice for bringing our organization into view; in doing this, art reorganizes us."[163] In the case of dance, for example, the loop starts with an instinctive set of motions, and then choreography reflects on those natural motions and reformats them, creating a new set of higher-order motions (giving rise to novel thoughts and reflections).

We agree with Noë's general point about the way cultural practices (e.g., art and religion) reorganize us, and we think that a major ingredient in the original or instinctive organization is affective. Yet, art and religion reflect upon that emotional organization—via disengaging and decoupling representational abilities (fairly unique to humans)—only to reorganize them into blended forms hitherto unseen (e.g., uncanny formulations, mixed emotions, etc.) or into new contexts (e.g., lust in wartime or horror, love in robots, anger in the ashram, etc.).

In this way, art and religion (as well as other cultural technologies like science and philosophy) shape a new layer of human emotion and cognition. Franz Kafka, for example, sensitizes us to and even creates a whole range of emotions related to the alienations of modernity, and, in doing so, he even reorganizes us as human beings to some extent.

But we can recognize some basic emotions or instinctual elements (i.e. fear, anger, etc.) in the deep organizational level of these reorganized cultural realities. As Noë says, a work of art is a "strange tool" in the sense that it has no obvious function but it has great significance, and "We make strange tools to investigate ourselves." Emotion, in our view, is a crucial part of the toolbox and the equipment (human nature) under investigation and reorganization; indeed, emotions are the strange tools of sentience.

This book has been a sustained study of the adaptive successes of sentience generally and human emotion specifically. We started with the primordial vertebrate drive of conatus, or seeking—an affective fire-starter for most animals. And we showed how this biological intentionality interacts with our content-rich direct perceptual experience, getting us to move as actors rather than as passive "knowers." The mind is a motivation; not a mirror.

Our mammalian emotional systems became increasingly flexible, open to conditioning and learning. The decoupling of emotions from simple stimulus-response pathways was as important as the decoupling of images and representations from perception. Cultural feedback loops, from stable families to large social structures, facilitated the adaptive management (including domestication) of human emotions.

We followed the trail of emotional evolution into the development of true symbol systems, like language, and even into the citadel of concepts, showing that affect is not a contamination of information but is the very core of informational salience. Finally, we explored the emotional aspects of cultural evolution, with special attention to religion, myth, and art.

We hope that the affective paradigm we have built is inviting and spacious enough for other researchers to move around in. Our expectation is that psychologists, philosophers, anthropologists, neuroscientists, and cultural theorists will put more detail and nuance into our theoretical armature. At the very least, through our exploration of the shadowy evolution of sentience and the emotional essence of mind, we hope to steer the conversation back toward how the human mind is a part of nature.

Notes

References

Acknowledgments

Index

Notes

Introduction

1. Darwin 1870.
2. Darwin 1871, 1872.
3. For example, see Willner et al. 2015.
4. See Solms and Friston 2018, which suggests a solution to the mind-body problem with which we concur (see Gabriel 2012), wherein the function of consciousness is interoceptive and basic selfhood is generated by the pursuit of homeostasis in goal-directed intentionality (see Damasio 2018). Also, see Gazzaniga 2018 on the layers of mind.
5. Panksepp 1998.
6. See Gould and Lewontin 1979.
7. Clark and Chalmers 1998. This work rests upon the foundations of ecological psychology; see Gibson 1979.
8. Evelyn Fox Keller, who spent her earlier career disentangling gender from sex, may now be showing us the way forward by re-entangling them. In her recent work, she stresses the plastic relationship between genes and environment and tries to counteract our tendency to privilege one cause over another by stressing "developmental pathways" rather than just bottom-up (molecular) explanations or top-down (structural) explanations. See Keller 2010.
9. See, for example, Johnson and Earle 2000; Richerson and Boyd 2008; Henrich and McElreath 2003.
10. Tooby and DeVore 1987; Pinker 2010.
11. Whiten and Erdal 2012.
12. Panksepp capitalizes the names to indicate that these specific affects are physio-logical / behavioral systems as well as correlated subjectively experienced

emotions. In this book, we will occasionally use the capitalized form to stress the bodily and subcortical aspects of an emotion. For readability, we will more often employ the lower case. But the psychological and physiological aspects are not separable in a dualistic or even epiphenomenal way.

13. De Waal 2003.
14. Panksepp 2011.
15. See Winkielman and Berridge 2004.
16. See Panksepp 2011. See also Izard 2011; Panksepp 1996.
17. See Hoffecker 2011.
18. See Barrett 2017a.
19. See the classic Tversky and Kahneman 1973. For how affect biases cognition in rats, compare Harding, Paul, and Mendl 2004.
20. Barrett 2017b. See also the earlier anthropologist constructionist theory, Lutz 1986, 1988.
21. Firestone and Scholl 2016. See also the classic discussion of interpenetrability of mental modules in Fodor 1983.
22. See Barrett 2014.
23. See James 1890, 141.
24. Sterelny 2012a; Narvaez et al. 2013.
25. Gamble 1998.
26. Johnson and Earle 2000.
27. See Pinker et al. 1997.
28. Asma 2018.
29. James 1902.

1. Why a New Paradigm?

1. Locke 1690; Hume 1779; Mill 1829; 1843; also, Hothersall 1984.
2. Boring 1929, 97.
3. Mill 1843; Bain 1868.
4. Mills 1998.
5. Watson 1930; Hebb 1944; Skinner 1974.
6. See, for example, Barrett 2011.
7. Schachtman and Reilly 2011.
8. de Houwer 2011; Olsson 2011; Von Overwalle 2011, contra Skinner 1953.
9. Barrett 2011; compare Dennett 1996b.
10. See Karl Lashley's definitive critique that set off cognitive science as discussed in Gardner 1983.
11. Mill 1998.
12. Panksepp 1998.
13. Mill 1998.
14. LeDoux 1996; Damasio 1994.
15. For example, Anderson and Bower 1972.
16. Gardner 1985.
17. Breedlove et al. 2015.

18. Chomsky 1956; Dreyfus 1972; Seed and Tomasello 2010. See, by way of comparison, Fodor 1983; Fodor and Pylyshyn 1988; Rorty 1979; Lee 1998.

19. See Bechtel 2008; Hutto and Myin, 2017; Chemero 2009.

20. Gibson 1966, 1979; M. Wilson 2002; see also Fodor and Pylyshyn 1981.

21. Goldstein 1981; Reed 1986.

22. Compare Barsalou 1999; Heft 2007; O'Regan and Noë 2001; Noë 2005.

23. On the resistance to cognitive interpretation, see Barrett 2011; Griffiths 2008; Solomon 2003. On successful behaviorist paradigms, see Davidson et al. 2009. On the role of phenomenology, see Damasio 1999, 2003; Dreyfus and Kelly 2007.

24. As cognition see Zajonc 1968, 1980. As conditioning, see Baeyens et al. 2001b. As conscious, see Ekman 2003; Solomon 2003. And as and generated unconsciously, see Bauer and Rubens 1985; Weiskrantz 1986; Gabriel 2008.

25. On human uniqueness, see Nussbaum 2003. On homology across mammals, see Panksepp 1998.

26. As appraisals, see James 1890; Lazarus 1966; Nussbaum 2003; Ekman 2003. As basic emotions, see Tomkins 1962. As core affect, see Russell (2003) and the vitality affects of Stern (1985). Compare Freeman 1999.

27. As intentional core, see Sartre 1939; Damasio 2010; Colombetti 2014.

28. Davidson and Irwin 1999.

29. We consider primary-level emotions to be tracking the situation of the organism in the environment at the behest of homeostatic processes. While researchers like Scherer (2009) claim this process to be a form of appraisal insofar as it instigates appraisal evaluations of the personal relevance of an event (i.e., its implications, the organism's coping potential, and the normative significance of the event), we do not see overwhelming evidence that these nonconscious relational processes necessitate the internal representations that are the key element in appraisal theories. Prinz's notion of embodied appraisal coheres more nicely in that he emphasizes the noncognitive nature of primary-level states that do not include judgment, which function instead as perceptions of bodily changes that track (or "represent," in Prinz's words, though our conceptions of representation diverge on this point; see Chapter 6) properties relevant to the well-being of the organism. See Prinz 2004.

30. Izard 2011.

31. Panksepp 1998, 2012.

32. See Damasio (1994) and work on the double fear pathway by LeDoux (1996).

33. Clark 2008; Lewontin 1984; Clark and Chalmers 1998. Also, Kono 2009.

34. On metaphors of mind, see Harre 2004; Black 1977; Bhaksar 1975; Gentner and Grudin 1985; and Gabriel (unpublished manuscript).

35. Porter 1997.

36. For primary-level, see Panksepp 1998. For secondary- and tertiary-level, see Millikan 1996; Damasio 1994; Ray and Heyes 2011.

37. Legrand 2010a, 2010b; Grammont et al. 2010; also see Merleau-Ponty 1942.

38. Jeannerod 1994, 2006.

39. Ungerleider and Mishkin 1982; Milner and Goodale 1993; compare Michaels et al. 2001.

40. Searle 1983.

41. McDowell 2011.

42. See the work of Edward S. Reed (for example, Reed and Jones 1982; Reed 1986, 1996). Also Heft (2007) on landscape affordances, Rietveld and Kiverstein (2014) on how the environment solicits affordances, and Ju and Takayama (2009) on how industrial design elicits action possibilities.

43. Pacherie 2000; compare Gibson 1979.

44. O'Regan and Noë 2001; see Barsalou 1999 on perceptual representations. See also Grammont et al. 2010; Colombetti 2014. While we have sympathies with an enactive interpretation of action, we are not convinced that representations, even so-called pragmatic representations, are necessary for many elements of sensorimotor behaviors. The work of Hutto and Myin (2012) on a content-less enactivism is more convincing to us, cf. Nishberg 2015.

45. Compare Carruthers 2008; Tomasello and Call 2008. Cf. Dennett 1996b on the Tower of Generate and Test. Additionally, there may be executive processes that function upon the decoupled action plan as mental simulation; see Barton 2012.

46. Klein 2014; see also 2017.

47. Dreyfus 2002; Dreyfus and Kelly (2007) further discuss the phenomenology of affordances.

48. For which we do not have overwhelming evidence of internal representations, see Barton 2012; Barrett and Rendall 2010.

49. Colombetti 2014.

50. Merleau-Ponty 1942, Dreyfus 2002; see also Hutto and Myin 2012.

51. Grammont 2010.

52. Allen 2010; Gallese 2010. Also see Heft (2007) on intentional attunements via socialization in the context of ecological psychology.

53. Gallese et al. 1996.

54. Gallese 2010, 207.

55. On the embodiment of intentionality, see Withagen et al. 2012.

56. Thompson and Zahavi 2007.

57. James 1899.

58. On background feeling state, see Damasio 2010; Russell 2003. On unconscious emotion, see Baeyens et al. 2001b; Bauer and Rubens 1985; Gabriel 2007.

59. See Phelps (2006) and consult proceedings from Society for Affective and Social Neuroscience (2012–2018).

60. Gabriel and Khooshabeh (unpublished manuscript) sought to determine whether unconscious emotion processes are subtended by associative or representational memory, with inconclusive results. We attempted to classically condition a prosopagnosic patient with positive and negative affect to "neutral" faces. We attempted using photographs and hand-drawn faces but could not escape the confound that no face is truly neutral. See also, Hammerl and Grabitz 2000, as well as Zanna et al. 1970.

61. Barton 2012.

62. Golonka 2015 claims affordances are dispositions but that the act of perceiving information specifying that affordance is relational. See also Damasio et al. 1982, 331.

63. This idea follows from a line of theorists who conceptualize the organism as a self-regulatory system; for example, Jennings 1906; Bandura 1988; Burgdorf and Panksepp 2006. On the global workspace theory of consciousness, see Baars 1988.

64. On nonverbal communication, see Ashenfelter et al. 2009; Mignault and Chaudhuri 2003; Hall et al. 2005; Keating et al. 1977; Cashdan 1998; Kraut and Johnston 1979. On motor commands, see Jeannerod 1994.

65. Katz and Fodor 1963. On images in the mind, see Kosslyn 2005. Although, see Hutto and Myin (2012) for a content-less framing of tertiary level processes.

66. For full details of summarized studies see Gabriel 2007; Gabriel et al. 2008. For more on prosopagnosia, see Bodamer 1947; and for covert recognition in prosopagnosic patients, see Bauer 1984; Bauer and Rubens 1985. In short, covert recognition in PA patients is a consistent and reliable finding across a wide range of indirect measures ranging from low-level responses, such as eye-tracking and skin conductance response (SCR), to higher-level responses such as semantic interference paradigms and facilitative effects. Experimental results on covert recognition suggest that PA patients are able to process and respond appropriately, to some degree, to overtly unrecognized faces (Sergent and Signoret 1992), suggesting that facial identity information may enter a PA patient's mind without their awareness while still having an effect on their behavior. This dissociation is in line with the distinction (Marcel 1983; Faulkner and Foster 2002) between phenomenology and information processing, since a PA patient is not consciously aware of facial identity information but nevertheless demonstrates responses that seem to reflect some processing.

We argue that affective information in the experiments functions as an unconscious form of recognition. Prosopagnosia in this regard is an absence of overt identification recognition of faces of people they are familiar with. It is in this unique situation that we investigate whether affective reactions remain as a type of secondary-level schema valence tag of the familiar face. This affective information is simply who that person is to *me;* that is, how does this person make *me* feel given life-historical factors; it is a strictly personal knowledge manifested in sentient affective markers.

The associationist / behaviorist learning mechanisms, *evaluative conditioning* and *attitudes* provide a route to valence acquisition based on the contiguous pairing of originally neutral stimuli with positively or negatively valenced events (Hermans et al. 2003; Hammerl 2000). Evaluative conditioning refers to mere spatiotemporal co-occurrence of a neutral stimulus X (CS) with a valenced stimulus Y (US), which results in the originally neutral stimulus X acquiring an evaluative meaning congruent with the valence of Y (Baeyens et al. 2001a). Some interesting characteristics of this type of conditioning are as follows: there is immunity to extinction, the CS valence tracks US valence when it changes, and awareness is not necessary. A consistent finding in this paradigm is that neutral pictures take on the valence of their respective US pairing; this suggests that a CS can be evaluatively conditioned. Similar results have been found using words, symbols, crossmodal stimuli, a single feature across different stimulus configurations, gustatory-olfactory stimuli, and ecologically valid situations like odors. Furthermore, it has been observed that evaluative conditioning may also occur in indirect observational situations where a subject becomes conditioned by simply watching a CS-US pairing occur on someone else (Baeyens et al. 2001b). This research paradigm is closely related to the methodology of *affective priming,* wherein a series of positive or negative stimuli are presented and must be evaluated by the subject as rapidly as possible; each target is preceded by a prime that is positive, negative, or neutral. The time needed to evaluate the target stimuli is consistently found to be mediated by the valence of the primes—for example, response latencies for evaluatively congruent trials are significantly shorter than on evaluative incongruent trials; this type of priming is said to be automatic in that it is fast, involuntary, and efficient (Herman et al. 1998).

Our adaptation of this paradigm for covert recognition in a PA patient is that a CS "neutral" face (but note that all faces are identity-neutral for a PA patient) gets paired with the US of the PA's affective reaction to the person. This affective reaction is established through interaction with, or knowledge of, the person. Henceforth, sight of the face (CS) is nonconsciously linked through evaluative conditioning to the particular affect garnered from interaction with, or knowledge of, the person (US).

A closely related construct for an affective mechanism that is more cognitive and representational (i.e. secondary-level unto tertiary-level processes) comes from research in social psychology. According to Fazio 1995, *attitudes* may be viewed as associations in memory between the representation of a stimulus or event and its summary evaluation (on a positive-negative continuum). In other words, attitudes are object-evaluation associations, or ready aids in sizing up objects and events in the environment (Fazio et al. 1983). Attitudes with more emotional bases appear to be characterized by stronger object-evaluation associations; individuals appear to trust their emotional reactions as diagnostic of their evaluations of an object (Fazio 1995). Automatic attitude activation refers to the situation in which the presentation of the object results in the activation of its memory representation and immediately leads to automatic activation of the related evaluation (Hermans et al. 2003).

In relation to evaluative conditioning, attitudes are thought to develop via classical conditioning through repeated pairings of potential attitude objects (CS) with positively and negatively valenced stimuli (US); this is referred to as attitude formation (Olsen and Fazio 2001), as tested in the Implicit Attitudes Task (IAT), for example. These mechanisms illustrate our conceptualization of secondary-level affective mechanisms as straddling a line between behaviorist and cognitive processes.

67. Gabriel et al. 2008. For diagnostic history of apperceptive PA patient MJH who was the subject in this series of experiments, see Biederman and Kalocsais 1997 and Duchaine et al. 2003.

68. The main results supporting this hypothesis are: (1) the correlation of likeability scores across conditions in the first experiment for faces of familiar people that MJH likes and (2) accurate affective reactions to faces of familiar people who MJH does not like in the next experiment.

69. As conceptualized by Mandler (1980), recognition includes: (1) retrieval of the familiarity value of an event and (2) a slower search-and-retrieval mechanism that attempts to identify the event. If MJH is in fact experiencing these two stages of recognition, then they must be functioning covertly because his subjective familiarity ratings are inaccurate and he is not able to explicitly identify the majority of target faces.

70. Pham et al. 2001.

71. Gardner 1985; Bechtel and Abrahamsen 1991; Sperber 2002.

72. Fodor 1983. Though there are many other interpretations, see Sperber and Wilson 2002.

73. Marr 1982.

74. See the appraisal of the critique of Maclean's theory in Cory 2002.

75. We see this layered version of brain functions as a helpful model or even metaphor for the way brains emerged in stages, not *in toto*. The actual brain is a tangled, beautiful mess of interconnecting circuits that evolution reorganizes slowly, but the triune model helps us see the major physiological categories of a central nervous system in action.

   The oldest brain, functionally speaking, is the reptilian complex of the brain stem, thalamus, cerebellum, and basal ganglia, which we share with all vertebrates. Here reside functions like circulation and respiration, motor control, and motivational or appetitive drive (in the higher regions of the basal ganglia). The limbic region (the amygdala, hippocampus, midbrain, and especially the periaqueductal gray [PAG]) includes the emotional and memory aspects of the mammal mind. And the third and final layer of the triune brain—immense in humans—is the neocortex. The cortex, which is deeply grooved in primates (to maximize surface area) contains various functional lobes: frontal (planning, calculating), parietal (touch, etc.), occipital (vision, etc.), and temporal (auditory, etc.). At the ontogenetic level, this part of the brain (especially the frontal areas) is the last to wire-up completely. See figure in Introduction.

76. Panksepp 2012; Davidson et al. 2009.

77. Charland 1996, 2005; Griffiths 2008; Fodor 1983. Although not everyone agrees as to the definition of a *module,* see the discussion in Barrett and Kurzban 2006, 628. Nor is the definition of a *basic affective mechanism* uncontroversial; see Davidson et al. 2009.

78. Alternatively, Charland (1996) and Griffiths (2008) argue *basic affective mechanisms* are themselves *modules,* with the difference that they function upon nonpropositional representations. Charland argues, tautologically, that these are symbolic because they can be used by *modules* and central processors in inferential processes.

79. Cannon 1926; Damasio 2010; Panksepp 1998.

80. Colombetti 2014. See also Wheeler 2005; Deacon 2011.

81. Damasio 2010; Solms and Friston 2018.

82. Bauer and Rubens 1985; Damasio 1996; Gabriel 2007. Also, on embodied intentions in the context of ecological psychology, see Withagen and Michaels 2005; Michaels et al. 2001.

83. Panksepp 1998, 15.

84. See Uttal (2001, 2005), a convincing series of books on the shortcomings of brain imaging and thus cognitive neuroscience.

85. That may be broached in coming years with neuroscientific research on the connectome; see http://www.humanconnectomeproject.org/.

86. Although Strawson has gone this improbable route in his argument for panpsychism. Strawson 2006; see also Rosenberg 2004.

87. Mountcastle and Mountcastle 1948. Similar results were demonstrated by Panksepp 1998; Damasio 2010 on a decorticated human. Furthermore, studies of subjects with spinal brain injuries report the intensity of feelings seems to depend on the location of the lesion: the higher up, the weaker the feeling (Colombetti 2007; Thompson 2007; Hohmann 1966; Chwalisz et al. 1988).

88. Gardner 1985; see Sober 2000. For a defense of the mechanistic view of mind, see Bechtel 2008.

89. Holbrook and Fessler 2015; Cosmides and Tooby 1992.

90. Proust 2007.

91. Compare Fodor 1974 and Gardner 1985.

92. See Carruthers 2008; Metcalf and Kober 2005.

2. Biological Aboutness

1. Spinoza 2001, part III, prop. 6.

2. Kant 1987, §75, 312.

3. Nagel 2012.

4. Craig and Moreland 2009.

5. Paley 2008. Kant argued that the human mind cannot avoid projecting purpose into nature, even when it produces Panglossian absurdities. But, contrary to many interpreters, this is not a free pass for Intelligent Design natural theology. Close study of his position reveals a nuanced alternative teleology. In addition to the instrumental teleology that seeks to link specific structures to functions (e.g., sharp teeth to carnivore diet, skin color to solar environment, sweat glands to thermoregulation, etc.), we must assume, he argues, a more universal teleology in all of nature in order to do science in the first place.

   How could we expect nature to give us answers to our questions unless there was some rational aspect or logic in nature that could be interpreted by our own rational minds? Science assumes some isomorphism between our rational faculties and nature's structure, otherwise the former could not limn the latter. This expectation of isomorphism is a kind of a priori projection that unifies all nature into a domain of possible exploration. Kant suggests that an encompassing "principle of purposiveness" (an expectation that we'll get answers to "What for?" questions) constitutes the unity of nature, and that "we must necessarily assume that there is such a unity without our comprehending it or being able to prove it" (Introduction, V).

   This is a fascinating thesis, but we shouldn't make too much of it. The religiously inclined would like to ride piggyback on this thesis in order to claim Kant on the side of Intelligent Design and natural theology—after all, one might read this isomorphism (between nature and our mind) as proof of God's overall beneficent design. Sociologist Steve Fuller, for example, makes this same confusion when he endorses Intelligent Design in books and courtrooms (e.g., see Fuller's testimony in *Kitzmiller v Dover Area School District,* 2005). Since natural religion predicts, via the divine design hypothesis, that we will decode nature's rational messages—and we do indeed succeed in limning these rational structures—then natural theology, according to Fuller, is corroborated as a bona fide research program. But design neither follows from the fact that nature is structured, nor from the isomorphism of mental faculties and nature. And Kant knew this, though he did not know the eventual Darwinian explanation of the isomorphism (evolved cognitive modules and adaptive general intelligence capacities). Even without the Darwinian naturalistic solution, Kant knew that our stubborn faith in a rational nature could not prove that it was designed. Kant's assumed cosmic teleology is unlike the religious tradition because his version must remain a vague requirement of the system, lacking any and all content.

This unity-of-nature tradition of teleology is necessary—when we expect answers to our scientific questions—but it will always be vague, lacking in predictive power, incapable of proof, and even incapable of true comprehension. In other words, this isomorphism assumption is another instrumental tool, albeit writ large across the whole of nature. The most that Kant can say about the content of this assumption is "There is in nature a subordination of genera and species comprehensible by us" (Introduction, V). And he adds that there is "a harmony of nature with our cognitive faculty." This paucity of content shows us that the *unity-of-nature* teleology is just the base-level assumption that *nature has predictability.* As a "kingdom of ends," nature can be anticipated and used. That is the broadest notion of teleology, but it can and should be conceptually detached from the design hypothesis of natural theology. We see this debate to be but a piece of the realism-versus-empirical-adequacy debate (van Fraassen 1980).

6. Hume 1779.
7. Behe 1996; Dembski 2002; Johnson 1991; Johnson 1989.
8. Dennett and Plantinga 2010.
9. After Thomas Nagel's *Mind and Cosmos,* Philip Kitcher published a rebuttal in the *New York Times* (September 8, 2013) on the grounds that Nagel was asking old-fashioned questions about the nature of life itself, values, mind, and teleology. If anything, Kitcher suggested, such grandiose general answers—about the unity of nature—may come at the very *end* of separate analytical strands in contemporary science. For now, Kitcher suggests, such grand unifying theories should be set aside. But it's interesting that Kant, perhaps equally skeptical about the content of such grandiose inquiry, nonetheless placed the most overarching unity (nature's teleological structure) at the very *beginning*—as a precondition of scientific inquiry—not at the end.

The important takeaway from this unity-of-nature tradition is that it is about the meeting point between nature's causal regularity and our cognitive tracking powers; that is its frustrating genius. Nature's "for the sake of" structure is a function of our hardwired minds. But if Kant is correct, then seeing nature as purposeful may be built into our cognitive faculties or, at the very least, our methodological norms. After Darwin, we acknowledge that such faculties were built slowly, as they served the survival of our ancestors. Thinking of nature as a kingdom of ends may fuel certain kinds of biological investigation into means / end relationships, but it remains agnostic about ultimate causes beyond material processes. And Ockham's razor preference for parsimonious explanations lead most scientists to rightly reject divine psychology as a cause of nature's structure.

Theist philosopher Alvin Plantinga thinks it unlikely that our minds would have evolved to match the structures of reality, if a benign deity had not aligned them. For Plantinga, we need a benevolent supernatural force outside the metaphysics of naturalism in order to ground our truth claims inside natural science—otherwise, all science is just instrumental, unreliable, and unfettered to reality. Our arriving at truths would be highly improbable, Plantinga thinks, without help from above.

Like other naturalists, we think Plantinga has gotten this whole issue backwards, and the isomorphism between our theories and reality is a result of an evolution of mind that selected for successful *prediction generators*. Of course, the best kind of prediction generator is the one that perceives and processes the actual causal pathways of physical nature (albeit indirectly and fallibly). In this sense, we are critical realists insofar as we think successful predictions, theories, and models take their success from the fact that they capture features of the real, even if this access is indirect and heavily mediated (Bhaskar 1998). And, with respect to prediction generation, we suggest that scientific models of the world are different from animal models of their worlds (life strategies) only by degree rather than kind. Yes, our linguistic symbol system allows for unparalleled abstraction and sophistication of modeling, but following John Dewey 1929, even science itself is an extension of the human organism's pragmatic attempt to navigate its own survival.

10. A process he referred to as kinetic teleology.
11. Aristotle, *Physics* II, *Metaphysics* V, and *On the Parts of Animals* I.
12. See Cleanthes' famous argument in Hume's *Dialogues Concerning Natural Religion*.
13. See Leroi 2014; Lloyd 1975; Balme and Gotthelf 1985.
14. See Carroll 2005.
15. Aristotle, *On the Parts of Animals,* book I.
16. Ibid., book III.
17. Dawkins 1986.
18. Hladký and Havlíček 2013.
19. Dennett 1996a.
20. See Russell (1905) for an influential statement on twentieth-century representational forms of intentionality.
21. Furthermore, among many other research programs, we see Konrad Lorenz's work on imprinting, J. J. Gibson's ecological psychology, and John Garcia's work on taste aversion to demonstrate meeting points between biological intentionality and "natural" behaviors in animals.

22. Gibson 1966; see also Fodor and Pylyshyn 1981.
23. See Tarski 1984; Lewis et al. 1991.
24. For example, see Cassirer 1950.
25. Ehrlich 2002.
26. Gould and Vrba 1982.
27. Aristotle, *De Anima* II; *Generation of Animals* II.
28. Eldredge and Grene 1992.
29. Sloan 2006.
30. Ghiselin 1997.
31. See Amundson 2007.
32. Wimsatt and Schank 2004.
33. Kauffman 1993.
34. For a helpful discussion of Andreas Wagner's work, see Ball 2015.
35. Wagner 2008.
36. Weiner 1948. For game theory economics, see Von Neumann and Morgenstern 1944.
37. Asma and Greif 2012.
38. Maturana and Varela 1980, 78.
39. See Asma 1996; Juarrero 1999; Thompson 2007; Deacon 2011.
40. Lin 2006.
41. Spinoza, *Ethics* III, prop. 9.
42. See Coleman 1977.
43. Many embryologists of the nineteenth century assumed a vital force because they didn't know how an undifferentiated organic blob could slowly become an articulated fetus. Push-and-pull physics did not normally turn unstructured mush into highly structured working parts and integrated wholes, so the fetus was either fully articulated (micro-size) inside the mother and just grew larger with nutrition (preformationism) or it was an amorphous blob that sequentially took on form—via a vital force (epigenesis).

      This vitalism tradition was very popular, even after Darwin's revolution. The idea of an occult invisible force that guides animal embryogenesis was congenial to solving the origin-of-life mystery too, and many used this stone to kill two birds. Darwin tried a mechanical replacement—with his maligned pangenesis theory—but, while it purported to account for heredity, it couldn't be corroborated, and it couldn't explain the goal-directed process of ontogenesis.
44. Dupré 2014.
45. Duncan and Wakefield 2011, 330.
46. Panksepp and Biven 2012.
47. Berridge 2009, 68.
48. Panksepp 1971.
49. Gould and Vrba 1982; see also Gould 1997.
50. See Marcus 2008.
51. See Bedau and Humphreys 2008.
52. Clayton and Davies 2006; see also Macdonald and Macdonald 2010.
53. Richards 1989; Morgan 1923.

54. Ryle 1949.
55. Searle 1992, 83.
56. See Merleau-Ponty 1962.
57. See Dretske 1995; Asma 1996; Millikan 1996.
58. Originally articulated by James J. Gibson 1969, 1977; Edward S. Reed 1987; and further developed by Ruth Millikan 2004 and Anthony Chemero 2009.
59. Chemero 2009.
60. Nagel 2012.
61. Fodor and Piattelli-Palmarini 2011.
62. For the original formulation of this conscious subjectivity, see Brentano 1995.
63. Arnold 2012a; see also 2012b. See Hutto and Myin (2012) for a content-less model of the mind.
64. See Barsalou 1999. On thinking with images, see Kosslyn et al. (2006) and our discussion in Chapter 6.
65. Deacon 2011.
66. Fodor 2012, 33.

3. Social Intelligence from the Ground Up

1. DeSalle and Tattersall 2012; Baron-Cohen 1999; Humphrey 1976; Dennett 1996b.
2. For Theory of Mind Mechanism, see Leslie 1987. For Bayesian inference models, see Griffiths et al. 2012.
3. Varela et al. 1991; M. Wilson 2002.
4. As opposed to, for example, Gallistel and King 2009.
5. For example, see recent work on how neurotransmitters in the striatum of primates relates to social conformity, social cuing within group aggression, and neuro-chemical origin of hominids; Raghanti et al. 2018.
6. Panksepp 1998; Gardner 1985; Goleman 1996.
7. Clark 2008, 219.
8. Carruthers and Ritchie 2012. On uncertain monitoring, see Smith et al. 2003. Also see the work of Kerry L. Marsh and collaborators in developing a social synergy model of social behavior.
9. Proust 2009.
10. Ibid., 178–180.
11. Call and Tomasello 2008.
12. Metcalfe 2008.
13. Barton 2012.
14. Cannon 1926, 91; Damasio 2018.
15. Palmer 1999.
16. See our Introduction.
17. Clark 2008; M. Wilson 2002.
18. Cussins 1992.
19. Gibson 1966; 1979, quoted in Withagen and Michaels 2005.
20. Fodor 1983.

21. For ecological psychology and action-oriented representations, see Gallagher and Zahavi 2012; Gallagher 2007; Clark 1997; O'Regan and Noë 2001. Here is a survey of the variants of alternative ontologies in cognitive science. An autopoietic enactivism (Varela et al. 1991). Also, Thompson 2007 claims mentality emerges from the self-organizing and self-creating activities of living organisms. A radical embodied cognitive science (Chemero 2009) claims that representational and computational views of the mind are mistaken and that a set of explanatory tools that do not posit mental representations is a more accurate descriptor of perception (also see Clark 2008; Beer 2003; 2014). A radicalized enactivism takes us further into the embodiment thesis, claiming that basic cognition is exhaustively constituted by patterns of environmentally situated organismic activity (see Hutto and Myin 2012). The latter theorists explain content-involving cognition via non-internally-generated cognitive gains brought on by engagement in shared practices that include external symbol systems, like natural language.
22. See Fodor 1990; Millikan 1984; Jacobson 2008.
23. Dietrich and Markman 2003.
24. Clark 2008.
25. Millikan 1996, 6.
26. Damasio 1996.
27. Millikan 1996, 6.
28. Strum 2001.
29. For Millikan's conception of PPRs, see Millikan 2009; also see Hutto and Myin 2017 on a content-less notion of enactive mind to which we are sympathetic and make some progress toward in this chapter.
30. Bermúdez 1998.
31. Fodor 1998; Rietveld and Kiverstein 2014.
32. For examples, see Costall 2010; Chemero 2009; Fiebich 2014.
33. Wheeler 2008.
34. Clark 2008; Wheeler 2005.
35. For defining characteristics of AORs, see Gallagher 2008.
36. See Barton 2012.
37. See Hurley 1998; O'Regan and Noë 2001; Noë 2009; Robbins and Aydede 2009.
38. Milner and Goodale 1995.
39. Colombetti 2014. Dynamic systems theory is a set of models where variables change continually and interdependently. Meanwhile, in the philosophy of perception literature, Heck 2000 made a helpful distinction between the content of a percept versus the conceptual state of a percept, while Dretske (1981) wrote on how perceptual content may more helpfully be described as analog, reflecting its fine-grainness and difference from the digital format favored by computational theorists of the mind. Finally, Raftopoulos and Müller (2006) suggest that what makes the content of a representation nonconceptual is precisely the fact that its content is insulated from other knowledge structures available to the person. We made similar arguments in Chapter 1 on the relation between modularity and affective mechanisms, and we discussed a set of prosopagnosic experiments

investigating implicit affective content, which is similar to research on subpersonal representation (in Burge 1986; Egan 1992; Bermúdez 1995) and to the analog magnitude representations in Beck 2012a and 2012b. On dynamic systems in social psychology, see Nowak and Vallacher 1998.

40. Sterelny 2003; Panksepp 2012.

41. Snowdon 2001.

42. For a review of affordances in social psychology, see Marsh et al. 2006. For coordination through interpersonal affordances, see Richardson et al. 2007. See the clear enunciation of the nature of affordance information in Turvey 1992.

43. Golonka (2015) claims that law-based information and information based on conventions, like language and culture, are distinct and that affordances only occur on the former; although she leaves room for this social intelligence model by emphasizing that conventional information can play a role in action selection. We see the social milieu of the creature as a part of the econiche and socializing as a form of life (Rietveld and Kiverstein 2014) that, while subject to change, offers a set of reliable-enough affordances across an individual lifetime.

44. McCall and Singer 2012.

45. Compare Proust 2009. For PPRs, see Millikan 1996; Gibson 1979; Reed 1996. For somatic marker encodings, see Damasio 1996.

46. Damasio 1996; but see Pessoa (2013), where the author claims that emotion and cognition are ultimately not distinguishable, and that affective motivations are embedded in perceptual and cognitive processes.

47. On how complex social thinking can emerge sans representational abilities, see Vallacher et al. 2002.

48. Marsh et al. 2006, 22.

49. See evaluative conditioning in Baeyens et al. 2001a; Gabriel 2007.

50. Gibson 1979. See also Chemero 2003a; Rietveld et al. 2013.

51. Millikan 1996; Gibson 1979; Cussins 1992. On social embodiment, see Barsalou et al. 2003. See also Pacherie 2015.

52. See Barrett and Rendall 2010. Despite Millikan's use of "representation" in PPR, these do not function like the traditional definition of representations, insofar as they cannot be manipulated in a separate inner space and their intrinsic purpose is more tied to real-time sensorimotor functions than to an internal space where symbols undergo information-processing. PPRs thus resemble a type of FBS or AOR. Alternatively, with some more philosophical work one could elaborate this characterization into a more radical enactive paradigm like that of Hutto and Myin 2017.

53. Hall 1966; Chemero 2003b.

54. On the ability to read sexual availability, see Gangestad et al. 1992.

55. Baron 1980.

56. Ashenfelter et al. 2009; Mignault and Chaudhuri 2003; Hall et al. 2005. See also Gifford 1994.

57. Keating et al. 1977; Cashdan 1998; Kraut and Johnston 1979.

58. Ellyson et al. 1980; Strongman and Champness 1968; Dovidio et al. 1988. On the universal ability to read conspecific body language, see Buck 1988.

59. Strum 2001.
60. Camp 2009; Cheney and Seyfarth 2007.
61. Snowdon 2001.
62. Bente et al. 2010.
63. Compare Barker 1968 and Costall 1995. See also Heft (2007), who claims socialization is the background condition from which ecological psychology operates.
64. See Marsh et al. 2009. See also Van Acker and Valenti (1989) on the possibilities of social interaction.
65. Millikan 1996.
66. Carruthers and Ritchie 2012, 83.
67. See the emulator system of Clark and Grush (1999), and subsequent discussion of the model.
68. Strum 2001.
69. See van de Waal et al. 2013.
70. Ekman 1999.
71. Sterelny 2012a.

4. Emotional Flexibility and the Evolution of Bioculture

1. Kronfeldner 2007.
2. Jablonka 2012.
3. Gowlett 2016.
4. See Bondariansky and Day 2009. Also see Strum 2014. Quite surprisingly, chimpanzees may need substantial social learning from bioculture in order to successfully engage in so "instinctual" a process as sex. It may not be an innately engraved skill set. Chimps in the wild observe sexual behavior and translate that information into successful copulation when their own post-puberty passions descend upon them. In captivity, without the model of adult sex, chimps become aroused but unable to engage the equipment effectively.

   Personal communication with Iner Peterson, who was the architect and physical engineer of the bioaeronautics chimp lab in the early 1960s at Holloman Air Force Base, reveals an interesting scenario. Peterson, who also worked with "Ham"—the first astrochimp in space—informed me that when the lab instituted a breeding program designed to create a pathogen-free chimp, they could not get any captive chimps to breed properly and had to import wild chimps from Gabon Africa to "demonstrate" the mounting techniques. This is a striking example of the way even fundamental survival skills can be offloaded into a species' biocultural reservoir.
5. Heyes 2012.
6. West-Eberhard 2003.
7. Hallson et al. 2012.
8. Sapolsky 2006.
9. van de Waal et al. 2013.
10. Hobaiter et al. 2014.

11. See Dennett 2017.

12. Panksepp 1998.

13. Klein 2009.

14. Arguably, Elias 1939 and Elias 1969 articulated a similar thesis—that changes in societal structures and changes in emotions co-evolved. But we are now in a privileged era of scientific data that can deliver evidence and fruitful research approaches for this general thesis.

15. Byrne et al. 2004.

16. Ibid.

17. It is often classed with the emotions, but it is really a master emotion, a motivational system that organisms enlist in order to find and exploit resources in their environment. In plain English, we call it *desire* or *craving*. It energizes mammals to pursue pleasures or satisfactions, but it is not the same as pleasure. It is that growing, intense sensation of heightened attention and the increasing feeling of anticipation—as if you are about to scratch a powerful itch. This desire or SEEKING system begins in the ventral tegmental area (VTA) of the midbrain (rising through the nucleus accumbens) and extends neurons up to the prefrontal cortex and down to the brain stem. Jaak Panksepp calls this phenomenon SEEKING (see Panksepp and Biven 2012 for an updated discussion). Kent Berridge has discovered a similar brain-based emotional system, which he calls "wanting." Berridge explains, "Usually a brain 'likes' the rewards that it 'wants.' But sometimes it may just 'want' them. Research has established that 'liking' and 'wanting' rewards are dissociable both psychologically and neurobiologically. By 'wanting' we mean *incentive salience,* a type of incentive motivation that promotes approach toward and consumption of rewards, and which has distinct psychological and neurobiological features." This incentive-salient form of wanting, or Panksepp's SEEKING, is below the neocortical circuits of conscious desire. Incentive salience and SEEKING are subcortical systems that seek to reset homeostasis, and while they often align with our neocortical conscious desires, they sometimes lead to irrational wanting—a want for what is not cognitively wanted.

18. Sapolsky 2004.

19. Eisenstein et al. 2012.

20. Stanford 2001.

21. de Waal 2013.

22. Gibson 1979, and further developed in Millikan 2004.

23. Narvaez et al. 2013.

24. Stanford 2001.

25. Kaplan et al. 2000.

26. Hawkes 1991.

27. Goldstone et al. 2016.

28. Bailey et al. 2013 suggests that many collaborative decisions may involve following simple rules and routines adapted to circumstance rather than involving anticipatory choice. Even complex cooperative behavior, like taking on helping roles in a hunt, can arise from simple mechanisms such as associative learning. Individuals, for example, follow their preferred stalking pattern (e.g., go straight

toward or circle around prey), but in a group-hunt context, only the timing of actions has to be adjusted relative to the actions of other group members in order to achieve a successful collaborative hunt. Bailey et al., drawing on Munro, points out that "coordinated hunting in wolves suggests that fanning out and encircling prey may simply arise as each wolf is attempting to (1) get to the closest safe distance from the prey and (2) the best possible view of the prey" (Munro et al. 2011). Individuals, thus, move toward the prey and then away from other individuals so those in front do not obstruct their view (ibid.). This picture of events is probably an oversimplification in many instances. However, it does illustrate how careful we should be when inferring the cognitive abilities necessary to perform apparently complex behaviours.

29. Barrett and Rendall 2010.
30. Panksepp 2011.
31. "Informavores" was coined by George A. Miller in 1983, but it has been used and further developed by evolutionary thinkers like Daniel Dennett and Steven Pinker.
32. Heyes 2012.
33. Cook et al. 2014.
34. Sterelny 2012a.
35. See Chapais 2008. Also see Palombit 1999.
36. Fortunato and Archetti (2010) argue that monogamy first arose in Eurasian societies just as true agriculture was taking hold—around 12,000 years ago. Since agriculture involves intensive cultivation of large tracts of land, often requiring ploughing, irrigation, fertilization, and so on, it demands long-term partnerships. Gavrilets (2012), on the other hand, using a mathematical model, places the transition much earlier, and we tend to agree. But notice that the logic of the argument is the same (ecology-based fitness selection for pair-bonding), while the precise historical trigger is debatable.
37. Excellent data has been collected showing the importance of tubers as fallback foods in contemporary Tanzanian Hadza peoples and, by extension, our *Homo* ancestors. Humans in the savanna compete with baboons for berries, small game, baobab, and some honey. But baboons can't get to the deep tubers in the way women can, so the tubers remain a very important fallback food when diet is strained (Marlowe and Berbesque 2009).
38. Dunbar 2001.
39. Berlyne 1954; Berlyne 1978.
40. Piaget and Inhelder 1969.
41. Lowenstein 1994.
42. Morgan and Sanz 2003.
43. See Panksepp and Biven 2012. Also see Karama et al. 2002 and Fisher 1998.
44. Bartels and Zeki 2000. Freud (1927) claimed love is nothing but lust infused with tenderness (cf. care).
45. Pusey 2001.
46. Harcourt and Stewart 2007.
47. Kingdon 1988.

48. Reno et al. 2003.
49. Dupanloup 2003.
50. Harcourt and Stewart 2007.
51. de Waal 2006.
52. Hohmann and Fruth 2003.
53. de Waal 2001.
54. Hrdy 1977; 2009.
55. Muller and Wrangham 2009.
56. Panksepp 1998.
57. Many mammals (and terrestrial vertebrates generally) imprint through the "Jacobson organ" (vomeronasal organ) at the roof of the mouth. Chemicals from offspring are taken up by the mother's Jacobson organ (through smell or oral contact), usually during the cleaning of the just-born baby, and signals are conducted upstream to target sites in the amygdala and hypothalamus. Humans have a vestigial Jacobson organ, observable in our embryonic phase but largely inactive in adulthood. Different species have diverse ways of harvesting the chemical information that bonds them together, but for mammals, the resulting brain chemistry of oxytocin is strikingly similar (Broad et al. 2006).
58. Simply introducing these neuropeptides in high doses in a nonpregnant female mammal will actually produce mothering behaviors. Nonpregnant female rats were given blood transfusions from females who had just given birth, and they immediately began engaging in new maternal behaviors (e.g., building nests, gathering another mother's dispersed pups together, hovering over them to provide warmth, etc.). The same triggering of maternal behaviors in non-mother rats can be achieved by directly injecting oxytocin (OT) into their brains (OT can't cross the blood-brain barrier). Studies like these have also established that OT is necessary for the onset of maternal behaviors but not the maintenance of mothering activities. Once OT flips the switch, mothering care is sustained on its own momentum, so to speak (Insel 2003).
59. Kendrick et al. 1987.
60. See chapter 12 of Panksepp 1998.
61. As mentioned earlier, the infanticide strategy is a common mechanism by which males enter a new territory and take over. When a male, whether it's a rat or a lion, enters new territory, it usually kills the babies of the group. This activity stops breastfeeding, which stops lactation in the mothers, which then restarts ovulation. This brutal pattern creates the opportunity for the new male to mate with the females and create a new gene line. In rats, the post-coital oxytocin spike correlates exactly with the gestation period (three weeks) of its own offspring. In other words, after sex, males calm down long enough to bond with their own offspring (not kill them), and then they slowly resume normal levels of pup-killing. Rodent mothers also engage in regular infanticide. For a discussion of aggression, parenting and oxytocin, see Oliver Bosch (2013).
62. Stanley and Adolphs 2013.
63. Hammock and Young 2006.

64. Kemp and Guastella 2011.
65. Hrdy 2009.
66. Fries and Pollak 2005.
67. See Spikins et al. 2014.
68. Clay and Zuberbuhler 2012.
69. Damasio 1996.
70. Heyes 2010.
71. LaBar and Cabeza 2006.
72. Ibid.
73. Barton and Venditti 2014.
74. Richerson and Boyd 2001. See also Boyd and Richerson 2005.
75. Henrich 2015.
76. Sapolsky 2006.
77. Byrne (2001) shows how primate social deception may not be as smart as it looks. Monkeys and apes have comparatively excellent memories of socially relevant information about conspecifics (e.g., kinship and rank). This is probably a genetically provisioned prerequisite for social sophistication. But they also acquire deception tactics by trial-and-error learning that links associations (e.g., whenever this crying sound is made, Mother appears with a reward); however, this does not bring understanding of "other minds." It is not possible to infer the Theory of Mind Mechanism from Machiavellian social deception behaviors. See also de Waal 2013.
78. Sterelny 2012a, 166.

5. The Ontogeny of Social Intelligence

1. This chapter is heavily indebted to the research and careful thinking of our friend and colleague Glennon Curran (associate fellow of the research group in Mind, Science, and Culture). We are grateful for our conversations and his perspectives on the ontogeny of social emotions and on many other topics over the years.
2. Sterelny 2012a; Klein 2009.
3. See Sterelny 2012a; Sterleny et al. 2013.
4. Sterelny 2012a, 123.
5. Ibid., 122–123.
6. Barker 1968.
7. For example, see Strum 2001 on Baboons; de Waal 2005 on Bonobos; Morris 1969.
8. Knickmeyer et al. 2008.
9. Lagercrantz and Ringstedt 2001.
10. Berridge 2003a.
11. Schore 2013.
12. See this chapter under "The Infant-Caregiver Relationship"; Schore 1994; Semrud-Clikeman and Hynd 1990.
13. Martin 1983.

14. Sakai et al. 2012, R792.
15. Rosenberg and Trevathan 2002.
16. Wong 2012; Portmann 1990.
17. Montagu 1989.
18. Wong 2012.
19. Portmann 1990.
20. On child socialization, see Cole et al. 1996; Cole 2006, as well as Ellis 1997, and Rogoff 2003.
21. See de Waal et al. 2001.
22. Sterelny 2012a, 61.
23. Ibid., xiv.
24. Sterelny 2012a; Klein 2009; de Waal 2001.
25. Humphrey (1976) and Dunbar (1998) emphasize the role of social manipulation, Machiavellian deception, and competition, while Potts (1996) emphasizes climate variability.
26. Hermann et al. 2007. Also Sterelny et al. (2013) discuss the exigencies of cooperative social strategies in early Pleistocene culture.
27. Andersen and Guerrero 1998.
28. "The PLAY urge may be one of the major emotional forces that coaxes children to insistently explore intersubjective space (Reddy 2008), which helps promote the cognitive and epigenetic construction of higher social brain areas" (Panksepp 2013).
29. *Evolution, Early Experience and Human Development* by Narvaez et al. (2013) presents an interdisciplinary compilation of pertinent research.
30. In their book *The Autistic Brain: Thinking across the Spectrum*, Temple Grandin and Richard Panek (2013) critique the DSM-V's decision to combine "social interaction" and "social communication" into one diagnostic criterion of autism. In so doing, they make a similar, though not identical, distinction regarding types of social intelligence. They state: "What isn't scientific about the DSM-5's handling of the diagnostic criteria, however, is collapsing together social interaction and social communication. Social interaction covers nonverbal *behavior* that involves being with another person—making eye contact, smiling and so on. Social communication covers the verbal or nonverbal *ability to converse*—sharing ideas and interests, for example. Do impairments in social communication and impairments in social interaction actually belong to one single domain?" (p. 108). We construe what they call "social interaction" to be a part of social intelligence. Also see Gardner 1985; Grandin and Panek 2013.
31. Herbert et al. 2002.
32. Carter and Porges 2013, 132.
33. Ibid.
34. Narvaez et al. 2013, 15.
35. See Bowlby 1969, 1973, 1980.
36. Ainsworth et al. 1978.
37. Bowlby 1951; Schore 2013.
38. Schore 2001.

39. For example, Van Ijzendoorn 1995, 1998; Heinrichs and Domes 2008. For a review of attachment theory, see Bretherton 1985.
40. Gunnar 2000; Weaver et al. 2004.
41. Panksepp 2013, 85.
42. Wismer-Fries et al. 2005.
43. Theodoridou et al. 2009.
44. Kosfeld et al. 2005.
45. Domes et al. 2007. For a skeptical critique of many intranasal oxytocin studies, see Walum et al. 2015.
46. Bartz et al. 2011.
47. Levy et al. 2015.
48. Carlson et al. 2009.
49. Vrtička and Vuilleumier 2012.
50. Zimmermann et al. 2009.
51. Heinrichs et al. 2003; Olff et al. 2013; Tops et al. 2014.
52. Alves et al. 2015, 48.
53. Young 2003, 4; see also Burkett et al. 2016.
54. Schore 2013.
55. Mammal touch is crucial in igniting the release of oxytocin. High levels of maternal licking and skin-to-skin contact produce increased levels of plasma oxytocin in neonatal rats. See Henriques et al. 2014.
56. We discuss how such information is represented, and its subsequent associational pathways, in Chapters 3 and 6. See also Schore 1994; Shapiro et al. 1997; Siegel 1999; Wang 1997.
57. Ryan et al. 1997.
58. Schore 2013, 40.
59. Ibid., 349.
60. Price et al. 1996.
61. Schore 2013.
62. Cerqueira et al. 2008.
63. Marquez et al. 2013.
64. See Sandi's interview at http://www.medicaldaily.com/childhood-abuse-changes -brain-treatment-might-change-it-back-244281.
65. Johnson et al. 1999. And on violent offenders, see Raine et al. 1997; Raine and Yang 2006.
66. Beer et al. 2003; Gross 1998; Saver and Damasio 1991; Rolls et al. 1994.
67. Beer et al. 2006.
68. Ibid.
69. Rolls et al. 1994.
70. Kringlebach and Rolls 2004.
71. Nakamura et al. 2008.
72. Larquet et al. 2010.
73. Pinkham et al. 2003.
74. Carbone et al. 2013. On eye contact, see Stern 1985.
75. Lee et al. 1998; Leekam et al. 1997; Tiegerman and Primavera 1984.

76. Arnold et al. 2000. On the relation to language acquisition, see Podrouzek and Furrow 1988.
77. Parker and Nelson 2005.
78. Schore 1994, 66.
79. Parker and Nelson 2005.
80. Baron-Cohen et al. 1992. On decreased white matter, see Girgis et al. 2007.
81. American Psychiatric Association 2000.
82. Schore 2013.
83. Grandin and Panek 2013, 36.
84. Heyes 2010.
85. Barrett and Rendall 2010.
86. See Allis et al. 2007. Indeed, according to Mark Solms, the task of development is to reconcile the needs of the seven affective systems in conjunction; affect thus guides behavioral learning. Paper presented at Jaak Panksepp memorial symposium, April 14, 2018, Washington State University.
87. Darwin 1859.
88. Dennett 1996a.
89. De Jong 2005; Crews 2003.
90. Belsky 1997, 2005.
91. Hartman and Belsky 2015.
92. Davidson and Begley 2012.
93. Jessee et al. 2010.
94. Aron et al. 2005.
95. Sturge-Apple and Cicchetti 2012.
96. Hartman and Belsky 2015, 20.
97. Pepper and Nettle 2017.
98. Frankenhuis and de Weerth 2013.
99. Nussbaum 2001; Deonni and Teroni 2012.
100. James 1983; Panksepp 1998; Damasio 1994; Prinz 2004.
101. Deonna and Teroni 2012; Nussbaum 2001; McGinn 2011 Solomon 2006, and most recently Barrett 2017a.
102. See Deonna and Teroni 2012, 3 and 7. Actually, we have never heard anyone speak as portrayed in this last sentence, but never mind, since the whole approach has bigger problems.
103. Perhaps an expert reporter of phenomenological nuance could capture some of the feelings of fear in a linguistic description, but that would be only half the story, given the Spinozistic monism we espouse. Moreover, even the richest propositional description of fear will be a retranslation of deeply embodied experiences that have more to do with homeostatic drives, and kinesthetic, haptic, proprioceptive experiences—all of which become mediated by passing into linguistic modes of expression.
104. Ibid., 81.
105. Grice 1961.
106. For a review, see Winkielman and Berridge 2004; also see the prosopagnosic experiments discussed in Chapter 1.

107. DeLancey 2001; Barrett 2017a; Nussbaum, 2001; Kenny 1963.
108. Griffiths et al. 2012.
109. DeLancey 2002.
110. Prinz 2004.
111. Frankland and Greene 2015.
112. Berwich and Chomsky 2016.
113. Kahneman 2011.
114. Davidson and Begley 2012.
115. Marsh et al. 2006; Colombetti 2007.

6. Representation and Imagination

1. Dennett (1996b) describes the shift from purely conditioned creatures, called Skinnerian, to modestly representational minds, or "Popperian" creatures. A Popperian creature is minimally embodied by a type of emulator system that takes input about the starting or current state of a system, along with control commands, and gives an output as a prediction of the next state of the system as a set of values for the future feedback that the new state should yield; it models the target system in real time (see also mechanisms in Clark and Grush 1999).

2. Marcus 2008.

3. Brooks 1991; Webb 2012 on cognition in insects; Beer 1998, 2000; Thompson 2007. Traditional theories of representation emphasize the constant causal connection between the object in the environment and the tracking relation that stands in for it in the mind. Representations are internal states that mediate between a system's inputs and outputs in virtue of that state's semantic content (Dietrich and Markman 2003). On the nature of representations, see also Fodor 1990; Millikan 1984; Jacobson 2008. We take representation to be a system that connects behavior with environmental features, representing an information state in some language of mind. According to Hutto and Myin (2017), representations both respond to and keep track of covariant information and make contentful claims and judgments that can be correct or incorrect.

4. Subpersonal behaviors include the following: (1) subintentional acts, such as non-intentional movements (e.g., of the tongue or fingers) that we are not aware of and that have no reason or purpose; (2) pre-intentional acts, or "deeds" that include such things as the positioning of your fingers in catching a ball or the movement of your fingers while playing a complicated piece on the piano; and (3) pre-intentional acts like "on-line, feedback-modulated adjustments that take place below the level of intention, but collectively promote the satisfaction of [an] antecedent intention"; for example, fine finger movements when playing a note on a fretless instrument like the oud. On sub-intentional acts, see O'Shaughnessy 1980; Pacherie 2008. On pre-intentional acts, see Rowland 2006. And, on pre-intentional acts, see Rowlands 2006, 103; quoted in Gallagher 2008. Also, cf. the notion of "absorbed coping" wherein an agent's body moves from a sense of deviation from satisfactory gestalt without the agent knowing what this moving will achieve (Dreyfus 2014; quoted in Hutto and Myin 2017).

5. Bermúdez 1998, 50.

6. Peacocke 1992, 1998. Previously, there were other indications that propositional conceptual content is not sufficient to cover all instances of mental content; for example, the work of Armstrong and Stalnaker (1998) on indexicals. At the same time, Cussins (1990) sought a way to tie up such content within a connectivist network, and Hurley (1989) expounded on the relation between perception and action in this regard, while Tye (2006) expanded the discussion of nonconceptual conscious experience.

7. Kelly 2001.

8. Ibid.

9. Bermúdez 1995; Griffiths and Scarantino 2008; Pacherie 2011. As we have discussed, somatic markers are activated in relation to elements of the environment; it may be that the medial orbitofrontal cortex matches affordances to these activations. See Damasio et al. 1996; Proust 2015.

10. In this regard, Bermúdez (1998) focuses on the fine-grainness and complexity of the somatic proprioceptive system as an example of nonconceptual content in the bodily self. This somatic information concerning a perceiver's position, movement, homeostatic state, muscular fatigue, limb disposition, etc. is a form of nonconceptual content available from birth and features in our description of the self as evolved from an initial bodily self. In a similar vein, Proust (2015) discusses how affordances resonate with the environment by carrying indexical and evaluative neural-somatic information.

11. Cf. Proust 2015, 2016.

12. Proust (2015) discusses affect as a nonconceptual, perceptual, and evaluative expressive mode of representation that expresses a subjective relation to the environment and a tendency to act. Feelings here express affordances that are a subjective appraisal of the environment. See also Bermúdez 1998; Scarantino 2003.

13. Drawn largely from Bermúdez and Cahen, "Nonconceptual Mental Content," *The Stanford Encyclopedia of Philosophy* (Fall 2015), edited by Edward N. Zalta. Retrieved from http://plato.stanford.edu/archives/fall2015/entries/content-nonconceptual/.

    On the non-conceptual dorsal stream, see Norman 2002. A crucial aspect of the debate for philosophers remains whether and how such mental functions can further account for the rational role of perceptual states in belief formation, a topic we will not directly confront.

14. On the perceptual given, see O'Shaughnessy 2003. On the study of perception, see Palmer 1999. On the philosophy of perception, see Thompson 1995; Anscombe 1965; and the work of Burge 1986 and Dretske 1981, 1995. We explore a quasi-pictorial view of perception in regards to imagistic thinking, cf. Thomas 1999; Neisser 1976.

15. Costall 1995.

16. Turvey 1992. Compare direct perception with our discussion of modes of perception in Chapter 3 with the notion of "seeing as" in Thomas 1999, which may be a form of involuntary imagination.

17. Clancey 1997. Also see Hutto and Myin (2017) for a radical enactive position wherein mental states lack content entirely.
18. Tye 1984; quote from Marcel 1983. Cf. recent predictive processing models of mind.
19. Reed 1996; Costall 1995.
20. See Millikan (1984) on teleosemantics and Lewontin (1984) on niche symbiosis.
21. Bermúdez 1998.
22. Gibson 1979. Also, see Chemero (2009) for a discussion of the kinds of information relayed by features of the environment; see Fiebich 2014 on the types of affordances.
23. Chemero 2009.
24. Warren 1984.
25. See Loomis and Beall 1998.
26. Young and Wasserman 1997.
27. Carello and Turvey 2000. One more important research program is Jiang and Mark (1994), investigating the relationship between eye height and the perception of gap-crossing as affordances. The Center for the Ecological Study of Perception and Action at the University of Connecticut continues a tradition of incisive empirical and philosophical study on the mechanisms of direct perception.
28. Reed 1996. Affordances may become decoupled in the linguistic representational mind as explicit choices to act. In fact, as we discuss, perceptual affordances may be the space where executive function developed within symbolic imagination. Proust (2015) explores similar territory but differentiates between the roles of affordances and PPRs.
29. Gallagher 2008.
30. According to Bermúdez (1998), this cannot occur at the nonconceptual level; a creature must have the cognitive ability to represent before decoupled thought is possible. We disagree because we claim decoupling can occur in body grammar (see section V).
31. Cf. Griffiths and Scarantino in Robbins and Aydede 2009.
32. Clark and Grush 1999.
33. Jeffery et al. 2013.
34. An egocentric map is a map based around the body axis of one's self, and allocentric maps encode information about objects relative to each other. On the bodily self, see Panksepp's core SELF in Panksepp 1998b; Asma and Greif 2012; Damasio 2010.
35. Rescorla 2009; Millikan 2004; Sloman 1978. Although we would like to open dialogue on the question of nested affordances as an incipient form of predication.
36. Bennett 1996. We are making the case that these maps are in between cognition and proto-representational affective systems; nevertheless. "cognitive maps" is the common term used in the literature, and we will use it herein. Other kinds of cognitive maps can be delineated: paradigmatic cognitive maps seem to integrate both a bearing map (a multicoordinate grid that is both directional and distributed) and a sketch map that encodes and stores fine-grained topographical data con-

structed from memory of positions of unique discrete cues that are positional. There are also strip or broad maps, mosaic or grid maps, sketch or bearing maps, and integrated cognitive and navigational maps. Hutto and Myin (2017, 156; quoting Gladziejewski 2016) recount that maps represent by (1) structurally resembling features of some domain; (2) guiding the actions of their users; (3) doing so in detachable (i.e., offline) ways; and (4) allowing their users to detect representational errors.

37. Tolman 1948.

38. Gallistel 1990.

39. Localization includes dead-reckoning, the mechanism of integrating distance and direction when moving, such that animals (including ants) can make a straight-line return to the starting point (see Gallistel 1990). See Rescorla (2009) on the distinction between localization and piloting.

40. The underlying direct perception mechanisms that furnish information to create the cognitive maps are optic flow (*tau*) and internal proprioceptive acceleration detectors in the semicircular canals.

41. For more on the evolution and function of memory as short-form valence summary, see Klein et al. 2004; Gabriel 2007 and Russell 2003.

42. See Clayton and Dickinson 1998. Though the scrub jay is not a mammal, we think analogous processes arose in the mammalian clade. Also see Gallistel (1990) for provocative work on insect navigation.

43. Gallagher 2008.

44. Kensinger et al. 2007a, 2007b.

45. Luria 1972.

46. Panksepp 1998, 125. Though see Solms (2000) on the role of forebrain ventromedial structures in activating the dream state.

47. Other forms of involuntary imagination include perseverative intrusive thoughts (as observed in OCD and paranoid schizophrenia, for example) and drug-induced hallucinations. Another interesting variant is involuntary episodic memories as observed in PTSD. All of these examples are involuntarily affect-inducing or affect-induced.

48. Solms 2000.

49. Panksepp 1999.

50. O'Keefe and Nadel 1978; Ledoux 1996; Tulving 1983; Panksepp and Biven 2012. Furthermore, according to Solms (2000), normal spatial cognition has been found to be essential for dreaming.

51. Cf. Nader and Hardt 2009.

52. For fear conditioning, see Kindt et al. 2009; for procedural memory, see Walker et al. 2003; for episodic memory, see Hardt et al. 2010; Hupbach et al. 2013.

53. Damasio 1994.

54. Hume 1748; Freud 1900; Solms 2000; Asma 2017.

55. Maren 2001; Baddeley 1990; Tulving 2005; Klein et al. 2008; Szpunar 2010.

56. Cohen and Eichenbaum 1993.

57. Damasio et al. 1994; Damasio 2003.

58. Tronson and Tayor 2007; Hupbach et al. 2013. There is also a popular argument that memory and imagination share a common cognitive basis, see the work of K. K. Szpunar.

59. Fox et al. 2013.

60. *Theaetetus,* 152c1.

61. See Plato, *Sophist,* 263d6–8, for imagination containing true and false qualities; see 264b2 regarding a mixture of sense-perception and opinion.

62. Grönroos 2013.

63. See Aristotle, *De Anima,* 428a5–16.

64. See ibid., 428aa1–2.

65. In Kant's *Critique of Pure Reason* (1781), he suggests that imagination is the cognitive synthesizer. It is "the act of putting different representations together, and of grasping (*begreifen*) what is manifold in them in one [act of] knowledge" (CPR A 77; B 103). In his *Critique of Judgment* (1790), imagination takes a central role in aesthetic judgments. Additionally, the productive imagination actually generates new content in the form of novel creations (part 1, sect. 49).

66. Quoted in *The Descent of Man,* chap. III.

67. Asma 2017.

68. Barton 2012.

69. Ibid.

70. Gallagher and Zahavi 2012; Gallagher 2007; Clark 1997; Noë and O'Regan 2001.

71. Barton 2012.

72. Ibid.

73. Byrne et al. 2011.

74. Barsalou 1999, 2005, 2010.

75. Cf. Pacherie 2011.

76. Baars 1988.

77. The specificity of mental images is an old debate in philosophy. Without the advantage of contemporary experimental work, philosopher George Berkeley (1685–1753) made up his mind about mental images. Berkeley had seen a theory similar to Barsalou's in John Locke's philosophy, but he strenuously objected:

> Whether others have this wonderful Faculty of *Abstracting their Ideas,* they best can tell: For myself I find indeed I have a Faculty of imagining, or representing to myself the Ideas of those particular things I have perceived and of variously compounding and dividing them. I can imagine a Man with Two Heads or the upper parts of a Man joined to the Body of a Horse. I can consider the Hand, the Eye, the Nose, each by it self abstracted or separated from the rest of the Body. But then whatever Hand or Eye I imagine, it must have some particular Shape and Color. Likewise the Idea of Man that I frame to my self, must be either of a White, or a Black, or a Tawny, a Straight, or a Crooked, a Tall, or a Low, or a Middle-sized Man. (*Principles of Human Knowledge,* introduction, part X)

In short, Berkeley introspects his own consciousness and finds that he cannot picture abstract images, only specific ones. He can compose novel hybrids, but their parts are also highly specific. Therefore, he concludes, human beings cannot imagine general man, or general circle, or a generic prototype bird.

In response to Berkeley's skepticism, however, having an abstracted general image in one's mind is an empirical issue—Berkeley cannot picture an image generally, but many people think they can. Not much follows about the human mind from Berkeley's inability.

78. Cf. Biederman 1987 on geons as shape prototypes.

79. Kossyln 1973; Kosslyn et al. 2006; Pearson and Kosslyn 2013.

80. Carey (2011a) also suggests that core cognition "primitives" are likely iconic or imagistic in structure. Subsequent bootstrapping extrapolates the primitive images into linguistic domains, where they receive additional semantic and syntactical possibilities. We differ with Carey's interpretation of these primitives as "innate," although her use of that terminology is unclear. Her claim is that a representational *capacity* is innate, and not any specific content, but then builds certain contents into her examples. For some primitives, like "agency" and "number," she suggests that the input analyzers that identify their respective referents are not products of learning. She bases this on the fact that infants seem to have primitive representations (e.g., objects, numbers, and agents) before language and before nonlinguistic learning. We remain formally agnostic on this, but we suspect that even primitive representational templates emerge from a combination of experiential (ecological) stability in perception or from some form of iconic induction.

81. See Hume's *Dialogues* V, 170.

82. "Many have assumed that native mechanisms interpret and organize images (e.g., Kant, Reid). All [pre-computationalists] have assumed that images can be dynamic, not just static, representing events as well as snapshots of time." Barsalou 1999, 581.

83. *Daodejing,* chap. 78.

84. Bergen 2012.

85. Merleau-Ponty 1962.

86. Ibid., 124.

87. Bergen 2012, 13.

88. Asma 2017.

89. Cf. Mithen 2006; for conditioned gestural mimicry, see Sterelny (2012a) and on dreams see Panksepp (1998).

90. Marcus 2008.

91. *Descent of Man,* chap. 3.

92. Rao and Ballard 1999; Friston 2010; Seth et al. 2011; Clark 2016.

7. Language and Concepts

1. Human language, however, has a seemingly infinite potential for meaningful constructions because, as famously pointed out in Chomsky (1957), clauses can be added indefinitely in linear sequence, and clauses can be embedded inside other clauses. Humans can express highly complex states of affairs like, "I played a guitar

that was handmade in Texas, for my grandmother who lives in New York, in order that she might have some enjoyment, because her dog is dying." Animals cannot do that level of grammatically organized, meaningful information transmission. Clauses are embedded inside clauses, and yet it makes sense. The great apes that learned sign language, like Washoe and Nim Chimpsky, acquired and understood many signs, but they were never very competent with grammar.

Chomsky's explanation of human linguistic complexity is that a universal grammar is innate in humans. How else, he argues, can children acquire an infinitely workable system from finite childhood language stimulation? According to this innate theory, even the most language-rich childhood environments cannot give a child the staggering amount of associations necessary for competent language use, not to mention the syntactical rules for novel compositions. But even children in poor language-stimulation environments manage to acquire language competence. This "poverty of stimulus argument" leads nativists like Chomsky and Pinker (1994) to posit the existence of a genetically endowed brain capacity for shuffling and ordering signs into meaningful composites.

2. Most recently, see Berwick and Chomsky 2016, 11. In this book that summarizes the field, the authors break the problem of language into three parts: (1) an internal computational system for building hierarchically structured expressions with systematic interpretations, (2) a sensorimotor system for externalization, and (3) a conceptual system for inference, interpretation, planning, and the organization of thought. Our focus is largely on the third problem.

3. Anthropological linguists have noted how nativists tend to exaggerate the universal aspects of human languages, when, in fact, the speech sounds and, importantly, the grammars of world languages are more diverse than presumed. Additionally, competing empirical theories of how children acquire language from the bottom up (induction) are coming into greater focus now and hold out a nontrivial alternative to the nativist tradition (see Cowie 1999; Deutscher 2010).

Empirical research reveals a surprising ability for prelinguistic eight-month-olds (see Bates and Elman 1996). Babies are able to discriminate word patterns from the wash of parental auditory speech noise by sheer statistical repetition of sounds. We know that emotional emphasis on words can help infants hone in on words and relevant speech sounds, but even if you take away the affective component (as researchers did, using a monotone computer), babies are able to parse "words" from sheer frequency in the otherwise buzzing confusion of language sounds. Even when the "language" is nonsensical and artificial, babies will start to "master it" if the stimuli have some regularity of patterning. This throws doubt on the nativist theory, because it shows empirically how babies can quickly acquire, store, and maybe predict, language fundamentals through experiences of the rate of sound recurrence.

Finally, we may also note that we are sympathetic to other responses to the problem of language; for example, scientists like Svante Paabo and Michael Tomasello do not think of language as *sui generis*, but rather as one aspect of cultural learning—that set of abilities and propensities to share attention and learn complex things from others (Paabo 2014).

4. Bloom 2000; Carey 2011a, 2011b; Spelke 2010.

5. Spelke and Kinzler 2007.

6. Bloom further claims that these cognitive presets include a psychological tendency toward "essentialism"—the view that an object of experience has a defining essential nature that remains constant throughout qualitative changes and makes a thing what it is. And, Bloom suggests, we have a natural cognitive bias toward a metaphysical or Cartesian "dualism" that carves the world into physical, tangible stuff, and unseen immaterial souls or minds, inhabiting those physical forms. We're not convinced of either of these strong claims for innate cognitive biases, and we think Bloom may be misinterpreting his empirical findings (Bloom 2000; cf. Gabriel 2013), but the general thesis is intriguing and worth exploration: language is not an innate modular adaptation, but a byproduct of other innate cognitive tendencies.

7. Heidegger 1927.

8. Strum 2001; De Waal 1982.

9. Berwick and Chomsky claim that the purpose of language is not, in fact, communication, but rather as a tool for thought (Jerison 1973). While we agree that language is the crucial cognitive organ, its communicative functions remain central for the affective nature of the social mind. In addition to theories that describe the purpose of language as being for social grooming, gossip, and an internal mental tool, other major alternative theories explain the emergence of human language as an outgrowth of hunting cooperation, an outcome of motherese, a mode of sexual selection, a requirement of exchanging status information, a song, a requirement for or an outcome of toolmaking, an outgrowth of gestural systems, and a device for deception (Szamado and Szathmary 2006).

10. As we discussed in Chapter 5, *motherese*—cooing, babbling, and sing-song vocalizations—are universal human bonding methods. Babies learn early that some parent vocalizations are positive, some negative. We all spontaneously produce vocalizations that are emotionally loaded, like laughing or crying, but we are also conditioned to associate some vocalizations with extrinsic positive and negative accompaniments and consequences, like eating, sympathetic cooing, smiling, shouting, approval, disapproval, and so on.

   Positively charged affective vocalizations, like motherese, release oxytocin and generally improve the moods of the senders and receivers. An interesting study shows that increases in the mother's oxytocin accompany motherese verbalization encounters with the baby, whereas increases in the father's oxytocin accompany tactile and proprioceptive contact with the baby. This suggests a neurochemical correlation of positive bonding with some gender stereotypes of human parenting. See Gordon et al. 2010. Unlike other vertebrates, mammals care extensively for their young and other kin. As primates, we share important attachment mechanisms in the brain. Mammal mothers have a distinctive circuit from the hypothalamus, through the stria terminalis, to the ventral tegmental area (VTA), in which the neurotransmitter oxytocin travels. Damage to this system destroys maternal feeling and behavior. See this remarkable Youtube video for an example

of conversation before word acquisition: https://www.youtube.com/watch?v
=lih0Z2IbIUQ.

11. Dunbar 1998.

12. Berwick and Chomsky (2016) insist on a system called Merge as the locus for
recursion and push for a strong minimalist theory that describes the most efficient computational scenario underlying Merge.

13. Mithen 2006, 17.

14. Berwick and Chomsky 2016, 90.

15. Noble and Davidson 1994.

16. Berwick and Chomsky 2016.

17. Coen 2006.

18. Pagani et al. (2015) place the appearance of anatomically modern humans in
southern Africa around 200 KYA, up until before the last exodus from Africa,
approx. 60 KYA, but before 80 KYA.

19. Dennett 1991.

20. Lakoff 1987, 162. See also Dennett 1991 and Rorty 1979.

21. Baron-Cohen et al. 2013.

22. Jablonka et al. 2012, 2157.

23. Kahneman 2011.

24. For a poignant dramatization of what early hominid communication might have
been like, see *The Inheritors* by William Golding (1955).

25. One way to make this into an empirical question could be to study the ontogenetic development of spontaneous naming sounds in children. But even here, the
natural experiment would be distorted by the rich linguistic environment in
which every child already exists. Rhodes 1994.

26. Jablonka et al. 2012.

27. Dennett 1991.

28. The exception to this rule may be poetry, which returns to the emotionally rich,
imagistic use of language.

29. See the linguistic turn and a great deal of Continental philosophy in the twentieth century.

30. Ramnani 2006, 2012.

31. Baddeley et al. 1996.

32. A true teacher satisfies three criteria. First, the agent modifies its behavior in the
presence of a naïve observer. Second, the agent incurs some cost to this behavioral adjustment; and third, the naïve observer acquires the skill more rapidly than
in scenarios without the agent's example. See Caro and Hauser 1992.

33. Wittgenstein 1953; Quine 1960; Davidson 1967.

34. Putnam 1975, 269.

35. We agree with Laurence and Margolis (1999) that most concepts are sub-
propositional mental representations. But concepts should not be assumed to be
essentially lexical (i.e., corresponding to natural language terms), nor should they
be assumed to have definitional structure (i.e., having constitutive necessary and
sufficient conditions).

Besides this "classical theory of concepts"—in which a list of necessary and sufficient criteria form a containing set of tokens—most philosophers recognize four other competing theories of concepts: prototype concepts ("Structured mental representations that encode the properties that objects in their extension tend to possess" [Laurence and Margolis 1999, 31]); the theory-theory of concepts ("Concepts are representations whose structure consists in their relations to other concepts as specified by a mental theory" [ibid., 47]); the neo-classical theory ("Structured mental representations that encode partial definitions; i.e., necessary conditions for their application" [ibid., 54]); and conceptual atomism ("Lexical concepts are primitive; they have no structure" [ibid., 62]).

Conceptual atomism holds some attraction in the sense that it rejects a definitional structure for concepts and treats concepts as primitive (Fodor 1998). Some concepts will be composed or be the result of combinations of more elemental concepts, but these elements take their content from the world (the referent) directly, and not a relationship to other concepts. In this view, concepts are more like proper names that tag their objects or sets and lack other definitional structure. There is something attractive about this approach, especially if it can reconcile with the nonhuman animal mind. But again, proponents of conceptual atomism have worked on the problem with little consideration of biology or developmental learning. How unlearned non-inductive concepts appear in the mind—and properly reflect their referents—sounds Platonic at best and incoherent at worst. If conceptual atomism requires an extreme nativism (of innate ideas), then we cannot take it seriously. If conceptual atomism can be articulated in a developmental paradigm without such nativism, then we can appreciate the pre-propositional aspects of such a theory.

36. Our preference for prototypes hinges partly on the philosophical critiques of definitional concepts (e.g., Wittgenstein's famous point that "games" and many other concepts have no exceptionless definitions). Other philosophical critiques, like the semantic holism of Putnam (1975), as well as the ontological arguments of Dupré (1995) and others, have also rendered definitional concepts provincial.

Moreover, we are additionally swayed toward prototype concepts (and certain aspects of conceptual atomism), because definitional concepts (classical criteria models) lack psychological and phenomenological corroboration and appear to be discounted in some empirical studies. For example, Kintsch (1974) measured the time it took for subjects to process the meaning of simple versus complex concepts in sentence format. No processing lag was measured when subjects encountered concepts with greater definitional complexity. Complex concepts like "bachelor" or "convinced" (each containing definitional parts) were processed at the same speeds as those constituent parts. This may be because "definitions aren't psychologically real: The reason definitions don't affect processing is that they're not there to have any effect" (Laurence and Margolis 1999, 18).

37. Aristotle, *Rhetoric* II, chap. 2.

38. Seneca, *De Ira*.

39. Izard 2011.

40. Lurz 2006.
41. See Barrett 2006, 2014.
42. Barrett 2014.
43. Also, as we have argued in Chapter 1, affective systems are not modules (Gabriel 2012), and contrary to Barrett's assumption, arguments that throw doubt on the modular TOM do not discredit an affective TOM.
44. Adolphs (2017) shares some of our suspicions about Barrett's Conceptual Act Theory. In particular, he suggests that Barrett lacks a coherent concept of concepts.
45. Fodor 1975.
46. The spark of the prototype concept can be found in Wittgenstein (1953), then Rosch (1978), but it has been well-championed by Johnson and Lakoff (1980).
47. Piaget 1923; Mandler 1984. The schemata theory is helpful and compelling in general. Consider Rumelhart 1980, 41: "schemata can represent knowledge at all levels—from ideologies and cultural truths to knowledge about the meaning of a particular word, to knowledge about what patterns of excitations are associated with what letters of the alphabet. We have schemata to represent all levels of our experience, at all levels of abstraction. Finally, our schemata are our knowledge. All of our generic knowledge is embedded in schemata." Schemata are secondary-level processes par excellence, they serve as conditioned frames for thinking. Social psychology research on attitudes is also relevant in this context (Olsen and Fazio 2001; Gabriel 2007).
48. Johnson 1987; Rohrer 2006.
49. Boyer 2002.
50. Additionally, the example offered by Boyer (2002) is confusing, because "walrus" is not so much a concept, as an elaborate network of concepts.
51. Gobet and Lane 2012.
52. Hofstadter 2001.
53. Also see our discussion in Chapter 6 of the affordance of visual entropy in pigeons as a form of analogical reasoning embedded in perception; Young and Wasserman 1997.
54. Truppa et al. 2011. Also see Fagot and Thompson 2011.
55. Behrens et al. 2008; Heyes 2012.
56. Lakoff and Johnson 1980.
57. Hofstadter and Sander 2013.
58. Miller 1956.

8. Affect in Cultural Evolution

1. Bogucki 1999; Okasha and Binmore 2012.
2. Kahneman and Tversky 2000; Kahneman 2011; Sterelny et al. 2013.
3. Johnson and Earle 2000; Neitzel and Earle 2014; Carballo et al. 2014.
4. Silberbauer 1981.
5. Marshall 1976.
6. Read 2016.

7. Malinowski 1935; Weiner 1976.

8. Johnson and Earle 2000.

9. Meggitt 1965.

10. Boas 1927.

11. Abrutyn and Lawrence 2010. See also Yoffee 2005.

12. Abrutyn and Lawrence 2010.

13. Liverani 2006.

14. Rowe 1946.

15. Ibid.; Johnson and Earle 2000.

16. Rostovtzeff 1960.

17. Moffett 2013; Johnson and Earle 2000.

18. Barofsky 2013.

19. Johnson 1975; Engels 2010.

20. See the discussion of religion in Chapter 9.

21. DeVries 2008.

22. See Johnson and Earle 1987, diagram p. 31.

23. Earle 1998. See also Dye 2010.

24. See, for example, Suetonius's *Twelve Caesars,* 121 CE.

25. Ayling 2009, 2011.

26. For Pleistocene origins of mind, see Cosmides and Tooby 1995. For EEA, Environment of Evolutionary Adaptedness, see Bowlby 1969.

27. Panksepp 1998.

28. Gamble 1998.

29. Weber 1921.

30. Johnson and Earle 1987, 365.

31. *Republic* II, 359a. One of the "proofs" of this view is Glaucon's famous myth of the "ring of Gyges," later reconfigured and retold by J. R. R. Tolkien in the *Lord of the Rings* series. If you had a ring that turned you invisible, you would devolve, he suggests, into the worst sort of "free-rider." If the threat of detection and punishment is removed, one's naked self-interest ascends to dominance. You would be able to attain the best of all possible worlds (ibid., 359d).

32. Freud 1929.

33. *Republic* II, 369c–d.

34. Ibid., 372d.

35. Olson 1965.

36. See Axelrod 1984.

37. Ostrom 2000.

38. Ostrom 2015.

39. Ibid., 138.

40. Schwartz et al. 2002. See also Gigerenzer and Goldstein 1996.

41. Cockburn 2013, 223.

42. Sterelny 2013.

43. Kongzi / Confucius 2008, *The Analects,* 1.2.

44. Okasha and Binmore 2012.

45. Gintis 2012.

46. See Ospovat 1995.

47. Hutchinson and Gigenrenzer 2005; Houston 2009; McNamara and Houston 2009.

48. Asma 2013.

49. Frank 1988, 2005.

50. Gintis 2009.

51. Frank 1988.

52. See Güth et al.1982; Kahnemann et al. 1990.

53. Warnecken 2013. See also Richerson and Boyd (2001) on tribal instincts built on reciprocal altruism and kin selection.

54. See, for example, "Iceland Celebrations vs England," accessed September 14, 2017, from https://www.youtube.com/watch?v=PVq0MrmezpI.

55. An early but still trenchant discussion of multiple levels of nested biological causation can be found in Grene 1987.

56. Jensen et al. 2007.

57. Lieberman et al. 2007; Cosmides and Tooby 1992.

58. Hauser 2006.

59. See Haidt 2001; Greene and Haidt 2002.

60. Tomasello 2009.

61. For curiosity's cognitive closure impetus, see Loewenstein, 1994.

62. See Sayer 1992; Archer et al. 1998; Quine 1960; Grice 1961; Dewey 1938; Rosenberg 2004; Brown 1992.

63. Sterelny 2012b.

64. Ibid., 263.

65. Humphrey 1976; Dunbar 2001.

66. See Tomasello 2009; Jensen 2007.

67. Hrdy 2009; Panksepp 1998.

68. Kahneman 2011; Kosslyn and Miller 2013.

69. Zelazo and Cunningham 2007.

70. An interesting dissenting argument is made in Rosenthal (2008), who is unconvinced by the various adaptation theories of consciousness. The operations of executive control, enhanced planning, and so on, can go on just fine without a conscious light bulb over the top. Rosenthal concludes that consciousness is not an adaptation at all, but an epiphenomenal construction (or byproduct) of language. Social events and communication lead us to posit the existence of an inner consciousness to "explain" actions and speech by ourselves and others. According to Rosenthal, even the developmental Theory of Mind Mechanism—which purports to recognize the other's consciousness—is merely a projection of the capacity for talking-to-yourself into the head of another (but this is not consciousness either).

71. Adolphs 2009.

72. Kaplan et al. 2009.

73. Richerson and Boyd 2001.

74. Classic formulations of kin selection can be found in Westermarck 1894 and Hamilton 1963. For group selection, see Boyd and Richerson 1985; Batson 1991;

compare multilevel selection. For reciprocal altruism, see Trivers 1971, 2005. Also see Axelrod and Hamilton 1981.

Among other models, strong reciprocity has been put forward in Gintis et al. 2008 to accommodate laboratory findings that people tend to behave prosocially and punish antisocial behavior, at a cost to themselves, even when the probability of future interactions is extremely low, or zero. Cultural influence on social behavior is also a factor, as sociocultural arrangements provide a secondary niche (see Alexander 1987). Another factor is indirect reciprocity in dyadic relationships, where third parties hear about or observe an individual's level of cooperation (see Nowak and Sigmund 1998). In well-mixed populations, which are admittedly rare, network reciprocity is another possible mechanism (Nowak 2006). Sachs et al. 2004 provide a comprehensive model that includes directed reciprocation, shared genes, and byproduct cooperation.

75. Pagel 2012.
76. Nowak and Sigmund 1998.
77. Leimar and Hammerstein 2001.
78. Boyd and Richerson 1997.
79. See our discussion of social intelligence in Chapter 3; see also Bechara et al. 1994.
80. For emotions and the prisoner's dilemma, see Fehr and Gächter 2002.
81. McElreath et al. 2003.
82. Ibid.
83. Sanfey et al. 2003.
84. Pagel 2012; Leith and Baumeister 1998.
85. See Axelrod and Hamilton 1981; Trivers 1971.
86. Tomasello and Call 1997.
87. de Waal 1982.
88. Byrne and Whiten 1988; Humphrey 1976.
89. See our Introduction and Pagel 2012.
90. Panksepp 1998.
91. Damasio 2010.
92. James 1899; Mead 1913.
93. Panksepp 1998.
94. Dennett 1996b.
95. Panksepp 1998, 312.
96. Asma and Greif 2012.
97. Llinas 2001
98. Klein et al. 2004. For episodic memory, see Tulving 1983.
99. Also see Smaldino and Epstein 2015.
100. Klein 2009.
101. Boyd and Richerson 1997.
102. Ibid.
103. Festinger 1954. See also Tajfel 2003; Erickson and Nosanchuk 1998.
104. Wiessner 1983.
105. Wiessner 1984.

106. See Gabriel 2013; McCracken 1988; Leiss et al. 2005. Indeed, objects have some-what promiscuous social lives of their own (Appadurai 1986).
107. Mauss 1954.
108. Calasso 1988.
109. Calasso 1998, 2015.
110. Norwich 1997.
111. Finkel 2005.
112. Chagnon and Irons 1979; Borgerhoff and Mulder 1988.
113. Darwin 1871.
114. Austen 1813, chap. V.
115. Bonabeau et al. 1999; Chase 1974; Wilson 1975. Also see Bergman et al. 2003.
116. Chase et al. 2002.
117. There is evidence in mammals that the winner effect is related to elevated testosterone (Oyegbile and Marler 2005).
118. de Waal 1997; McElreath et al. 2003.
119. Henazi and Barrett 1999.
120. Tomasello 2000.
121. For chickens, see Landau 1951; for primates, see de Waal 1986; for human adolescents, see Savin-Williams 1979.
122. Kummer 1978; Mason 1978; Seyfarth et al. 1978; Stammback 1978; Vaitl 1978.
123. Chase 1982. Though see Bonabeau et al. 1999 for a revitalization of the intrinsic initial differences between individuals (i.e., the correlational model of Chase 1974).
124. Chase et al. 2002.
125. Mauss 1954.
126. Sahlins 1972; Frank 1988.
127. Malinowshi 1926; Thurnwald 1932; Gouldner 1960; Leakey and Lewin 1978.
128. Burger et al. 2009.
129. For social value exchanges, see Brown 1986; for reciprocity's function contra antisocial actions, see Keysar et al. 2008.
130. Cialdini 2001.
131. Diekmann 2004. Anterior cingulate cortex and striatal activations are correlated with cooperative social interactions in humans (Rilling et al. 2002), but fMRI does not penetrate to many parts of the brain implicated in the primary affective systems we have been discussing (see Panksepp 2012).
132. For the function of reciprocity in group-cooperation context, see Gouldner 1960; Wilson and Sober 1994; for reciprocity in an individual context, see Perugini et al. 2003.
133. Ridley 1996. Although there have been studies of social-grooming reciprocity in baboons—see Strum 2001; for bonobos, see de Waal 1989.
134. Wilson 1975, 562.
135. For reciprocity and morality, see Gintis et al. 2008; for reciprocity's connection to reasoning, see Pizarro and Bloom 2003; for reciprocity and empathy in primates, see de Waal 2006.
136. Haidt 2001; Wheatley and Haidt 2005; Knobe and Leiter 2006.

137. Geertz 1973.
138. Tomasello 2014.
139. See Bermudez 2003.
140. Tomasello 2014.
141. Compare Dennett 1991.
142. Grafenhain et al. 2009.
143. Tomasello 2014, 152.
144. See Ekman and the Dalai Lama's Atlas of Emotions website, accessed September 14, 2017: http://www.paulekman.com/atlas-of-emotions/#states:enjoyment.
145. Ferrada and Camarinha-Matos 2012.
146. Tomasello 2014.
147. Harris 1975, 1991.
148. See Johnson and Earle 2000; Diamond 1997; Sokoloff and Engerman 2000.
149. Darwin 1871, chap. 4.
150. Cook et al. 2014.
151. de Waal 2006.
152. Philosopher Jesse Prinz (2011) offers a compelling conceptual clarification of empathy. It is often assumed to be a species of prosocial care, but it may be an intrinsically neural mirroring ability. Perhaps empathy is not an emotion per se, but an ability to share or mirror an emotion. On this account, empathy is the mirror itself and it simply reproduces whatever comes before it. It comes with no preset value settings or normative content. As such, empathy would be a precondition for downstream evaluations (e.g., moral reasoning and behavior), but not the font of good will.
153. Darwin 1871, chap. 4.
154. See Hoogland 1995 for a dissenting view.
155. Gould (1979, 255) continues, "For example, in most sexually reproducing organisms, an individual shares (on average) one-half the genes of his sibs and one-eighth the genes of his first cousins. Hence, if faced with a choice of saving oneself alone or sacrificing oneself to save more than two sibs or eight first cousins, the Darwinian calculus favors altruistic sacrifice; for in so doing, an altruist actually increases his own genetic representation in future generations."
156. See the helpful discussion of group selection, especially as it helped to create the human cooperation explosion, in Chapter 7 (Sterelny 2003).
157. See O'Wolff and Sherman 2007. Also see Hoogland 1995.
158. In the spring of 2011, a debate emerged in the pages of *Nature* surrounding a recent volte-face by Edward O. Wilson, one of the earliest and best popularizers of kin selection. Wilson has changed his mind about the power of kin selection to explain altruism. He now favors a distinct causal mechanism called "group selection" (sometimes confused with kin selection) that operates in social creatures who are not necessarily related. In other words, he reverses the order of selection. Creatures that live in collectives evolved cooperative behaviors first (based only on group membership), and later this cooperation intensified for genetically related kin groups. The majority of the scientific community has roundly rejected Wilson's new suggestion, but it is still a live issue. For Wilson's controversial paper,

see Martin A. Nowak, Corina E. Tarnita, and Edward O. Wilson, "The Evolution of Eusociality," *Nature* 466 (August 26, 2010). For scathing critiques, see *Nature* 471 (March 24, 2011).

159. One thing that seems clear from the biology of bias is this: we don't come into the world as selfish Hobbesian mercenaries. Contrary to the usual pessimistic contract theory, we mammals don't start out as self-serving egotistical individuals who then need to be socialized (through custom, reason, and law) to endure the compromises of tribal living. Rather, we start out in a sphere of emotional-chemical values—created by family care—in which feelings of altruistic bonding are preset before the individual ego even extricates itself. Matt Ridley assumes the more pessimistic social-contract view of "egocentric individual first" and social cohesion later, in his *The Origins of Virtue* (Ridley 1996). He recognizes an alternative "origin story"—a communist utopia of "all for all" (instead of "all against all"). In this collectivist utopian view (still popular on the far left), selfishness is a fall from grace, brought on by protocapitalist inequities in the distribution of goods. Ridley lampoons this, and maybe he's right to do so. But our view is different. The family is the small tribe at the very origin of each human's development, so it is before *both* the egotist and the altruist (i.e., it is not derivative). Family is the original tribe that acts as the nursery for hatching out the subsequent values (selfishness and selflessness).

160. Freud's famous assumption *Homo homini lupus* ("Man is a wolf to other men") tries to posit not only the isolated aggressive interests of the early ego phase but also the artificial nature of morality itself. Biologist Frans de Waal calls this the "veneer theory" because "it sees morality as a thin layer barely disguising less noble tendencies." De Waal contrasts this veneer theory with the Darwinian view that morality is a "natural outgrowth of the social instincts, hence continuous with the sociality of other animals." The story we've been telling about ancient emotional kin bonding dovetails with this Darwinian view that moral sentiments are early and foundational for both individual and species ethics. See Frans de Waal's Tanner Lecture, included as part 1 of De Waal 2006.

161. The Romans used to say "*do ut des*" or "I give, so that you will give." But they also argued that giving (at least for humans) is its own reward. The discussion of the intrinsic aspect of altruistic behavior is not very popular in contemporary ethical discussions, but it was extremely important in earlier ethical philosophy. The ancients would not be able to go from monkey justice to egalitarian laws without some serious consideration of the soul's or the psyche's health. The cost-benefit analysis of benevolent action was well known to them, but they argued for a more primordial reason for good deeds—namely, that they are pleasing to the soul (they are their own reward). Seneca says that he doesn't give favor to his friends because he wants favors returned "in a circle." Echoing Greeks like Plato and Aristotle, Seneca says that virtuous giving is its own reward. He explains that "the wages of a good deed is to have done it. I am grateful, not in order that my neighbor, provoked by the earlier act of kindness, may be more ready to benefit me, but simply in order that I may perform a most pleasant and beautiful act;

I feel grateful, not because it profits me, but because it pleases me." See Seneca 1962.

162. Rappaport 1979.

163. Darwin, 1871, chap. 4.

164. Boyer 2002, 187.

165. Tangney and Dearing 2003.

166. Tronick and Cohn 1989.

167. Gilbert 1998.

168. Compare Thaler and Sunstein 2008.

169. Mauss 1930.

170. Sahlins 1972.

171. Wiessner 2005.

172. Ibid., 122.

173. Gintis et al. 2001.

174. See de Quervainet al. 2004.

175. Brosnan and de Waal 2003.

176. Also see Darby Proctor's experiments with the ultimatum game, which are fascinating but over-interpreted to be indicators of primate fairness. We could just as easily read these experiments as evidence of Confucian magnanimity or Aristotelian megalopsuchia (or even Nietzschean pagan virtues!) in chimpanzees. See Proctor et al. 2013.

177. Panksepp and Biven 2012; Berridge 2003b.

178. Wiessner 2014.

179. See video "CARTA: Violence in Human Evolution—Polly Wiessner: Violence: What's Culture Got to Do with It?," https://www.youtube.com/watch?v=_ITnD 8oz0sI, accessed September 12, 2017.

180. Wiessner 2014, 14027

181. Kaplan et al. 2009.

182. Ibid.

183. The Massai Association, an NGO based in Bellevue, Washington, and Kajiado Central District of Kenya offers a description of the enkipaata and emuratare rituals at: http://www.maasai-association.org/ceremonies.html, accessed September 9, 2017.

184. Llewellyn-Davis 1981; see also Talle 1988.

185. See Kibutu 2006.

186. Casebeer 2003.

187. Damasio 1994, 2003.

188. Satpute and Leiberman 2006; Lebreton et al. 2009.

189. Hare et al. 2009; Plassman et al. 2012.

190. On the restraint model, see Lohse 2017; Martinson et al. 2012; on "deliberating," see Capraro et al. 2016, 2017.

191. See Wills et al. 2018. Also see Wout et al. 2005 and Knoch et al. 2006.

192. Rand et al. 2012.

193. Camus et al. 2009.

194. See Ravosa and Dagosta 2007.

195. TOM is no doubt a crucial aspect of our contemporary moral lives, but its presence in Pleistocene normative life is unclear. Baron-Cohen (1999) speculates that TOM emerges around the time of the phylogenetic split of great apes from monkeys (circa 6 MYA). We suspect the origins of true TOM are much more recent. Baron-Cohen ascribes many downstream capacities to TOM, including communication repair, teaching, intentional deception, pretending, moral empathy, and more. We suspect that an earlier associational system or mirror neuron system (like those found in the PFC region of macaque monkeys) may be the system that allowed us to create inner simulations of other's behaviors, eventually giving rise to full-blown TOM during the Pleistocene. Others point out the importance of Machiavellian cognitive concern for reputation as a prerequisite for moral motivations (Sperber and Baumard 2012).

Last on the scene would have been abstract reasoning. Abstract inductions can help us create normative rules, and individual cases can be compared against such rules. Deductions and new insights can derive from careful consideration of nemocentric principles. But all this is a relatively recent type of cognition, and is unhelpful in understanding the evolution of social life. Additionally, psychologists and philosophers have also recently thrown doubt on such cognition to explain even contemporary moral judgment (Haidt 2001; Asma 2012). It is possible, however, that life-boat scenarios, trolley car dilemmas, and such can be imagined and help us guide our normative commitments. And these forms of artificial contextualized normative thinking may be closer to the kinds of ancient imaginative simulations that our ancestors engaged in during firelight talk and other cultural rituals.

9. Religion, Mythology, and Art

1. Tylor 1899, 1.
2. James 1902, 51.
3. Kant 1790.
4. For interested readers, there is no better place to start than William James's chapter on mysticism in his *Varieties of Religious Experience* (1902). See also Saarinen 2015.
5. Koestler 1954, 350.
6. Nielsen et al. 2013; Vollenweider et al. 1999; Vollenweider 2001.
7. Ehrenzweig 1967; Storr 1989; Fauteux 1995; Rooney and McKenna 2007; Comte-Sponville 2007; Harrison 2006.
8. Vollenweider et al. 1999.
9. Vollenweider 2001.
10. Francis and Loendorf 2002.
11. Csikszentmihályi 1990.
12. Ehrenzweig 1967; Milner 1987; Newton 2008; Krausz 2009.
13. Malraux 1964.
14. Gombrich 1951, 1; Dutton 2009, 51–58.
15. Boas 1927.
16. Leder and Nadal 2004, 2014.

17. Leder and Nadal 2014, 445. Also see the following reviews of the neuropsychological literature on aesthetic production and appreciation: Bäzner and Hennerici 2006; Bogousslavsky 2005; Chatterjee 2003; Miller and Hou 2004; Zaidel 2005.
18. Konecni 2005, 31.
19. Barker (1968) and Heft (2007) emphasize the importance of place as a behavioral setting that relates actions to milieu in an ecological psychology framework. Rietveld and Kiverstein (2014) and Withagen et al. (2012) expand on the importance of design for soliciting action in our sociocultural niche.
20. Dutton 2009.
21. Tooby and Cosmides 2001; Carroll 1999.
22. Boyer 2002.
23. Chatterjee 2003; Davies 2012.
24. Andersen 1979.
25. Compare to Hertzberger 1991, lessons for students in architecture.
26. Schiller 1971; Smith 1973.
27. Tylor 1899.
28. Harris 1975, 583.
29. Zaidel 2007.
30. Stout 1971, 32. See also Heft (2010) on the idiosyncratic and intrapsychic nature of environmental experience.
31. d'Azevedo 1958; Durkheim 1938; Radcliffe-Brown 1935; Boas 1927.
32. Geertz 1983, 119–120.
33. White 2003.
34. Flood 1990.
35. Lewis-Williams and Challis 2011.
36. Andersen 1979.
37. McBrearty and Brooks 2000.
38. Clottes 1996; Vialou 1996.
39. Pettitt and Bader 2000.
40. Clottes 1996.
41. White 2003.
42. Leroi-Gourhan 1968; Clottes 1996; Vialou 1996. Adornment, which we discussed in Chapter 8, as a form of social identity construction in such local-level groups in the Upper Paleolithic, is usually made of materials like ivory, soapstone, mother-of-pearl, or dental enamel that are distinct from the bone and antler used for practical functions and thus may have been perceived as having spiritual qualities (White 2003).
43. Lewis-Williams and Challis 2011. Notably, Eland blood was a highly desired substance for mixing paints used in the rock art, due to its sparkling appearance (How 1962). Also, cf. sympathetic magic below.
44. Flood 1990.
45. Lewis-Williams and Challis 2011.
46. Lewis-Williams 2002.
47. Boyer and Ramble 2010.

48. Boas 1927.
49. Bataille 2005.
50. Bégouen 1929.
51. Boas 1927; Wingert 1962.
52. Wingert 1962.
53. Lewis-Williams and Challis 2011.
54. As opposed to the conscious elision of the two during the Italian Renaissance (Burckhardt 1860).
55. Lorblanchet 1995; Dauvois and Boutillon 1994.
56. Whitely 2009; Leroi-Gourhan 1968.
57. Leroi-Gourhan 1968; Vialou 1996.
58. Lewis-Williams and Challis 2011.
59. Bégouen 1929; Lewis-Williams et al. 1988; Clottes and Lewis-Williams 1998.
60. Whitley 2009.
61. Ibid.
62. Lewis-Williams 1995, 65.
63. Conkey 2009.
64. Clottes and Lewis-Williams 1998.
65. Lewis-Williams and Challis 2011.
66. Clottes 1996.
67. Ibid.
68. Ibid.
69. Lewis-Williams and Dowson 1988.
70. Lewis-Williams 2010.
71. Clottes and Lewis-Williams 1998.
72. Orpen 1874; Lewis-Williams 2010.
73. See Lewis-Williams 2010.
74. Whitley 2000.
75. Eliade 1964.
76. Whitley 2009.
77. Ibid.
78. Ibid.
79. Andreasen 1987; Richards et al. 1988; Simmora et al. 2005.
80. Hershman and Lieb 1994.
81. Boas 1927.
82. Boyer 2002.
83. See Haidt 2003.
84. Bell 1914; Carroll 2005.
85. Barbara Stoler Miller, trans. 1986.
86. Ibid.
87. Cinema takes on these themes very effectively, for example in Michael Haneke's *Amour* (2012) or Alain Resnais's *Hiroshima, mon amour* (1960).
88. Both poems translated by Robin Skelton.
89. Elizabeth T. Gray, trans., *Ghazal* 1995, 18.
90. Vygotsky 1962; Mitchell 1981.

91. Bruner 1991.

92. Booker 2004.

93. Leach 1969; Lévi-Strauss 1967.

94. Lewis-Williams and Challis, 2011.

95. Norenzayan et al. 2016.

96. White 2003; Gray et al. 2001; Tramo 2001.

97. See Snyder 2000.

98. See Sachs et al. 2015.

99. Boyer 2002; Atran 2002a; Norenzayan 2013.

100. Slingerland and Sullivan 2016.

101. Mumford 1961.

102. Panchanathan and Boyd 2003.

103. Diamond 1997.

104. Norenzayan et al. 2016.

105. D. Wilson 2002.

106. Soltis and Richerson 1995.

107. Barrett 2012; Atran 2002; Boyer 2002.

108. Boyer and Ramble 2010.

109. Sosis and Alcorta 2003. And see D. Wilson 2002.

110. For example, Boyer 2002.

111. Diamond 1997.

112. Asma 2018; Gabriel 2017; De Superstitione (unpublished manuscript).

113. Hurley and Chater 2005.

114. James 1879, 319.

115. Compare to Sterelny 2013.

116. Hrdy 2009.

117. See Wynn and Bloom 2014. Also see Warneckan 2013 and F. Warnecken and Tomasello 2006.

118. Asma 2012. Laura Hillenbrand's book *Unbroken* (Hillenbrand 2010) gives us a glimpse into the very local bonds of affection that fueled WWII American soldiers as they fought overseas—or in the case of Louie Zamperini (the subject of *Unbroken*), as they languished in prisoner of war camps. "On an October afternoon, Louie stepped out of an army car and stood on the lawn at 2028 Gramercy Avenue, looking at his parents' house for the first time in more than three years. 'This, this little home,' he said, 'was worth all of it.'"

119. Norenzayan et al. 2016; Boyer 2002; Atran 2002; Norenzayan 2013.

120. Buddhism as a factor of culture and development in Cambodia, in *Cambodia Report* (a publication of the Center for Advanced Study) II (2) March–April, 1996.

121. Sosis 2000.

122. See "Statistics" at the Philanthropy Roundtable website: at http://www.philan thropyroundtable.org/almanac/statistics/, accessed September 9, 2017.

123. Verification in religion should not be understood on the scientific model. The religious believer makes a sacrifice to the spirits, for example, motivated by hope or fear. If the wish is granted, it stands as verification. If the wish is not granted, it still stands as verification, only the failure is accounted for by the devotee's

impurity or error. The devotee believes in the existence of the unverified spirit in the same way that some of us believe in bacterial food-poisoning or quarks. We experience the effects but not the responsible entities. But, one protests, a scientist (a specialized member of the community) can verify the existence of salmonella and quarks. The devotee agrees that his shaman (another specialized member of the community) can also verify the existence of the spirits. Granted, these are unimpressive forms of "verification" when compared with the hypothetico-deductive model, but for the religious community, they do act as daily warrants or justifications for supernatural beliefs.

Even members of prosperous, well-educated communities commit regularly to unverifiable entities and pseudo-scientific systems, devoting real resources and time. Half of all Americans, for example, believe in extra-sensory perception (ESP) and ghosts. All this suggests that it is extremely easy to commit to unverifiable entities, and religion is only a subspecies of this tendency.

Additionally, there is another kind of "verification" that makes special sense in supernatural domains; in particular, "emotional verification." When the suffering believer desperately needs psychological consolation and engages in prayer or devotional rites that resemble adjunctive behaviors and then feels better, this stands as confirmatory evidence or verification of the entities in question.

124. Dijksterhuis 1961.

125. Religion helps people manage their emotional lives. While it may not do very much for skeptics, we would be very inattentive if we failed to notice how much relief and comfort religion gave to other people. No amount of scientific explanation or sociopolitical theorizing is going to console the mother of a murdered boy, for example. But the irrational hope that she will see her murdered son again can sustain her, and such an emotional belief may give her the energy and vitality to continue caring for her other children. Effectively securing the success (i.e., health, wealth, and procreative eligibility) of one's offspring (via elaborated CARE processes) is the very condition of evolutionary fitness for humans, so we can imagine a selective pressure for such emotional beliefs.

People who critique such emotional responses and strategies with the refrain "But is it true?" are missing the point. Most religious beliefs are not true. But the emotional brain does not care about verisimilitude. The limbic system does not operate on the grounds of true and false. An emotion is not a representation or a judgment, so it cannot be evaluated like a theory. Emotions are not true or false. Even a terrible fear inside a dream is still a terrible fear. This means that the criteria for measuring a healthy theory is not the criteria for measuring a healthy emotion. Unlike a healthy theory—which must correspond to empirical facts—a "healthy emotion" might be one that contributes to neurochemical homeostasis or other affective states that promote biological flourishing. The intellectual life answers to the all-important criterion: Is this or that claim *accurate?* But the emotional life has a different master. It answers to the more ancient criterion: Does this or that feeling help the organism *thrive?* Often, an accurate belief also produces thriving (How else could intelligence be selected for?), but frequently there is no such happy correlation. Mixing up these criteria is a common category mistake that fuels a lot of the theist / atheist debate.

126. Asma 2006.

127. Darwin 2004, 682.

128. Barnes and Bloor 1982; Feyerabend 1999.

129. Scruton 2004, 10.

130. Grandin, "Why Do Kids with Autism Stim?" *Autism / Asperger's Digest,* November / December 2011, http://autismdigest.com/why-do-kids-with-autism -stim/.

131. Fibiger and Phillips 1986.

132. Brown and Jenkins 1968.

133. See Sheffield 1965; Williams and Williams 1969.

134. Panksepp 2012, 115; Burghardt 1973, 1999.

135. Phelps and Gazzaniga 1992; Gabriel 2007.

136. Panksepp 1998, 116.

137. Traditional: *A Shelter in the Time of Storm.*

138. Cieri et al. 2014.

139. Hare et al. 2012.

140. Tan and Hare 2013.

141. Strum 2001.

142. The Sapolsky quotes are drawn from the interview "When Monkeys Make Up," http://incharacter.org/archives/forgiveness/when-monkeys-make-up-chimps -are-from-mars-bonobos-are-from-venus/, accessed September 1, 2017.

143. Ibid.

144. See, for example, Bosch and Young 2017.

145. DeSteno 2018.

146. Casey et al. 2011.

147. See, for example, Bataille 1957.

148. Bellah 2011.

149. Piaget and Inhelder 1969.

150. Ibid., 12.

151. James 1902.

152. Davidson 2002. Also see Pollan (2018) for an exploration of microdosing and the oceanic feeling for a popular audience.

153. Damasio 2010.

154. Dennett 1992.

155. Wright 2018.

156. Freud 1927.

157. Seabright 2013.

158. Norenzayan et al. 2016.

159. Obviously in Christianity, there is a parallel narrative that sees the suffering of Jesus as the ultimate enemy-killer—redeeming the world from sin through his suffering. But, while this is a familiar version to us now, we must recognize that it made little sense to the pagan cultures (of the Mediterranean and Northern Europe). The idea that one "wins" (righteousness) by "losing" (undergoing

suffering) was and remains very paradoxical to the cultures of strength, loyalty, and power.

160. See Orchard 1995, chap. IV, in *Pride and Prodigies: Studies in the Monsters of the Beowulf-Manuscript* (D. S. Brewer, 1995).

161. From "Alexander the Great's Journey to Paradise" in Richard Stoneman's edited volume *Legends of Alexander the Great* (The Everyman Library, 1994).

162. We are skeptical about a universal tendency in human history toward peace and progressive liberalism (Shermer 2015; Pinker 2011). Hierarchy is deeply rooted in our phylogeny, and egalitarianism may be the recent evolutionary arrival (Boehm 1993, 1999). Given our deep Enlightenment cultural values of liberal individualism, it is difficult to imagine the larger good that comes from non-egalitarian social hierarchies. But we should not confuse the universal human aversion to bullying as tantamount to universal egalitarian preference (Boehm 1999). Cultures with emphasized rank structures do provide benefits to their members and not just to the elites. "The existence of hierarchy may reduce competition so that those lower down in the hierarchy can devote their energies to enterprises other than social or political competition. Hierarchies may thus benefit the majority" (Wiessner 2009, 197). Big Man chief systems and timocracies can also mediate internal conflict well, provide social stability, and respond quickly to emergencies. Hierarchic groups also function well as a collective when competition with other outgroups becomes intense.

Both egalitarian and hierarchic cultures can be adaptive and provide fitness, but both are fraught with constant internal tensions. Elites on the top of a ranked society can abuse their power and hoard resources, but egalitarian societies often punish ambitious and exceptional individuals with harsh scrutiny and leveling (Kelly 1993; Lee 1993; Wiessner 2009). It is important that we do not romanticize early forms of human society and culture as "noble savage" egalitarians, with hierarchies supposed to emerge only later.

163. Noë 2005, 29 and 30.

References

Abrutyn, Seth and Kirk Lawrence. 2010. From chiefdom to state: Toward an integrative theory of the evolution of polity. *Sociological Perspectives* 53 (3): 419–442.

Adolphs, Ralph. 2009. The social brain: Neural basis of social knowledge. *Annual Review of Psychology* 60: 693–716.

———. 2017. Reply to Barrett: Affective neuroscience needs objective criteria for emotions. *Social Cognitive and Affective Neuroscience* 12 (1): 32–33.

Ainsworth, M. D. S., M. Blehar, E. Waters, and S. Wall. 1978. *Patterns of attachment: A psychological study of the strange situation.* Oxford: Lawrence Erlbaum.

Alexander, R. 1987. *The biology of moral systems.* New York: Aldine de Gruyter.

Allis, C. D, T. Jenuwein, D. Reinberg, and M. L.Caparros, eds. 2007. *Epigenetics.* Cold Spring Harbor, NY: Cold Spring Harbor Laboratory Press.

Alves, Emily, Andrea Fielder, Nerelle Ghabriel, Michael Sawyer, and Femke T. A. Buisman-Pijlman. 2015. Early social environment affects the endogenous oxytocin system: A review and future directions. *Frontiers in Endocrinology* 6: 32.

American Psychiatric Association. 2000. *Diagnostic and statistical manual of mental disorders,* 4th ed., text rev. Washington, DC: American Psychiatric Association.

Amundson, Ron. 2007. *The changing role of the embryo in evolutionary thought: Roots of evo-devo.* New York: Cambridge University Press.

Andersen, Peter A. and Laura K. Guerrero. 1998. Principles of communication and emotion in social interaction. In P. A. Andersen and L. K. Guerrero, eds., *Handbook of communication and emotion: Research, theory, applications, and contexts,* 49–96. Cambridge, MA: Academic Press.

Andersen, Richard L. 1979. *Art in primitive societies.* New York: Prentice-Hall.

Anderson, John R., and Gordon H. Bower. 1972. Recognition and retrieval processes in free recall. *Psychological Review* 79 (2): 97–123.

365

Andreasen, Nancy C. 1987. Creativity and mental illness: Prevalence rates in writers and their first-degree relatives. *American Journal of Psychiatry* 144 (10): 1288–1292.

Anscombe, G. E. M. 1965. The intentionality of sensation: A grammatical feature. In Ronald J. Butler, ed., *Analytic philosophy,* 158–180. Hoboken, NJ: Blackwell.

Appadurai, Arjun. 1988. *The social life of things: Commodities in cultural perspective.* Cambridge: Cambridge University Press.

Aristotle. 1942. *Generation of animals.* Loeb Classical Library. A. L. Peck, trans. Cambridge, MA: Harvard University Press.

———. 1987. *De anima.* Penguin Classics. Hugh Lawson-Tancred, trans. London: Penguin.

———. 1999. *Metaphysics.* Penguin Classics. Hugh Lawson-Tancred, trans. London: Penguin.

———. 2002. *On the parts of animals.* Clarendon Aristotle Series. James G. Lennox, ed. New York: Oxford University Press.

———. 2008. *Physics.* Oxford World's Classics. David Bostock, ed., Robin Waterfield, trans. New York: Oxford University Press.

Arnold, Angela, Randye J. Semple, Ivan Beale, and Claire M. Fletcher-Flinn. 2000. Eye contact in children's social interactions: What is normal behaviour? *Journal of Intellectual and Developmental Disability* 25 (3): 207–216.

Arnold, Dan. 2012a. *Brains, buddhas and believing: The problem of intentionality in classical Buddhist and cognitive scientific philosophy of mind.* New York: Columbia University Press.

———. 2012b. Reaching bedrock: Buddhism and cognitive-science. *Berfrois,* April 27. www.berfrois.com/2012/04/dan-arnold-buddhism-cognitive-science/.

Aron, Elaine N., Arthur Aron, and Kristin M. Davies. 2005. Adult shyness: The interaction of temperamental sensitivity and an adverse childhood environment. *Personality and Social Psychology Bulletin* 31 (2): 181–197.

Ashenfelter, Kathleen T., Steven M. Boker, Jennifer R. Waddell, and Nikolay Viranov. 2009. Spatiotemporal symmetry and multifractal structure of head movements during dyadic conversation. *Journal of Experimental Psychology* 35 (4): 1072–1091.

Asma, Stephen T. 1996. *Following form and function: A philosophical archaeology of life science.* Evanston, IL: Northwestern University Press.

———. 2006. *The gods drink whiskey.* San Francisco: HarperOne.

———. 2012. *Against fairness.* Chicago: University of Chicago Press.

———. 2013. Families made us human: The evolution of emotional modernity. *Aeon Digital Magazine,* November 7.

———. 2017. *The evolution of imagination.* Chicago: University of Chicago Press.

———. 2018. *Why we need religion.* New York: Oxford University Press.

Asma, Stephen, and Thomas Greif. 2012. Affective neuroscience and the philosophy of the self. *Journal of Consciousness Studies* 19 (3 / 4): 6–48.

Atran, Scott. 2002a. *In gods we trust: The evolutionary landscape of religion.* New York: Oxford University Press.

―――. 2002b. Modest adaptationism: Muddling through cognition and language. *Behavioral and Brain Sciences* 25 (4): 504–506.

Austen, Jane. 1813. *Pride and prejudice.* Dover Thrift ed.

Axelrod, Robert. 1984. *The evolution of cooperation.* New York: Basic Books.

Axelrod, Robert, and William D. Hamilton. 1981. The evolution of cooperation science. *New Series* 211 (4489): 1390–1396.

Ayling, Julie. 2009. Criminal organizations and resilience. *International Journal of Law, Crime and Justice* 37 (4): 182–196.

―――. 2011. Gang change and evolutionary theory. *Crime, Law and Social Change* 56 (1): 1–26.

Baars, Bernard. 1988. *A cognitive theory of consciousness.* Cambridge: Cambridge University Press.

Baddeley, Alan D. 1990. *Human memory: Theory and practice.* Needham Heights, MA: Allyn and Bacon.

Baddeley, Alan, Sergio Della Sala, and T. W. Robbins. 1996. Working memory and executive control [and discussion]. *Philosophical Transactions of the Royal Society London B* 351 (1346): 1397–1404.

Baeyens, Frank, Debora Vansteenwegen, Dirk Hermans, and Paul Eelen. 2001a. Chilled white wine, when all of a sudden the doorbell rings: Mere reference and evaluation versus expectancy and preparation in human Pavlovian learning. In F. Columbus, ed., *Advances in Psychology Research,* 4:241–277. Huntington, NY: Nova Science.

―――. 2001b. Human evaluative flavor-taste conditioning: Conditions of learning and underlying processes. *Psychologica Belgica* 41 (4): 169–186.

Bailey, Ida E., Julia P. Myatt, and Alan M. Wilson. 2013. Group hunting within the carnivora: Physiological, cognitive and environmental influences on strategy and cooperation. *Behavioral Ecology and Sociobiology* 67 (1): 1–17.

Bain, Alexander. 1868. *Mental science: A compendium of psychology, and the history of philosophy.* New York: D. Appleton and Co.

Ball, Philip. 2015. The strange inevitability of evolution. *Nautilus* 20, January 8. http://nautil.us/issue/20/creativity/the-strange-inevitability-of-evolution.

Bandura, Albert 1988. Self-regulation of motivation and action through goal systems. In V. Hamilton, G. H. Bower, and N. H. Frijda, eds., *Cognitive Perspectives on Emotion and Motivation,* 37–61. Dordrecht: Kluwer Academic.

Barker, Roger G. 1968. *Ecological psychology: Concepts and methods for studying the environment of human behavior.* Stanford, CA: Stanford University Press.

Barnes, Barry, and David Bloor. 1982. Relativism, rationalism and the sociology of knowledge. In Martin Hollis and Steven Lukes, eds., *Rationality and Relativism.* Cambridge, MA: MIT Press.

Barofsky, N. 2013. *Bailout: How Washington abandoned Mainstreet while rescuing Wall Street.* New York: Free Press.

Baron, Rueben M. 1980. Contrasting approaches to social knowing: An ecological perspective. *Personality and Social Psychology Bulletin* 6 (4): 591–600.

Baron-Cohen, Simon. 1999. The evolution of a theory of mind. In Michael Corballis and Stephen E. G. Lea, eds., *The descent of mind: Psychological perspectives on hominid evolution,* 261–277. New York: Oxford University Press.

Baron-Cohen, Simon, Jane Allen, and Christopher Gillberg. 1992. Can autism be detected at 18 months?: The needle, the haystack, and the CHAT. *British Journal of Psychiatry* 161 (6): 839–843.

Baron-Cohen, Simon, Helen Tager-Flusberg, and Michael V. Lombardo, eds. 2013. *Understanding other minds: Perspectives from developmental social neuroscience,* 3rd ed. New York: Oxford University Press.

Barrett, H. Clark, and Robert Kurzban. 2006. Modularity in cognition: Framing the debate. *Psychological Review* 113 (3): 628–647.

Barrett, Justin L. 2012. *Born believers: The science of childhood religion.* New York: Free Press.

Barrett, Lisa Feldman. 2006. Are emotions natural kinds? *Perspectives on Psychological Science* 1 (1): 28–58.

———. 2014a. The conceptual act theory: A précis. *Emotion Review* 6 (4): 292–297.

———. 2014b. Why emotions are situated conceptualizations. Interview with Andrea Scarantino in *Emotion Researcher: International Society for Emotion Research's Sourcebook for Research on Emotion and Affect.* http://emotionresearcher.com/lisa-feldman-barrett-why-emotions-are-situated-conceptualizations/.

———. 2017. *How emotions are made: The secret life of the brain.* Boston: Houghton Mifflin Harcourt.

Barrett, Louise. 2011. *Beyond the brain: How body and environment shape animal and human minds.* Princeton, NJ: Princeton University Press.

Barrett, Louise, and Drew Rendall. 2010. Out of our minds: The neuroethology of primate strategic behavior. In M. L. Platt and A. A. Ghazanfar, eds., *Primate neuroethology,* 570–586. New York: Oxford University Press.

Barsalou, Lawrence W. 1999. Perceptual symbol systems. *Behavior and Brain Sciences* 22 (4): 577–660.

———. 2010. Grounded cognition: Past, present, and future. *Topics in Cognitive Science* 2: 716–724.

Barsalou, Lawrence W., Paula M. Niedenthal, Aron K. Barbey, and Jennifer A. Ruppert. 2003. Social embodiment. *Trends in Cognitive Sciences* 9: 309–311.

Bartels, Andreas and Semir Zeki. 2000. The neural basis of romantic love. *NeuroReport* 11 (17): 3829–3834.

Barton, Robert A. 2012. Embodied cognitive evolution and the cerebellum. *Philosphical Transactions of the Royal Society London B* 367 (1599): 2097–2107.

Barton, Robert A., and Chris Venditti. 2014. Rapid evolution of the cerebellum in humans and other great apes. *Current Biology* 24 (20): 2440–2444.

Bartz, Jennifer A., Jamil Zaki, Niall Bolger, and Kevin N. Ochsner. 2011. Social effects of oxytocin in humans: Context and person matter. *Trends in Cognitive Sciences* 15 (7): 301–309.

Bataille, Georges. 1957 / 2005. *Erotism: Death and sensuality.* Mary Dalwood, trans. San Francisco: City Lights.

Bates, Elizabeth and Jeffrey Elman. 1996. Learning rediscovered. *Science* 274 (5294): 1849–1850.

Batson, C. Daniel. 1991. *The altruism question: Toward a social-psychological answer.* New York: Lawrence Erlbaum.

Bauer, Russell M. 1984. Autonomic recognition of names and faces in Prosopagnosia: A neuropsychological application of the guilty knowledge test. *Neuropsychologia* 22 (4): 457–469.

Bauer, Russell M., and A. R. Rubens. 1985. Agnosia. In K. M. Heilman and E. Valenstein, eds., *Clinical Neuropsychology*, 2nd ed., 187–241. New York: Oxford University Press.

Bäzner, H., and Michael Hennerici. 2006. Stroke in painters. *International Review of Neurobiology* 74: 165–191. doi:10.1016/S0074-7742(06)74013-2.

Bechara, Antoine, Antonio R. Damasio, Hanna Damasio, and Steven W. Anderson. 1994. Insensitivity to future consequences following damage to human prefrontal cortex. *Cognition* 50: 7–15.

Bechtel, William. 2008. *Mental mechanisms: Philosophical perspectives on cognitive neuroscience*. New York: Taylor and Francis.

Bechtel, William, and Adele Abrahamsen. 1991. *Connectionism and the mind: An introduction to parallel processing in networks*. Cambridge, MA: Blackwell.

———. 2010. Dynamic mechanistic explanation: Computational modeling of circadian rhythms as an exemplar for cognitive science. *Studies in History and Philosophy of Science Part A* 1: 321–333.

Beck, Jacob. 2012a. Do animals engage in conceptual thought? *Philosophy Compass* 7 (3): 218–229.

———. 2012b. The generality constraint and the structure of thought. *Mind* 121 (483): 563–600.

Bedau, Mark A., and Paul Humphreys, eds. 2008. *Emergence: Contemporary readings in philosophy and science*. Cambridge, MA: MIT Press.

Beer, Jennifer S., Erin A. Heerey, Dacher Keltner, Donatella Scabini, and Robert T. Knight. 2003. The regulatory function of self-conscious emotion: Insights from patients with orbitofrontal damage. *Journal of Personality and Social Psychology* 85 (4): 594–604.

Beer, Jennifer S., Oliver P. John, Donatella Scabini, and Robert T. Knight. 2006. Orbitofrontal cortex and social behavior: Integrating self-monitoring and emotion-cognition interactions. *Journal of Cognitive Neuroscience* 18 (6): 871–879.

Beer, Randall D. 2003. The dynamics of active categorical perception in an evolved model agent. *Adaptive Behavior* 11 (4): 209–243.

———. 2014. Dynamical systems and embedded cognition. In Keith Frankish and William M. Ramsey, eds., *The Cambridge handbook of artificial intelligence,* 128–148. Cambridge: Cambridge University Press.

Bégouen, H. 1929. The magic origin of prehistoric art. *Antiquity* 3 (9): 5–19.

Behe, Michael J. 1996. *Darwin's black box: The biochemical challenge to evolution*. New York: Free Press.

Behrens T. E., L. T. Hunt, M. W. Woolrich, and M. F. Rushworth. 2008. Associative learning of social value. *Nature* 456 (7219): 245–249.

Bellah, Robert. 2011. *Religion in human evolution: From the paleolithic to the axial age*. Cambridge, MA: Belknap.

Belsky, J. 2005. Differential susceptibility to rearing influence. In Bruce J. Ellis and David F. Bjorklund, eds., *Origins of the social mind: Evolutionary psychology and child development*, 139–163. New York: Guilford Press.

Belsky, Jay. 1997. Variation in susceptibility to environmental influence: An evolutionary argument. *Psychological Inquiry* 8 (3): 182–186.

Bente, Gary, Haug Leuschner, Ahmad Al Issa, and James L. Blascovich. 2010. The others: Universals and cultural specificities in the perception of status and dominance from nonverbal behavior. *Consciousness and Cognition* 19 (3): 762–777.

Bergen, Benjamin K. 2012. *Louder than words: The new science of how the mind makes meaning*. New York: Basic.

Bergman, Thore J., Jacinta C. Beehner, Dorothly L. Cheney, and Robert M. Seyfarth. 2003. Hierarchical classification by rank and kinship in baboons. *Science* 302 (5648): 1234–1236.

Berlyne, D. E. 1954. A theory of human curiosity. *British Journal of Psychology* 45 (3): 180–191.

———. 1978. Curiosity and learning. *Motivation and Emotion* 2 (2): 97–175.

Bermúdez, José Luis. 1995. Nonconceptual content: From perceptual experience to subpersonal computational states. *Mind and Language* 10 (4): 333–369.

———. 1998. *The paradox of self-consciousness*. Cambridge, MA: MIT Press.

———. 2003. *Thinking without words*. New York: Oxford University Press.

Bermúdez, José, and Arnon Cahen. 2015. Nonconceptual mental content. In Edward N. Zalta, ed., *The Stanford encyclopedia of philosophy,* Fall ed. http://plato.stanford.edu/archives/fall2015/entries/content-nonconceptual/.

Berridge, Kent C. 2003a. Comparing the emotional brains of humans and other animals. In R. Davidson, K. R. Scherer, and H. H. Goldsmith, eds., *Handbook of affective sciences,* 25–51. New York: Oxford University Press.

———. 2003b. Pleasures of the brain. *Brain and Cognition* 52 (1): 106–128.

———. 2009. Wanting and liking: Observations from the neuroscience psychology laboratory. *Inquiry* 52 (4): 378–398.

Berridge, Kent C., Terry E. Robinson, and J. Wayne Aldridge. 2009. Dissecting components of reward: "Liking," "wanting," and "learning." *Current Opinion in Pharmacology* 9 (1): 65–73.

Berwick, Robert, and Noam Chomsky. 2016. *Why only us: Language and evolution.* Cambridge, MA: MIT Press.

Bhaskar, Roy. 1998. Philosophy and scientific realism. In Margaret Archer, Roy Bhaskar, Andrew Collier, Tony Lawson, and Alan Norrie, eds. *Critical realism: Essential readings,* 16–47. New York: Routledge.

Biederman, Irving. 1987. Recognition-by-components: A theory of human image understanding. *Psychological Review* 94 (2): 115–147.

Biederman, Irving, and Peter Kalocsais. 1997. Neurocomputational bases of object and face recognition. *Philosphical Transactions of the Royal Society London B* 352 (1358): 1203–1219.

Black, M. 1977. More about metaphor. *Dialectica* 31: 431–457. doi:10.1111/j.1746-8361.1977.tb01296.x.

Bloom, Paul. 2000. *How children learn the meanings of words*. Cambridge, MA: MIT Press.

Boas, Franz. 1927 / 2010. *Primitive art*. Mineola, NY: Dover Publications.

Bodamer, J. 1947. Die Prosopagnosie. *Archiv fur Psychiatrie und Nervenkrankheiten* 179: 6–53.

Boehm, Christopher, Harold B. Barclay, Robert Knox Dentan, Marie-Claude Dupre, Jonathan D. Hill, Susan Kent, Bruce M. Knauft, Keith F. Otterbein, and Steve Rayner. 1993. Egalitarian behavior and reverse dominance hierarchy. *Current Anthropology* 34 (3): 227–254.

———. 1999. *Hierarchy in the forest: The evolution of egalitarian behavior*. Cambridge, MA: Harvard University Press.

Bogousslavsky, Julien. 2005. Artistic creativity, style and brain disorders. *European Neurology* 54: 103–111. doi:10.1159/000088645.

Bogucki, P. 1999. *The origins of human society*. Hoboken, NJ: John Wiley and Sons.

Bonabeau, Eric, Guy Theraulaz, and Jean-Louis Deneubourg. 1999. Dominance orders in animal societies: The self-organization hypothesis revisited. *Bulletin of Mathematical Biology* 61: 727.

Bonduriansky, Russell, and Troy Day. 2009. Nongenetic inheritance and its evolutionary implications. *Annual Review of Ecology, Evolution, and Systematics* 40: 103–125.

Booker, Christopher. 2004. *The seven basic plots: Why we tell stories*. London: Continuum.

Boring, Edwin G. 1929. The psychology of controversy. *Psychological Review* 36 (2): 97.

Bosch, O. 2013. Maternal aggression in rodents: brain oxytocin and vasopressin mediate pup defence. *Philosophical Transactions of the Royal Society of London B* 368 (1631).

Bosch, Oliver J., and Larry J. Young. 2017. Oxytocin and social relationships: From attachment to bond disruption. *Current Topics Behavioral Neurosciences* 35: 97–117.

Bowlby, John. 1951. *Maternal care and mental health: A report prepared on behalf of the World Health Organization as a contribution to the United Nations programme for the welfare of homeless children*. Geneva: World Health Organization.

———. 1969. *Attachment and Loss: Attachment* (vol. 1). New York: Basic Books.

———. 1973. *Attachment and Loss: Separation* (vol. 2). International Psycho-Analytical Library no. 95. London: Hogarth Press.

———. 1980. *Attachment and Loss: Sadness and Depression* (vol. 3). New York: Basic Books.

Boyd, Robert. 2017. *A different kind of animal: How culture transformed our species*. Princeton, NJ: Princeton University Press.

Boyd, Robert, and Peter J. Richerson. 1985. *Culture and the evolutionary process*. Chicago: University of Chicago Press.

———. 2001. *The origin and evolution of cultures*. Oxford: Oxford University Press.

Boyer, Pascal. 2002. *Religion explained: The evolutionary origins of religious thought*. New York: Basic Books.

Boyer, Pascal, and Charles Ramble. 2010. Cognitive templates for religious concepts: Cross-cultural evidence for recall of counter-intuitive representations. *Cognitive Science* 25 (4): 535–564.

Breedlove, Mark, Neil V. Watson, and Mark R. Rosenzweig. 2015. *Biological psychology: An introduction to behavioral, cognitive, and clinical neuroscience.* Sunderland, MA: Sinauer Associates, Inc.

Brentano, Franz. 1874 / 1995. *Psychology from an empirical standpoint,* 2nd ed. Oskar Kraus, ed., Antos C. Rancurello, D. B. Terrell, and Linda L. McAlister, trans. London: Routledge.

Bretherton, Inge. 1985. Attachment theory: Retrospect and prospect. *Monographs of the Society for Research in Child Development* 50 (1 / 2).

Broad, K., J. Curley, and E. Keverne. 2006. Mother-infant bonding and the evolution of mammalian social relationships. *Bulletin of Mathematical Biology* 361 (1476): 2199–2214.

Brosnan, Sarah, and Frans B. M. de Waal. 2003. Monkeys reject unequal pay. *Nature* 425 (6955): 297–299.

Brown, Harold I. 1992. Direct realism, indirect realism, and epistemology. *Philosophy and Phenomenological Research* 52 (2): 341–363.

Brown, Paul L., and Herbert M. Jenkins. 1968. Auto-shaping of the pigeon's key-peck. *Journal of the Experimental Analysis of Behavior* 11 (1): 1–8.

Brown, R. 1986. *Social psychology,* 2nd ed. London: Collier Macmillan.

Bruner, Jerome. 1991. The narrative construction of reality. *Critical Inquiry* 18 (1): 1–21.

Buck, Ross. 1988. *Human motivation and emotion,* 2nd ed. Oxford: John Wiley and Sons.

Burckhardt, Jacob. 1860 / 2002. *The civilization of the Rennaissance in Italy.* New York: Penguin Random House.

Burgdorf, Jeffrey, and Jaak Panksepp. 2006. The neurobiology of positive emotions. *Neuroscience and Biobehavioral Reviews* 30: 173–187.

Burge, Tyler. 1986. Intellectual norms and foundations of mind. *Journal of Philosophy* 83 (12): 697–720.

Burger, Jerry M., Jackeline Sanchez, Jenny E. Imberi, and Lucia R. Grande. 2009. The norm of reciprocity as an internalized social norm: Returning favors even when no one finds out. *Social Influence* 4 (1): 11–17.

Burghardt, Gordon M. 1973. Instinct and innate behavior: Toward an ethological psychology. In J. A. Nevin and G. S. Reynolds, eds., *The study of behavior: Learning, motivation, emotion, and instinct.* Glenview, IL: Scott, Foresman.

———. 1999. Conceptions of play and the evolution of animal minds. *Evolution and Cognition* 5 (2): 115–123.

Burkett, James P., Elissar Andari, Zachary V. Johnson, Daniel C. Curry, Frans B. M. de Waal, and Larry J. Young. 2016. Oxytocin-dependent consolation behavior in rodents. *Science* 22 (6271): 375–378.

Byrne, Richard, and Andrew Whiten. 1988. *Machiavellian intelligence: Social expertise and the evolution of intellect in monkeys, apes, and humans.* Oxford: Oxford University Press.

Byrne, Richard W. 2001. Social and technical forms of primate intelligence. In Frans B. M. de Waal, ed., *Tree of origin: What primate behavior can tell us about human social evolution,* 145–172. Cambridge, MA: Harvard University Press.

Byrne, Richard W., P. J. Barnard, I. Davidson, V. M. Janik, W. C. McGrew, A. Miklosi, and P. Wiessner. 2004. Understanding culture across species. *Trends in Cognitive Sciences* 8 (8): 341–346.

Byrne, Richard W., Catherine Hobaiter, and Michelle Klailova. 2011. Local traditions in gorilla manual skill: Evidence for observational learning of behavioral organization. *Animal Cognition* 14 (5): 683–693.

Calasso, Roberto. 1988. *The marriage of cadmus and harmony.* New York: Vintage Press.

———. 1998. *Ka: Stories of the mind and gods of India.* London: Jonathan Cape.

———. 2015. *Literature and the gods.* New York: Vintage Press.

Call, Josep, and Michael Tomasello. 2008. Does the chimpanzee have a theory of mind? 30 years later. *Trends in Cognitive Science* 12 (5): 187–192.

Camp, Elisabeth. 2009. A language of baboon thought? In Robert Lurz, ed., *Philosophy of animal minds,* 108–127. Cambridge: Cambridge University Press.

Camus, Mickael, N. Halelamien, H. Plassmann, S. Shimojo, J. O'Doherty, C. Camerer, and A. Rangel. 2009. Repetitive transcranial magnetic stimulation over the right dorsolateral prefrontal cortex decreases valuations during food choices. *European Journal of Neuroscience* 30 (10): 1980–1988.

Cannon, Walter B. 1926. Physiological regulation of normal states: Some tentative postulates concerning biological homeostatics. In A. Pettit, ed., *À Charles Richet: Ses amis, ses collègues, ses élèves.* Paris: Éditions Médicales.

Capraro, Valerio, Brice Corgnet, Antonia M. Espín, and Roberto Hernán-González. 2017. Deliberation favours social efficiency by making people disregard their relative shares: Evidence from USA and India. *Royal Society Open Science* 4, 160605.

Capraro, Valerio, Francesca Giardini, Daniele Vilone, and Mario Paolucci. 2016. Partner selection supported by opaque reputation promotes cooperative behavior. *Judgment and Decision Making* 11 (6): 589–600.

Carballo, David M., Paul Roscoe, and Gary M. Feinman. 2014. Cooperation and collective action in the cultural evolution of complex societies. *Journal of Archaeological Method and Theory* 21 (1): 98–133.

Carbone, Vincent J., Leigh O'Brien, Emily J. Sweeney-Kerwin, and Kristin M. Albert. 2013. Teaching eye contact to children with autism: A conceptual analysis and single case study. *Education and Treatment on Children* 36 (2): 139–159.

Carello, Claudia, and Michael T. Turvey. 2000. Rotational dynamics and dynamic touch. In M. Heller, ed., *Touch, representation and blindness,* 27–66. Oxford: Oxford University Press.

Carey, Susan. 2011a. *The origin of concepts.* Oxford: Oxford University Press.

———. 2011b. Précis of *The origin of concepts. Behavioral and Brain Sciences* 34 (3): 113–167.

Carlson, Elizabeth A., Byron Egeland, and L. Alan Sroufe. 2009. A prospective investigation of the development of borderline personality symptoms. *Development and Psychopathology* 21 (4): 1311–1334.

Caro, T. M., and M. D. Hauser. 1992. Is there teaching in nonhuman animals? *Quarterly Review of Biology* 67 (2): 151–174.

Carroll, Noel. 1999. *Philosophy of art: A contemporary introduction.* Routledge Contemporary Introductions to Philosophy. London: Routledge Press.

Carroll, Sean B. 2005. *Endless forms most beautiful: The new science of evo devo and the making of the animal kingdom*. New York: W. W. Norton.

Carruthers, Peter. 2008. Meta-cognition in animals: A skeptical look. *Mind and Language* 23 (1): 58–89.

Carruthers, Peter, and J. Brendan Ritchie. 2012. The emergence of metacognition: Affect and uncertainty in animals. In Michael J. Beran, Johannes L. Brand, Josef Perner, Joëlle Proust, eds., *Foundations of metacognition*, 191–234. New York: Oxford University Press.

Carter, C. Sue, and Stephen W. Porges. 2013. Neurobiology and the evolution of mammalian social behavior. In D. Narvaez, J. Panksepp, A. N. Schore, and T. R. Gleason, eds., *Evolution, early experience and human development: From research to practice and policy*, 132–151. New York: Oxford University Press.

Casebeer, W. D. 2003. Moral cognition and its neural constituents. *Nature Reviews Neuroscience* 10: 840–846.

Casey, B. J., Leah H. Somerville, Ian H. Gotlib, Ozlem Ayduk, Nicholas T. Franklin, Mary K. Askren, John Jonides, Marc G. Berman, Nicole L. Wilson, Theresa Teslovich, Gary Glover, Vivian Zayas, Walter Mischel, and Yuichi Shoda. 2011. Behavioral and neural correlates of delay of gratification 40 years later. *Proceedings of the National Academy of Sciences* 108 (36): 14998–15003.

Cashdan, Elizabeth. 1998. Smiles, speech, and body posture: How women and men display sociometric status and power. *Journal of Nonverbal Behavior* 22 (4): 209–228.

Cassirer, Ernst. 1950 / 1969. *The problem of knowledge: Philosophy, science, and history since Hegel*. William H. Woglom, trans. New Haven, CT: Yale University Press.

Cerqueira, João J., Osborne F. X. Almeida, and Nuno Sousa. 2008. The stressed prefrontal cortex. Left? Right! *Brain, Behavior, and Immunity* 22 (5): 630–638.

Chagnon, Napoleon, and William Irons. 1979. Cultural and biological success. In N. A. Chagnon and W. Irons, eds., *Evolutionary biology and human social behavior: An anthropological perspective*, 284–302. North Scituate, MA: Duxbury.

Chapais, Bernard. 2008. *Primeval linship: How pair-bonding gave birth to human society*. Cambridge, MA: Harvard University Press.

Charland, Louis C. 1996. Feeling and representing: Computational theory and the modularity of affect. *Synthese* 105: 273–301.

———. 2005. The heat of emotion: Valence and the demarcation problem. *Journal of Consciousness Studies* 12 (8–10): 82–102.

Chase, Ivan. 1974. Models of hierarchy formation in animal societies. *Behavioral Science* 19 (6): 374–382.

———. 1982. Dynamics of hierarchy formation: The sequential development of dominance relationships. *Behaviour* 80 (3 / 4): 218–240.

Chase, Ivan, Craig Tovey, Debra Spangler-Martin, and Michael Manfredonia. 2002. Individual differences versus social dynamics in the formation of animal dominance hierarchies. *Proceedings of the National Academy of Sciences* 99 (8): 5744–5749.

Chatterjee, Anjan. 2003. Prospects for a cognitive neuroscience of visual aesthetics. *Bulletin of Psychology and the Arts* 4: 55–60. doi: 10.1037/e514602010–003.

Chemero, Anthony. 2003a. An outline of a theory of affordances. *Ecological Psychology* 15 (2): 181–195.

―――. 2003b. Radical empiricism through the ages. *PsycCRITIQUES* 48 (1): 18–21.

―――. 2009. *Radical embodied cognitive science*. Cambridge, MA: MIT Press.

Cheney, Dorothy L., and Robert M. Seyfarth. 2007. *Baboon metaphysics: The evolution of a social mind*. Chicago: University of Chicago Press.

Chomsky, Noam. 1956. Three models for the description of language. *IRE Transactions on Information Theory* 2 (3): 113–124.

―――. 1957. *Syntactic structures*. New York: Walter de Grutyer.

Chwalisz, Kathleen, Ed Diener, and Dennis Gallagher. 1988. Autonomic arousal feedback and emotional experience: Evidence from the spinal cord injured. *Journal of Personality and Social Psychology* 54 (5): 820–828.

Cialdini, Robert B. 2001. *Influence: Science and practice*. New York: Allyn and Bacon.

Cieri, Robert L., Steven E. Churchill, Robert G. Franciscus, Jingzhi Tan, and Brian Hare. 2014. Craniofacial feminization, social tolerance, and the origins of behavioral modernity. *Current Anthropology* 55 (4): 419–443.

Clancey, William J. 1997. *Situated cognition: On human knowledge and computer representations*. Cambridge: Cambridge University Press.

Clark, Andy. 1997. *Being there: Putting brain, body, and world together again*. Cambridge, MA: MIT Press.

―――. 2008. *Supersizing the mind: Embodiment, action, and cognitive extension*. New York: Oxford University Press.

―――. 2016. *Surfing uncertainty: Prediction, action, and the embodied mind*. Oxford: Oxford University Press.

Clark, Andy, and David J. Chalmers. 1998. The extended mind. *Analysis* 58 (1): 7–19.

Clark, Andy, and Rick Grush. 1999. Towards a cognitive robotics. *Adaptive Behavior* 7 (1): 5–16.

Clayton, N. S., and A. Dickinson. 1998. Episodic-like memory during cache recovery by scrub jays. *Nature* 395 (6699): 272–274.

Clayton, Philip, and Paul Davies, eds. 2006. *The re-emergence of emergence: The emergentist hypothesis from science to religion*. Oxford: Oxford University Press.

Clottes, Jean. 1996. Thematic changes in upper palaeolithic art: A view from the Grotte Chauvet. *Antiquity* 70 (268): 276–288.

Clottes, Jean, and J. David Lewis-Williams. 1998. *The shamans of prehistory: Trance and magic in the painted caves*. New York: Harry N. Abrams.

Cockburn, Andrew. 2013. Cooperative breeding in birds: Toward a richer conceptual. In Kim Sterelny, R. Joyce, B. Calcott, B. Fraser, eds., *Cooperation and its evolution,* 223–246. Cambridge, MA: MIT Press.

Cohen N. J., and H. Eichenbaum. 1993. *Memory, amnesia, and the hippocampal system*. Cambridge, MA: MIT Press.

Cole, Pamela. 2006. Cultural variations in the socialization of young children's anger and shame. *Child Development* 77: 1237–1251.

Cole, Pamela, Carolyn Zahn-Waxler, Nathan A. Fox, Barbara A. Usher, and Jean D. Welsh. 1996. Individual differences in emotion regulation and behavior problems in preschool children. *Journal of Abnormal Psychology* 105: 518–529.

Coleman, William. 1977. *Biology in the nineteenth century: Problems of form, function and transformation.* Cambridge: Cambridge University Press.

Colombetti, Giovanna. 2007. Enactive appraisal. *Phenomenology and the Cognitive Sciences* 6 (4): 527–546.

———. 2014. *The feeling body: Affective science meets the enactive mind.* Cambridge, MA: MIT Press.

Comte-Sponville, André. 2007. *The little book of atheist spirituality.* Nancy Huston, trans. New York: Viking.

Conkey, Margaret W. 2009. Materiality and meaning-making in the understanding of the Paleolithic "arts." In C. Renfrew and I. Morley, eds., *Becoming human: Innovation in prehistoric material and spiritual culture,* 179–194. Cambridge: Cambridge University Press.

Cook, Richard, Geoffrey Bird, Caroline Catmur, Clare Press, and Cecilia Heyes. 2014. Mirror neurons: From origin to function. *Behavioral and Brain Sciences* 37 (2): 177–192.

Cory, Gerald, and Gardner Russell, eds. *The evolutionary neuroethology of Paul Maclean: Convergences and frontiers.* Westport, CT: Praeger.

Cosmides, Leda, and John Tooby. 1992. Cognitive adaptations for social exchange. In J. Barkow, L. Cosmides, and J. Tooby, eds., *The adapted mind: Evolutionary psychology and the generation of culture,* 163–228. New York: Oxford University Press.

———. 1995. From function to structure. In M. Gazzaniga, ed., *The cognitive neurosciences.* Cambridge, MA: MIT Press.

Costall, Alan. 1995. Socializing affordances. *Theory and Psychology* 5 (4): 467–481.

———. 2010. The future of experimental psychology. *The Psychologist* 23 (12): 1022–1023.

Cowie, Fiona. 1999. *What's within: Nativism reconsidered.* New York: Oxford University Press.

Craig, William Lane, and J. P. Moreland, eds. 2009. *Blackwell Companion to Natural Theology.* Malden, MA: Blackwell, 2009.

Crews, David. 2003. The development of phenotypic plasticity: Where biology and psychology meet. *Developmental Psychobiology* 43 (1): 1–10.

Csikszentmihályi, Mihály. 1990. The domain of creativity. In M. A. Runco and R. S. Albert, eds., *Theories of creativity,* vol. 115 of *Sage Focus Editions.* Thousand Oaks, CA: Sage Publications.

Cussins, Adrian. 1990. The connectionist construction of concepts. In Margaret A. Boden, ed., *The philosophy of AI.* Oxford: Oxford University Press.

———. 1992. Content, embodiment and objectivity: The theory of cognitive trails. *Mind* 101 (404): 651–688.

Damasio, Antonio R. 1994. *Descartes' error: Emotion, reason, and the human brain.* New York: Putnam.

———. 1996. Executive and cognitive functions of the prefrontal cortex: The somatic marker hypothesis and the possible functions of the prefrontal cortex. *Philosophical Transactions of the Royal Society London B* 351 (1346): 1413–1420.

———. 1999. *The feeling of what happens: Body and emotion in the making of consciousness.* New York: Harcourt Brace.

————. 2003. *Looking for Spinoza: Joy, sorrow, and the feeling brain*. Orlando, FL: Harcourt.

————. 2010. *Self comes to mind: Constructing the conscious brain*. New York: Random House.

————. 2018. *The strange order of things: Life, feeling, and the making of cultures*. New York: Pantheon Press.

Damasio, Antonio R., Hanna Damasio, and Gary W. Van Hoesen. 1982. Prosopagnosia: Anatomic basis and behavioral mechanisms. *Neurology* 32 (4): 331.

Darwin, Charles. 1859. *On the origin of species by means of natural selection*. London: John Murray.

————. 1870. *On the origin of species by means of natural selection*. New York: D. Appleton and Company.

————. 1871. *Descent of man, and selection in relation to sex*. London: John Murray.

————. 1872. *Expression of the emotions in man and animals*. London: John Murray.

————. 2004. *Descent of man*. Penguin Classics.

Dauvois, Michel, and Xavier Boutillon. 1994. Caractérisation acoustique des grottes ornées paléolithiques et de leurs lithophones naturels. In Catherine Homo-Lechner, Annie Bélis, eds., *La pluridisciplinarité en archéologie musicale*, vol. 1, 209–251. Paris: Les Editions de la MSH.

Davidson, D. 1967. Truth and meaning. *Synthese* 17: 304–323.

Davidson, Richard J. 2002. Toward a biology of positive affect and compassion. In Richard J. Davidson and Anne Harrington, eds., *Visions of compassion: Western scientists and Tibetan Buddhists examine human nature*, 107–130. Oxford: Oxford University Press.

Davidson, Richard J., and Sharon Begley. 2012. *The emotional life of your brain: How its unique patterns affect the way you think, feel, and live—and how you can change them*. New York: Hudson Street Press.

Davidson, Richard J., and William Irwin. 1999. The functional neuroanatomy of emotion and affective style. *Trends in Cognitive Sciences* 3 (1): 11–21.

Davidson, Richard J., Klaus R. Scherer, and H. Hill Goldsmith, eds. 2009. *Handbook of affective sciences*. Oxford: Oxford University Press.

Dawkins, Richard. 1986. *The blind watchmaker: Why the evidence of evolution reveals a universe without design*. New York: W. W. Norton.

Deacon, Terrence W. 2011. *Incomplete nature: How mind emerged from matter*. New York: W. W. Norton.

De Houwer, J. 2011. Why the cognitive approach in psychology would profit from a functional approach and vice versa. *Perspectives on Psychological Science* 6 (2): 202–209.

De Jong, Gerdien. 2005. Evolution of phenotypic plasticity: Patterns of plasticity and the emergence of ecotypes. *New Phytologist* 166 (1): 101–118.

DeLancey, Criag. 2002. *Passionate Engines: What emotions reveal about the mind and artificial intelligence*. New York: Oxford University Press.

Dembski, William A. 2002. *Intelligent design: The bridge between science and theology*. Westmont, IL: InterVarsity Press.

Dennett, Daniel C. 1991. *Consciousness explained*. Boston: Little, Brown and Company.

———. 1992. The self as the center of narrative gravity. In Frank S. Kessel, Pamela M. Cole, and Dale L. Johnson, eds., *Self and consciousness: Multiple perspectives,* 103–115. New York: Taylor and Francis.

———. 1996a. *The intentional stance.* Cambridge, MA: MIT Press.

———. 1996b. *Kinds of minds: Toward an understanding of consciousness.* New York: Basic Books.

———. 2017. *From bacteria to Bach and back: The evolution of minds.* New York: W. W. Norton.

Dennett, Daniel C., and Alvin Plantinga. 2010. *Science and religion: Are they compatible?* New York: Oxford University Press.

Deonna, Julien A., and Fabrice Teroni. 2012. *The emotions: A philosophical introduction.* New York: Routledge.

de Quervain, D. J., F. U. Fischbacher, V. Treyer, M. Schellhammer, U. Schynder, A. Buck, et al. 2004. The neural basis of altruistic punishment. *Science* 305: 1254–1258.

DeSalle, Rob, and Ian Tattersall. 2012. *The brain: Big bangs, behaviors, and beliefs.* New Haven, CT: Yale University Press.

DeSteno, David. 2018. *Emotional success: The power of gratitude, compassion, and pride.* Boston: Houghton Mifflin Harcourt.

Deutscher, Guy. 2010. *The unfolding of language: An evolutionary tour of mankind's greatest invention.* New York: Henry Holt.

DeVries, J. 2008. *The industrious revolution: Consumer behavior and the household economy, 1650 to the present.* Cambridge: Cambridge University Press.

de Waal, Frans. 1982. *Chimpanzee politics: Power and sex among apes.* New York: Harper and Row.

———. 1986. The integration of dominance and social bonding in primates. *Quarterly Review of Biology* 61 (4): 459–479.

———. 1989. Food sharing and reciprocal obligations among chimpanzees. *Journal of Human Evolution* 18 (5): 433–459.

———. 1997. *Bonobos: The forgotten ape.* Berkeley: University of California Press.

———, ed. 2001. *Tree of origin: What primate behavior can tell us about human social evolution.* Cambridge, MA: Harvard University Press.

———. 2003. *Morality and the social instincts: Continuity with the other primates.* The Tanner Lectures on Human Values. Princeton University, November 19–20, 2003.

———. 2005. *Our inner ape: A leading primatologist explains why we are who we are.* New York: Riverhead Books.

———. 2006. *Primates and philosophers: How morality evolved.* Princeton, NJ: Princeton University Press.

———. 2013. *The bonobo and the atheist: In search of humanism among the primates.* New York: W. W. Norton.

Dewey, John. 1929 / 1960. *The quest for certainty. The Gifford lectures.* New York: Capricorn Books.

———. 1938. *Logic: The theory of inquiry.* New York: Holt, Rinehart and Winston.

Diamond, Jared. 1997. *Guns, germs, and steel: The fates of human societies.* New York: W. W. Norton.

Diekmann, Andreas. 2004. The power of reciprocity: Fairness, reciprocity, and stakes in variants of the dictator game. *Journal of Conflict Resolution* 48 (4): 487–505.

Dietrich, Eric, and Arthur B. Markman. 2003. Discrete thoughts: Why cognition must use discrete representations. *Mind and Language* 18 (1): 95–119.

Dijksterhuis, Eduard Jan. 1961. *The mechanization of the world picture.* Oxford: Clarendon Press.

Domes, G., M. Heinrichs, A. Michel, C. Berger, and S. C. Herpertz. 2007. Oxytocin improves "mind-reading" in humans. *Biological Psychiatry* 61: 731–733.

Dovidio, John F., Clifford E. Brown, Karen Heltman, Steve L. Ellyson, and Caroline F. Keating. 1988. Power displays between women and men in discussions of gender-linked tasks: A multichannel study. *Journal of Personality and Social Psychology* 55 (4): 580–587.

Dretske, Fred I. 1981. *Knowledge and the flow of information.* Cambridge, MA: MIT Press.

———. 1995. *Naturalizing the mind. The Jean Nicod lectures.* Cambridge, MA: MIT Press.

Dreyfus, Hubert L. 1972. *What computers can't do: A critique of artificial reason.* Cambridge, MA: MIT Press.

———. 2002. Intelligence without representation. *Phenomenology and the Cognitive Sciences* 1: 367–383.

———. 2014. *Skillful coping: Essays on the phenomenology of everyday perception and action.* Oxford: Oxford University Press.

Dreyfus, Hubert, and Sean D. Kelly. 2007. Heterophenomenology: Heavy-handed sleight-of-hand. *Phenomenology and the Cognitive Sciences* 6 (1–2): 45–55.

Duchaine, Bradley C., Holly Parker, and Ken Nakayama. 2003. Normal recognition of emotion in a prosopagnosic. *Perception* 32 (7): 827–838.

Dunbar, Robin I. M. 1998. The social brain hypothesis. *Evolutionary Anthropology* 6 (5):178–190.

———. 2001. Brains on two legs: Group size and the evolution of intelligence. In Frans B. M. de Waal, ed., *Tree of origin: What primate behavior can tell us about human social evolution,* 173–191. Cambridge, MA: Harvard University Press.

Duncan, Tommy, and James G. Wakefield. 2011. Fifty ways to build a spindle: the complexity of microtubule generation during mitosis. *Chromosome Research* 19 (3): 321–333.

Dupanloup I., L. Pereira, G. Bertorelle, F. Calafell, M. J. Prata, A. Amorim, and G. Barbujani. 2003. A recent shift from polygyny to monogamy in humans is suggested by the analysis of worldwide Y-chromosome diversity. *Journal of Molecular Evolution* 57 (1): 85–97.

Dupré, John. 2012. *Why philosophy of biology?* Public lecture at the Center for Science and Philosophy, University of Bristol. December 4, 2012, 23:7.

———. 2014. The role of behaviour in the recurrence of biological processes. *Biological Journal of the Linnean Society* 112 (2): 306–314.

Dye, Thomas S. 2010. Social transformation in old Hawai'i: A bottom-up approach. *American Antiquity* 75 (4): 727–741.

Earle, Timothy. 1998. Property rights and the evolution of Hawaiian chiefdoms. In Robert C. Hunt and Antonia Gilman, eds., *Property in economic context,* 89–118. Lanham, MD: University Press of America.

Egan, Frances. 1992. Individualism, computation, and perceptual content. *Mind* 101 (403): 443–459.

Ehrenzweig, Anton. 1967. *The hidden order of art: A study in the psychology of artistic imagination.* Berkeley: University of California Press.

Ehrlich, Melanie. 2002. DNA methylation in cancer: Too much, but also too little. *Oncogene* 21 (35): 5400–5413.

Eisenstein, Edward M., Doris L. Eisenstein, Jonnalagedda S. M. Sarma, Herschel Knapp, and James C. Smith. 2012. Some new speculative ideas about the "behavioral homeostasis theory" as to how the simple learned behaviors of habituation and sensitization improve organism survival throughout phylogeny. *Communicative and Integrative Biology* 5 (3): 233–239.

Ekman, Paul. 1999. Basic emotions. In Tim Dalgleish and Mick J. Power, eds., *Handbook of cognition and emotion,* 45–60. Sussex: John Wiley and Sons.

———. 2003. Darwin, deception, and facial expression. In P. Ekman, J. J. Campos, R. J. Davidson, and F. B. M. de Waal, eds., *Emotions inside out: 130 years after Darwin's* The expression of the emotions in man and animals, 205–221. New York: New York Academy of Sciences.

Eldredge, Niles, and Marjorie Grene. 1992. *Interactions: The biological context of social systems.* New York: Columbia University Press.

Eliade, Mircea. 1964. *Shamanism: Archaic techniques of ecstasy.* Princeton, NJ: Princeton University Press.

Elias, Norbert. 1939 / 1969. *The civilizing process: Sociogenetic and psychogenetic investigations.* Edmund Jephcott, trans. Oxford: Blackwell.

Ellis, P. 1997. The pathology of fatal child abuse. *Pathology* 29 (2): 113–121.

Ellyson, Steve L., John F. Dovidio, Randi L. Corson, and Debbie L. Vinicur. 1980. Visual dominance behavior in female dyads: Situational and personality factors. *Social Psychology Quarterly* 43 (3): 328–336.

Engels, F. 2010 / 1884. *The origin of the family, private property and the state.* London: Penguin.

Erickson, Bonnie H., and T. A. Nosanchuk. 1998. Contact and stereotyping in a voluntary association. *Bulletin of Sociological Methodology / Bulletin de Méthodologie Sociologique* 60 (1) : 5–33.

Fagot, Joël, and Roger K. R. Thompson. 2011. Generalized relational matching by Guinea baboons (papio papio) in two-by-two-item analogy problems. *Psychological Science* 22 (10): 1304–1309.

Faulkner, D., and J. K. Foster. 2002. The decoupling of "explicit" and "implicit" processing in neuropsychological disorders: Insights into the neural basis of consciousness? *Psyche* 8 (2).

Fauteux, Kevin. 1995. Regression and redemption in religious experience. *Journal of Pastoral Care and Counseling* 49 (1): 48–58.

Fazio, R. H. 1995. Attitudes as object-evaluation association: Determinants, consequences, and correlates of attitude accessibility. In Richard E. Petty and Jon A.

Krosnick, eds., *Attitude strength: Antecedents and consequences,* 247–282. Mahwah, NJ: Lawrence Erlbaum.

Fazio, R. H., M. C. Powell, and P. M. Herr. 1983. Toward a process model of the attitude–behavior relation: accessing one's attitude upon mere observation of the attitude object. *Journal of Personality and Social Psychology* 44 (4): 723–735.

Fehr, Ernst, and Simon Gächter. 2002. Altruistic punishment in humans. *Nature* 415: 137–140. doi: 10.1038/415137a.

Ferrada, Filipa and Luis M. Camarinha-Matos. 2012. Emotions in collaborative networks: A monitoring system. Doctoral Conference on Computing, Electrical and Industrial Systems, February 27–29, 2012, 9–20. Costa de Caparica, Portugal.

Festinger, Lionel. 1954. A theory of social comparison process. *Human Relations* 7: S.117–140.

Feyerabend, Paul. 1999. *Knowledge, science, and relativism: 1960–1980.* Cambridge: Cambridge University Press.

Fibiger, H. C., and A. G. Phillips. 1986. Reward, motivation, cognition: Psychobiology of mesotelencephalic dopamine systems. In F. E. Bloom and S. D. Geiger, eds., *The handbook of physiology: The nervous system IV,* 647–675. Bethesda, MD: American Physiology Society.

Fiebich, Anika. 2014. Perceiving affordances and social cognition. In Mattia Gallotti and John Michael, eds., *Perspectives in social ontology and social cognition,* vol. 4, 149–166. New York: Springer.

Finkel, Caroline. 2005. *Osman's dream: The history of the Ottoman Empire.* New York: Basic Books.

Firestone, Chaz, and Brian J. Scholl. 2016. Cognition does not affect perception: Evaluating the evidence for "top-down" effects. *Behavioral and Brain Sciences* 39: 1–72.

Fisher, Helen E. 1998. Lust, attraction and attachment in mammalian reproduction. *Human Nature* 9 (1): 23–52.

Flood, Josephine. 1990. *The riches of ancient Australia: A journey into prehistory.* Queensland: Queensland University Press.

Fodor, Jerry A. 1974. Special sciences: Or the disunity of science as a working hypothesis. *Synthese* 28: 97–115.

———. 1983. *The modularity of mind: An essay on faculty psychology.* Cambridge, MA: MIT Press.

———. 1990. Information and representation. In Philip P. Hanson, ed., *Information, language and cognition,* 175–190. Vancouver: University of British Columbia Press.

———. 1998. *Concepts: Where cognitive science went wrong.* New York: Oxford University Press.

———. 2012. What are trees about? *London Review of Books* 34 (10): 34.

Fodor, Jerry A., and Massimo Piattelli-Palmarini. 2011. *What Darwin got wrong.* New York: Farrar, Straus and Giroux.

Fodor, Jerry A., and Zenon W. Pylyshyn. 1981. How direct is visual perception? Some reflections on Gibson's "ecological approach." *Cognition* 9 (2): 139–196.

———. 1988. Connectionism and cognitive architecture. *Cognition* 28 (1–2): 3–71.

Fortunato, Laura, and Marco Archetti. 2010. Evolution of monogamous marriage by maximization of inclusive fitness. *Journal of Evolutionary Biology* 23 (1): 149–156.

Fox, Kieran C. R., S. Nijeboer, E. Solomonova, G. Domhoff, and K. Christoff. 2013. Dreaming as mind wandering: Evidence from functional neuroimaging and first-person content reports. *Frontiers in Human Neuroscience* 7, article 412.

Francis, Julie E., and Lawrence L. Loendorf. 2002. *Ancient visions: Petroglyphs and pictographs of the Wind River and Bighorn Country, Wyoming and Montana.* Salt Lake City: University of Utah Press.

Frank, Robert H. 1988. *Passions within reason: The strategic role of emotions.* New York: W. W. Norton.

———. 2005. Departures from rational choice: With and without regret. In Francesco Parisi and Vernon L. Smith, eds., *The law and economics of irrational behavior,* 13–36. Stanford, CA: Stanford University Press.

Frankenhuis, Willem E., and Carolina de Weerth. 2013. Does early-life exposure to stress shape or impair cognition? *Current Directions in Psychological Science* 22 (5): 407–412.

Frankland, Steven M., and Joshua D. Greene. 2015. An architecture for encoding sentence meaning in left mid-superior temporal cortex. *Proceedings of the National Academy of Sciences* 112 (37): 11732–11737.

Freud, Sigmund. 1900. *The interpretation of dreams.* Leipzig, Vienna: Franz Deuticke.

———. 1927. *The future of an illusion.* Reprinted (1953–1974) in J. Strachey, ed. and trans., *The standard edition of the complete psychological works of Sigmund Freud,* vol. 21. London: Hogarth Press.

———. 2005. *Civilization and its discontents.* New York: W. W. Norton

Friston, K. J. 2010. The free-energy principle: A unified brain theory? *Nature Reviews Neuroscience* 11 (2):127–138.

Gabriel, Rami. 2007. *Affective reactions in a prosopagnosic patient.* Doctoral dissertation. Retrieved from ProQuest Dissertations and Theses. (UMI No.: 3274419).

———. 2013. *Why I buy: Self, taste, and consumer society in America.* Bristol, UK: Intellect Press.

Gabriel, Rami, Stanley B. Klein, and Cade McCall. 2008. Affective reactions to facial identity in a prosopagnosic patient. *Cognition and Emotion* 22 (5): 977–983.

Gabriel, Rami, with Jaak Panksepp, Stephen Asma, Glennon Curran, Rami Gabriel, and Thomas Greif. 2012. Modularity in affective neuroscience and cognitive psychology. *Journal of Consciousness Studies* 19 (3 / 4): 19–25.

Gallagher, Shaun. 2007. Review of *Body language: Representation in action,* by Mark Rowlands *Notre Dame Philosophical Reviews* 9. https://ndpr.nd.edu/news/body-language-representation-in-action/.

———. 2008. Are minimal representations still representations? *International Journal of Philosophical Studies* 16 (3): 351–369.

Gallagher, Shaun, and Dan Zahavi. 2012. *The phenomenological mind.* New York: Routledge.

Gallese, V., and C. Sinigaglia. 2010. The bodily self as power for action. *Neuropsychologia* 48: 746–755.

Gallese, Vittorio, Luciano Fadiga, Leonardo Fogassi, and Giacomo Rizzolatti. 1996. Action recognition in the premotor cortex. *Brain* 119 (2): 593–609.

Gallistel, C. R. 1990. *Learning, development, and conceptual change. The organization of learning.* Cambridge, MA: MIT Press.

Gallistel, C. R., and Adam Philip King. 2009. *Memory and the computational brain: Why cognitive science will transform neuroscience.* New York: Wiley-Blackwell.

Gamble, Clive. 1998. Palaeolithic society and the release from proximity: A network approach to intimate relations. *World Archaeology* 29 (3): 426–449.

Gangestad, Steven W., Jeffry A. Simpson, Kenneth DiGeronimo, and Michael Biek. 1992. Differential accuracy in person perception across traits: Examination of a functional hypothesis. *Journal of Personality and Social Psychology* 62 (4): 688–698.

Gardner, Howard. 1983. *Frames of mind: Theory of multiple intelligences.* New York: Basic Books.

Gavrilets, Sergey. 2012. Human origins and the transition from promiscuity to pair-bonding. *Proceeding of the National Academy of Sciences* 109 (25): 9923–9928.

Gazzaniga, Michael. 2018. *The consciousness instinct: Unraveling the mystery of how the brain makes the mind.* New York: Farrar, Straus, and Giroux.

Geertz, Clifford. 1973. Thick description: Toward an interpretive theory of culture. In *The interpretation of cultures: Selected essays,* 3–32. New York: Basic Books.

Gentner, Deirdre, and J. Grudin. 1985. The evolution of mental metaphors in psychology: A 90-year retrospective. *American Psychologist* 40: 181–192.

Ghiselin, Michael T. 1997. *Metaphysics and the origin of species.* Albany, NY: SUNY Press.

Gibson, James J. 1966. *The senses considered as perceptual systems.* Oxford: Houghton Mifflin.

———. 1977. The theory of affordances. In R. Shaw and J. Bransford, eds., *Perceiving, acting, and knowing: Toward an ecological psychology,* 127–143. Oxford: Lawrence Erlbaum.

———. 1979. *The ecological approach to visual perception.* Boston: Houghton Mifflin.

Gifford, Robert. 1994. A lens-mapping framework for understanding the encoding and decoding of interpersonal dispositions in nonverbal behavior. *Journal of Personality and Social Psychology* 66 (2): 398–412.

Gigerenzer, Gerd, and Daniel G. Goldstein. 1996. Reasoning the fast and frugal way: Models of bounded rationality. *Psychological Review* 103 (4): 650–669.

Gilbert, P. 1998. What is shame? Some core issues and controversies. In P. Gilbert and B. Andrews, eds., *Shame: Interpersonal behavior, psychopathology, and culture,* 3–38. New York: Oxford University Press.

Gintis, H., J. Henrich, S. Bowles, R. Boyd, and E. Fehr. 2008. Strong reciprocity and the roots of human morality. *Social Justice Research* 21 (2): 241–253.

Gintis, H., Eric A. Smith, and Samuel Bowles. 2001. Costly signaling and cooperation. *Journal of Theoretical Biology* 213 (1): 103–119.

Gintis, Herbert. 2009. *The bounds of reason: Game theory and the unification of the behavioral sciences.* Princeton, NJ: Princeton University Press.

———. 2012. An evolutionary perspective on the unification of the behavioral sciences. In S. Okasha and K. Binmore, eds., *Evolution and rationality: Decisions, cooperation and strategic behavior,* 213–245. Cambridge: Cambridge University Press.

Girgis Ray R., Nancy J. Minshew, Nadine M. Melhem, Jeffrey J. Nutche, Matcheri S. Keshavan, and Antonio Y. Hardan. 2007. Volumetric alterations of the orbitofrontal cortex in autism. *Progress in Neuro-Psychopharmacology and Biological Psychiatry* 31 (1): 41–45.

Gobet, Fernand, and P. C. R. Lane. 2012. Chunking mechanisms and learning. In *Encyclopedia of the sciences of learning,* 541–544. New York: Springer.

Golding, William. 1955. *The inheritors.* London: Faber and Faber

Goldstone, Lucas G., Volker Sommer, Niina Nurmi, Colleen Stephens, and Barbara Fruth. 2016. Food begging and sharing in wild bonobos (*pan paniscus*): Assessing relationship quality? *Primates* 57 (3): 367–376.

Goleman, Daniel. 1996. Emotional intelligence: Why it can matter more than IQ. *Learning* 24 (6): 49–50.

Golonka, Sabrina. 2015. Laws and conventions in language-related behaviors. *Ecological Psychology* 27 (3): 236–250.

Gombrich, Ernst H. 1951. Hypnerotomachiana. *Journal of the Warburg and Courtauld Institutes* 14 (1 / 2): 119–125.

Gordon, Ilanit, Orna Zagoory-Sharon, James F. Leckman, and Ruth Feldman. 2010. Oxytocin and the development of parenting in humans. *Biological Psychiatry* 68 (4): 377–382.

Gotthelf, Allan, and D. M. Balme. 1985. *Aristotle on nature and living things: Philosophical and historical studies: Presented to David M. Balme on his seventieth birthday.* Pittsburgh, PA: Mathesis Publishing.

Gould, Stephen Jay. 1979. Biological potentiality vs. biological determinism. In *Ever since Darwin: Reflections in natural history,* 251–260. New York: W. W. Norton.

Gould, Stephen Jay, and Richard C. Lewontin. 1979. The spandrels of San Marco and the Panglossian paradigm: A critique of the adaptationist programme. *Proceedings of the Royal Society London B* 205 (1161): 581–598.

Gould, Stephen Jay, and Elizabeth S. Vrba. 1982. Exaptation—a missing term in the science of form. *Paleobiology* 8 (1): 4–15.

Gouldner, Alvin W. 1960. The norm of reciprocity: A preliminary statement. *American Sociological Review* 25 (161). doi: 10.2307/2092623.

Gowlett, John A. J. 2016. The discovery of fire by humans: A long and convoluted process. *Philosophical Transactions of the Royal Society London B* 371 (1696): 20150164.

Gräfenhain M., T. Behne, M. Carpenter, and M. Tomasello. 2009. Young children's understanding of joint commitments. *Development Psychology* 45: 1430–1443.

Grammont, Frederic D. Legrand, and P. Livet, eds. 2001. *Naturalizing intention in action,* 3–17. Cambridge, MA: MIT Press.

Grandin, Temple. 2011. Why do kids with autism stim? *Autism Asperger's Digest,* November / December 2011.

Grandin, Temple, and Richard Panek. 2013. *The autistic brain: Thinking across the spectrum.* Boston: Houghton Mifflin Harcourt.

Gray, Patricia M., Bernie Krause, Jelle Atema, Roger Payne, Carol Krumhansl, and Luis Baptista. 2001. The music of nature and the nature of music. *Science* 291 (5501): 52–54.

Greene, J., and J. Haidt. 2002. How and where does moral judgment work? *Trends in Cognitive Sciences* 6 (12).

Grene, Marjorie. 1987. Hierarchies in biology. *American Scientist* 75 (5): 504–510.

Grice, H. P. 1961. The causal theory of perception. *Proceedings of the Aristotelian Society.* Supplementary volume (35): 121–152.

Griffiths, Paul E. 2008. *What emotions really are: The problem of psychological categories.* Chicago: University of Chicago Press.

Griffiths, Paul E., and Andrea Scarantino. 2005. Emotions in the wild: The situated perspective on emotion. In *Cambridge Handbook of Situated Cognition.* Cambridge: Cambridge University Press.

Griffiths, Thomas L., Joshua B. Tenenbaum, and Charles Kemp. 2012. Bayesian inference. In Keith J. Holyoak and Robert G. Morrison, eds., *The Oxford handbook of thinking and reasoning,* 22–35. New York: Oxford University Press.

Grönroos, Gösta. 2013. Two kinds of belief in Plato. *Journal of the History of Philosophy* 51 (1): 1–19.

Gross, James J. 1998. The emerging field of emotional regulation: An integrative review. *Review of General Psychology* 2 (3): 271–299.

Gunnar, Megan R. 2000. Early adversity and the development of stress reactivity and regulation. In Charles A. Nelson, ed., *The effects of early adversity on neurobehavioral development* (vol. 31 of *Minnesota symposium on child psychology*). New York: Lawrence Erlbaum.

Güth, Werner, Rolf Schmittberger, and Bernd Schwarze. 1982. An experimental study of ultimatum bargaining. *Journal of Economic Behavior and Organization* 3 (4): 367–388.

Haidt, Jonathan. 2001. The emotional dog and its rational tail: A social intuitionist approach to moral judgment. *Psychological Review* 108: 814–34. doi: 10.1037//0033 –295X.108.4.814.

———. 2003. The moral emotions. In R. Davidson, K. Scherer, and H. Goldsmith, eds., *Handbook of affective sciences,* 852–870. Oxford: Oxford University Press.

Hall, Edward T. 1966. *The hidden dimension.* New York: Anchor Books.

Hall, Judith A., Erik J. Coats, and Lavonia Smith LeBaeau. 2005. Nonverbal behavior and the vertical dimension of social relations: A meta-analysis. *Psychological Bulletin* 131 (6): 898–924.

Hallson, L. R., S. Chenoweth, and R. Bonduriansky. 2012. The relative importance of genetic and nongenetic inheritance in relation to trait plasticity in *C allosobruchus maculatus. Journal of Evolutionary Biology* 25 (12): 2422–2431.

Hamilton, W. D. 1963. The evolution of altruistic behavior. *American Naturalist* 97 (896): 354–356

Hammerl, M. 2000. I like it, but only when I'm not sure why: Evaluative conditioning and the awareness issue. *Consciousness and Cognition 9:* 37–40.

Hammerl, M., and H. J. Grabitz. 2000. Affective-evaluative learning in humans: A form of associative learning or only an artifact. *Learning and Motivation* 31: 345–363.

Hammock, Elizabeth A. D., and Larry J. Young. 2006. Oxytocin, vasopressin and pair-bonding: Implications for autism. *Philosophical Transactions of the Royal Society London B* 361 (1476): 2187–2198.

Harcourt, Alexander H., and Kelly J. Stewart. 2007. *Gorilla society: Conflict, compromise, and co-operation between sexes.* Chicago: University Press of Chicago.

Harding, Emma J., Elizabeth S. Paul, and Michael Mendl. 2004. Animal behaviour: Cognitive bias and affective state. *Nature* 427 (6972): 312.

Hardt, Oliver, Einar Örn Einarsson, and Karim Nader. 2010. A bridge over troubled water: Reconsolidation as a link between cognitive and neuroscientific memory research traditions. *Annual Review of Psychology* 61: 141–167.

Hare, Brian, Victoria Wobber, and Richard Wrangham. 2012. The self-domestication hypothesis: Evolution of bonobo psychology is due to selection against aggression. *Animal Behaviour* 83 (3): 573–585.

Hare, Todd A., Colin F. Camerer, and Antonio Rangel. 2009. Self-control in decision-making involves modulation of the vmPFC valuation. *Science* 324 (5927): 646–648.

Harre, Rom. 2004. *Modeling: Gateway to the unknown*. New York: Elsevier.

Harris, Marvin. 1975 / 1991. *Cows, pigs, wars and witches: The riddles of culture*. London: Hutchinson and Co.

Harrison, Peter. 2006. "Science" and "religion": Constructing the boundaries. *Journal of Religion* 86 (1): 81–106.

Hartman, Sarah, and Jay Belsky. 2015. An evolutionary perspective on family studies: Differential susceptibility to environmental influences. *Family Process* 55 (4): 700–712.

Hauser, Marc. 2006. *Moral minds: How nature designed our universal sense of right and wrong*. New York: Ecco.

Hawkes, Kristen. 1991. Showing off: Tests of an hypothesis about men's foraging goals. *Ethology and Sociobiology* 12 (1): 29–54.

Hebb, Donald. 1944. *The organization of behavior: A neuropsychological theory*. New York: John Wiley and Sons.

Heck, Richard. 2000. Nonconceptual content and the "space of reasons." *Philosophical Review* 109 (4): 483–523.

Heft, Harry. 2007. The social constitution of perceiver-environment reciprocity. *Ecological Psychology* 19 (2): 85–105.

———. 2010. Affordances and the perception of landscape: An inquiry into environmental perception anaesthetics. In C. W. Thompson, P. Aspinall, and S. Bell, eds., *Innovative approaches to researching landscape and health*, 9–32. London: Routledge.

Heinrichs, Markus, Thomas Baumgartner, Clemens Kirschbaum, and Ulrike Ehlert. 2003. Social support and oxytocin interact to suppress cortisol and subjective responses to psychosocial stress. *Biological Psychiatry* 54 (12): 1389–1398.

Heinrichs, Markus, and Gregor Domes. 2008. Neuropeptides and social behaviour: effects of oxytocin and vasopressin in humans. *Progress in Brain Research* 170: 337–350.

Henazi, Peter S., and Louise Barrett. 1999. The value of grooming to female primates. *Primates* 40: 47–59. doi: 10.1007/BF02557701.

Henrich, Joseph. 2015. *The secret of our success: How culture is driving human evolution, domesticating our species and making us smarter*. Princeton, NJ: Princeton University Press.

Henrich, Joseph, and Richard McElreath. 2003. The evolution of cultural evolution. *Evolutionary Anthropology* 12 (3): 123–135.

Henriques, T. P., R. E. Szawka, L. A. Diehl, M. A. de Souza, C. N. Corrêa, B. C. C. Aranda, V. Sebben, et al. 2014. Stress in neonatal rats with different maternal care backgrounds: Monoaminergic and hormonal responses. *Neurochemical Research* 39 (12): 2351–2359.

Herbert, James D., Ian A. Sharp, and Brandon A. Gaudiano. 2002. Separating fact from fiction in the etiology and treatment of autism: A scientific review of the evidence. *Scientific Review of Mental Health Practice* 1 (1): 23–43.

Hermans, D., F. Baeyens, and P. Eelen. 2003. On the acquisition and activation of evaluative information. In J. Mushc and K. C. Klauer, eds., *The psychology of evaluation: Affective processes in cognition and emotion,* 139–168. New York: Lawrence Erlbaum.

Herrmann, Esther, Josep Call, María Victoria Hernandez-Lloreda, Brian Hare, and Michael Tomasello. 2007. Humans have evolved specialized skills of social cognition: The cultural intelligence hypothesis. *Science* 317 (5843): 1360–1366.

Hertzberger, Herman. 1991. *Lessons for students in architecture.* Ina Rike, trans. Rotterdam: 010.

Heyes, Cecilia. 2010. Where do mirror neurons come from? *Neuroscience and Biobehavioral Reviews* 34 (4): 575–583.

———. 2012. Grist and mills: On the cultural origins of cultural learning. *Philosophical Transactions of the Royal Society London B* 367 (1599): 2181–2191.

Hillenbrand, Laura. 2010. *Unbroken.* New York: Random House.

Hladký, Vojtěch, and Jan Havlíček. 2013. Was Tinbergen an Aristotelian? Comparison of Tinbergen's four whys and Aristotle's four causes. *Human Ethology Bulletin* 28 (4): 3–11.

Hobaiter, Catherine, Timothée Poisot, Klaus Zuberbühler, William Hoppitt, and Thibaud Gruber. 2014. Social network analysis shows direct evidence for social transmission of tool use in wild chimpanzees. *PLoS Biology* 12 (9): e1001960.

Hoffecker, John F. 2011. *Landscapes of the mind: Human evolution and the archaeology of thought.* New York: Columbia University Press.

Hofstadter, Douglas R. 2001. Analogy as the core of cognition. In Dedre Gentner, Keith J. Holyoak, and Boiko N. Kokinov, eds., *The analogial mind: Perspectives from cognitive science,* 499–538. Cambridge, MA: MIT Press.

Hofstadter, Douglas R., and Emmanuel Sander. 2013. *Surfaces and essences: Analogy as the fuel and fire of thinking.* New York: Basic Books.

Hohmann, George W. 1966. Some effects of spinal cord lesions on experienced emotional feelings. *Psychophysiology* 3 (2): 143–156.

Hohmann, Gottfried, and Barbara Fruth. 2003. Intra- and inter-sexual aggression by bonobos in the context of mating. *Behavior* 140 (11): 1389–1413.

Holbrook, Colin, and Daniel M. T. Fessler. 2015. The same, only different: Threat management systems as homologues in the tree of life. In P. J. Carroll, R. M. Arkin, and A. L. Wichman, eds., *Handbook of personal security,* 95–109. New York: Psychology Press.

Hoogland, John. 1995. *The black-tailed prairie dog: Social life of a burrowing animal.* Chicago: University of Chicago Press.

Hothersall, David. 1984. *History of psychology,* 2nd ed. New York: McGraw-Hill.

Houston, A. I. 2009. Flying in the face of nature. *Behavioural Processes* 80: 295–305.

Hrdy, Sarah Blaffer. 1977. *The langurs of Abu: Female and male strategies of reproduction.* Cambridge, MA: Harvard University Press.

———. 2009. *Mothers and others: The evolutionary origins of mutual understanding.* Cambridge, MA: Harvard University Press.

Hume, David. 1779 / 1998. *Dialogues concerning natural religion.* Richard Popkin, ed. Indianapolis: Hackett Publishing.

Humphrey, Nicholas K. 1976. The social function of intellect. In P. P. G. Bateson and R. A. Hinde, eds., *Growing points in ethology,* 303–317. Cambridge: Cambridge University Press.

Hupbach, Almut, Rebecca Gomez, and Lynn Nadel. 2013. Episodic memory reconsolidation: An update. In Cristina M. Alberini, ed., *Memory reconsolidation,* 233–247. San Diego, CA: Academic Press.

Hurley, Susan L. 1989. *Natural reasons.* New York: Oxford University Press.

———. 1998. *Consciousness in action.* Cambridge, MA: Harvard University Press.

Hurley, Susan L., and Nick Chater, eds. 2005. *Perspectives on imitation: From neuroscience to social science,* 2:55–77. Cambridge, MA: MIT Press.

Hutchinson, J., and G. Gigerenzer. 2005. Simple heuristics and rules of thumb: Where psychologists and behavioural biologists might meet. *Behavioural Processes* 69: 97–124

Hutto, Daniel D., and Erik Myin. 2012. *Radicalizing enactivism: Basic minds without content.* Cambridge, MA: MIT Press.

———. 2017. *Evolving enactivism: Basic minds meet content.* Cambridge, MA: MIT Press.

Insel, Thomas R. 2003. The neurobiology of affiliation: Implications for autism. In R. Davidson, K. R. Scherer, and H. H. Goldsmith, eds., *Handbook of affective sciences,* 1010–1020. New York: Oxford University Press.

Izard, Carroll E. 2011. Forms and functions of emotions: Matters of emotion–cognition interactions. *Emotion Review* 3 (4): 371–378.

Jablonka, Eva. 2012. Epigenetic inheritance and plasticity: The responsive germline. *Progress in Biophysics and Molecular Biology* 111 (2–3): 99–107.

Jablonka, Eva, Simona Ginsburg, and Daniel Dor. 2012. The co-evolution of language and emotions. *Philosophical Transactions of the Royal Society London B* 367 (1599): 2152–2159.

Jacobson, Anne J. 2008. Empathy, primitive reactions and the modularity of emotion. In Luc Faucher and Christine Tappolet, eds., *The modularity of emotions,* 95–113. Calgary, AB: University of Calgary Press.

James, William. 1879. The sentiment of rationality. *Mind* 4 (15): 317–346.

———. 1890 / 1899. *Principles of psychology,* 2 vols. London: Dover Publications.

———. 1902. *Varieties of religious experience.* London: Longmans Green.

———. 1983. What is an emotion? In *Essays in psychology.* Cambridge, MA: Harvard University Press.

Jeannerod, Marc. 1994. The representing brain: Neural correlates of motor intention and imagery. *Behavioral and Brain Sciences* 17: 187–245. doi: 10.1017/S0140525 X00034026.

———. 2006. *Motor cognition: What actions tell the self.* Oxford: Oxford University Press.

Jeffery, Kathryn J., Aleksandar Jovalekic, Madeleine Verriotis, and Robin Hayman. 2013. Navigating in a three-dimensional world. Target article plus responses in *Behavioral and Brain Sciences* 36 (5): 523–543.

Jennings, H. S. 1906. *Behavior of the lower organisms.* New York: Columbia University Press.

Jensen, K., and M. Tomasello. 2007. Chimpanzees are rational maximizers in an ultimatum game. *Science* 318 (5847): 107–109.

Jensen, Keith, Josep Call, and Michael Tomasello. 2007. Chimpanzees are vengeful but not spiteful. *Proceedings of the National Academy of Sciences* 104 (32): 13046–13050.

Jerison, Harry J. 1973. *Evolution of the brain and intelligence.* New York: Academic Press.

Jessee, Allison, Sarah C. Mangelsdorf, Geoffrey L. Brown, Sarah J. Schoppe-Sullivan, Aya Shigeto, and Maria S. Wong. 2010. Parents' differential susceptibility to the effects of marital quality on sensitivity across the first year. *Infant Behavior and Development* 33 (4): 442–452.

Jiang, Y., and L. S. Mark. 1994. The effect of gap depth on the perception of whether a gap is crossable. *Perception and Psychophysics* 56: 691.

Johnson, Allen W., and Timothy Earle. 2000. *The evolution of human societies: From foraging group to agrarian state,* 2nd ed. Stanford, CA: Stanford University Press.

Johnson, Jeffrey G., Patricia Cohen, Jocelyn Brown, Elizabeth M. Smailes, and David P. Bernstein. 1999. Childhood maltreatment increases risk for personality disorders during early adulthood. *Archives of General Psychiatry* 56 (7): 600–606.

Johnson, M. 1987. *The body in the mind: The bodily basis of meaning, imagination, and reason.* Chicago: University of Chicago Press.

Johnson, Orna R., and Allen Johnson. 1975. Male / Female relations and the organization of work in a machiguenga community. *American Ethnologist* 2 (4): 634–648.

Johnson, Phillip E. 1989. *Of pandas and people: The central question of biological origins.* Richardson, TX: Foundation for Thought and Ethics.

———. 1991. *Darwin on trial.* Washington, DC: Regnery Gateway Publishing.

Ju, Wendy, and Leila Takayama. 2009. Approachability: How people interpret automatic door movement as gesture. *International Journal of Design* 3 (2).

Juarrero, Alicia. 1999. *Dynamics in action: Intentional behavior as a complex.* Cambridge, MA: MIT Press.

Kahneman, Daniel. 2011. *Thinking, fast and slow.* New York: Farrar Straus and Giroux.

Kahneman, Daniel, Jack L. Knetsch, and Richard H. Thaler. 1990. Experimental tests of the endowment effect and the coase theorem. *Journal of Political Economy* 98 (6): 1325–1348.

Kahneman, Daniel, and Amos Tversky, eds. 2000. *Choices, values and frames.* New York: Cambridge University Press.

Kant, Immanuel. 1790 / 1987. *Critique of judgment,* §75, 312. Werner S. Pluhar, trans. Indianapolis: Hackett Publishing.

Kaplan, Hillard, Kim Hill, Jane Lancaster, and A. Magdalena Hurtado. 2000. A theory of human life history evolution: Diet, intelligence, and longevity. *Evolutionary Anthropology* 9 (4): 156–185.

Kaplan, Hillard, Paul L. Hooper, and Michael Gurven. 2009. The evolutionary and ecological roots of human social organization. *Philosophical Transactions of the Royal Society B* 364 (1533): 3289–3299.

Karama, Sherif, André Roch Lecours, Jean-Maxime Leroux, and Pierre Bourgouin, et al. 2002. Areas of brain activation in males and females during viewing of erotic film excerpts. *Human Brain Mapping* 16 (1): 1–13.

Katz, Jerrold J., and Jerry A. Fodor. 1963. The structure of a semantic theory. *Language* 39 (2): 170–210.

Kauffman, Stuart A. 1993. *The origins of order: Self-organization and selection in evolution.* Oxford: Oxford University Press.

Keating, Caroline F., Allan Mazur, and Marshall H. Segall. 1977. Facial gestures which influence the perception of status. *Sociometry* 40 (4): 374–378.

Keller, Evelyn Fox. 2010. *The mirage of a space between nature and nurture.* Durham, NC: Duke University Press.

Kelly, R. 1993. *Constructing inequality.* Ann Arbor: University of Michigan Press.

Kelly, Sean D. 2001. The non-conceptual content of perceptual experience: Situation dependence and fineness of grain. *Philosophy and Phenomenological Research* 62 (3): 601–608.

Kemp, Andrew H., and Adam J. Guastella. 2011. The role of oxytocin in human affect: A novel hypothesis. *Current Directions in Psychological Science* 20 (4): 222–231.

Kendrick, Keith M., Eric B. Keverne, and Basil A. Baldwin. 1987. Intracerebroventricular oxytocin stimulates maternal behaviour in the sheep. *Neuroendocrinology* 46 (1): 56–61.

Kenny, Anthony. 1963 / 2003. *Action, emotion and will.* New York: Routledge.

Kensinger, Elizabeth A., Rachel J. Garoff-Eaton, and Daniel L. Schacter. 2007a. Effects of emotion on memory specificity: Memory trade-offs elicited by negative visually arousing stimuli. *Journal of Memory and Language* 56 (4): 575–591.

———. 2007b. How negative emotion enhances the visual specificity of a memory. *Journal of Cognitive Neuroscience* 19 (11): 1872–1887.

Keysar, Boaz, Benjamin A. Converse, Jiunwen Wang, and Nicholas Epley. 2008. Reciprocity is not give and take: Asymmetric reciprocity to positive and negative acts. *Psychological Science* 19 (12): 1280–1286.

Kibutu, Thomas Njuguna. 2006. *Development, gender and the crisis of masculinity among the Maasai people Ngong, Kenya.* Doctoral Dissertation. University of Leicester.

Kindt, Merel, Marieke Soeter, and Bram Vervliet. 2009. Beyond extinction: Erasing human fear responses and preventing the return of fear. *Nature Neuroscience* 12: 256–258.

Kingdon, Jonathan. 1988. *East African mammals: An atlas of evolution in Africa,* vol. 3, part A. Chicago: University of Chicago Press.

Kintsch, W. 1974. *The representation of meaning in memory.* New York: Lawrence Erlbaum.

Kitcher, Philip. 2013. Things fall apart. The Stone (blog). *New York Times.* September 8, 2013. https://opinionator.blogs.nytimes.com/2013/09/08/things-fall-apart/.

Klein, Malcolm, Hans-Jürgen Kerner, Cheryl Maxson, and Elmar Weitekamp, eds. 2004. *The Eurogang paradox: Street gangs and youth groups in the U.S. and Europe.* Dordrecht: Kluwer Academic.

Klein, Richard G. 2009. *The human career: Human biological and cultural origins,* 3rd ed. Chicago: University of Chicago Press.

Klein, Stanley B. 2014. What can recent replication failures tell us about the theoretical commitments of psychology? *Theory and Psychology* 24 (3): 326–338.

———. 2017. The unplanned obsolescence of psychological science and an argument for its revival. *Psychology of Consciousness: Theory, Research, and Practice* 3 (4): 357–379.

Klein, Stanley B., Rami H. Gabriel, Cynthia E. Gangi, and Theresa E. Robertson. 2008. Reflections on the self: A case study of a prosopagnosic patient. *Social Cognition* 26 (6): 766–777.

Knickmeyer, Rebecca C., Sylvain Gouttard, Chaeryon Kang, Dianne Evans, Kathy Wilber, J. Keith Smith, Robert M. Hamer, Weili Lin, Guido Gerig, and John H. Gilmore. 2008. A structural MRI study of human brain development from birth to 2 years. *Journal of Neuroscience* 28 (47): 12176–12182.

Knobe, Joshua, and Brian Leiter. 2006. The case for Nietzschean moral psychology. *SSRN Electronic Journal.* doi: 10.2139/ssrn.816224.

Knoch, Daria, Alvaro Pascual-Leone, Kaspar Meyer, Valerie Treyer, and Ernst Fehr. 2006. Diminishing reciprocal fairness by disrupting the right prefrontal cortex. *Science* 314 (5800): 829–832.

Kongzi / Confucius. 2008. *The analects.* Oxford: Oxford University Press.

Kosfeld, Michael, Markus Heinrichs, Paul J. Zak, Urs Fischbacher, and Ernst Fehr. 2005. Oxytocin increases trust in humans. *Nature* 435 (7042): 673–676.

Kosslyn, Stephen M. 1973. Scanning visual images: Some structural implications. *Perception and Amp; Psychophysics* 14: 90–94.

———. 2005. Mental images and the brain. *Cognitive Neuropsychology* 22 (3 / 4): 333–347.

Kosslyn, Stephen M., William L. Thompson, and Giorgio Ganis. 2006. *The case for mental imagery.* New York: Oxford University Press.

Krausz, Michael. 2009. Creativity and self-transformation. In Karen Bardsley, Denis Dutton, and Michael Krausz, eds., *The idea of creativity,* 191–204. Leiden: Brill.

Kraut, Robert E., and Robert E. Johnston. 1979. Social and emotional messages of smiling: An ethological approach. *Journal of Personality and Social Psychology* 37 (9): 1539–1553.

Kringelbach, Morten L., and Edmund T. Rolls. 2004. The functional neuroanatomy of the human orbitofrontal cortex: Evidence from neuroimaging and neuropsychology. *Progress in Neurobiology* 72 (5): 341–372.

Kummer, Hans. 1978. On the value of social relationships to nonhuman primates: A heuristic scheme. *Social Science Information* 17 (4 / 5): 687–705.

LaBar, Kevin S., and Roberto Cabeza. 2006. Cognitive neuroscience of emotional memory. *Nature Reviews Neuroscience* 7 (1): 54–64.

Lakoff, George. 1987. *Women, fire and dangerous things: What categories reveal about the mind.* Chicago: University of Chicago Press.

Lakoff, George, and Mark Johnson. 1980. *Metaphors we live by.* Chicago: University of Chicago Press.

Landau, H. G. 1951. On dominance relations and the structure of animal societies. *Bulletin of Mathematical Biophysics* 13: 1–19. doi: 10.1007/BF02478336.

Langercrantz, H., and T. Ringstedt. 2001. Organization of the neuronal circuits in the central nervous system during development. *Acta Paediatrica* 90 (7): 707–715.

Larquet, Marion, Giorgio Coricelli, Gaëlle Opolczynski, and Florence Thibaut. 2010. Impaired decision making in schizophrenia and orbitofrontal cortex lesion patients. *Schizophrenia Research* 116 (2–3): 266–273.

Laurence, S., and E. Margolis. 1999. Concepts and cognitive science. In E. Margolis and S. Laurence, eds., *Concepts: Core Readings,* 3–81. Cambridge, MA: MIT Press.

Leach, Edmund Ronald. 1969. *Genesis as myth and other essays,* vol. 39. London: J. Cape.

Leakey, Richard E., and Roger Lewin. 1978. *People of the lake: Mankind and its beginnings.* New York: Anchor Press.

Lebreton, Maël, Soledad Jorge, Vincent Michel, Bertrand Thirion, and Mathias Pessiglione. 2009. An automatic valuation system in the human brain: Evidence from functional neuroimaging. *Neuron* 64 (3): 431–439.

LeDoux, Joseph. 1996. *The emotional brain: The mysterious underpinnings of emotional life.* New York: Simon and Schuster.

Lee, D. N. 1976. A theory of visual control of braking based on information about time-to-collision. *Perception* 5: 437–459.

Lee, Kang, Michelle Eskritt, Lawrence A. Symons, and Darwin Muir. 1998. Children's use of triadic eye gaze information for "mind reading." *Developmental Psychology* 34 (3): 525–539.

Lee, R. 1993. *The Dobe Ju'hoansi.* New York: Harcourt Brace.

Leekam, Susan, Simon Baron-Cohen, Dave Perrett, Maarten Milders, and Sarah Brown. 1997. Eye-direction detection: A dissociation between geo-metric and joint attention skills in autism. *British Journal of Developmental Psychology* 15 (1): 77–95.

Legrand, Dorothee, and Marco Iacoboni. 2010. Intersubjective intentional actions. In F. Grammont, D. Legrand, and P. Livet, eds., *Naturalizing intention in action.* Cambridge, MA: MIT Press.

Leimar, Olof, and Peter Hammerstein. 2001. Evolution of cooperation through indirect reciprocity. *Proceedings of the Royal Society of London; Biological Sciences* 268: 745–753.

Leiss, William, Stephen Kline, Jackie Botterill, Jacqueline Botterill, and Sut Jhally. 2005. *Social communication in advertising: Consumption in the mediated marketplace.* New York: Psychology Press.

Leith, Karen P., and Roy Baumeister. 2008. Empathy, shame, guilt, and narratives of interpersonal conflicts: Guilt-prone people are better at perspective taking. *Journal of Personality* 66: 1–37. doi: 10.1111/1467-6494.00001.

Leroi, Armand Marie. 2014. *The lagoon: How Aristotle invented science.* New York: Viking.

Leroi-Gourhan, Andre. 1968. *The art of prehistoric man in Western Europe.* London: Thames and Hudson.

Leslie, Alan M. 1987. Pretence and representation: The origins of "theory of mind." *Psychological Review* 94 (4): 412–426.

Lévi-Strauss, Claude. 1967. Structure and dialectics. *Structural Anthropology,* 229–238.

Levy, Kenneth N., Benjamin N. Johnson, Tracy L. Clouthier, J. Scala, and Christina M. Temes. 2015. An attachment theoretical framework for personality disorders. *Canadian Psychology / Psychologie Canadienne* 56 (2): 197–207.

Lewis, David K., John P. Burgess, and A. P. Hazen. 1991. *Parts of classes.* Cambridge, MA: Blackwell.

Lewis-Williams, J. David. 2002. *The mind in the cave: Consciousness and the origins of art.* London: Thames and Hudson.

———. 2010. The imagistic web of San myth, art and landscape. *Southern African Humanities* 22 (1): 1–18.

Lewis-Williams, J. David, and Sam Challis. 2011. *Deciphering ancient minds: The mystery of San bushmen rock art.* London: Thames and Hudson.

Lewis-Williams, J. David, Thomas A. Dowson, Paul G. Bahn, H-G. Bandi, Robert G. Bednarik, John Clegg, Mario Consens, et al. 1988. The signs of all times: Entoptic phenomena in upper palaeotlithic art. *Current Anthropology* 29 (2): 201–245.

Lewontin, Richard. 1984. *Not in our genes: Biology, ideology, and human nature.* New York: Pantheon Books.

Lieberman, D., John Tooby, and Leda Cosmides. 2007. The architecture of human kin detection. *Nature* 445: 727–731.

Lin, Martin. 2006. Teleology and human action. In Spinoza, ed., *Philosophical Review* 115 (3): 317–354.

Liverani, Mario. 2006. *Uruk: The first city.* Z. Bahrani and M. Van De Mieroop, trans. London: Equinox.

Llewelyn-Davis, Melissa. 1981. Women, warriors and patriarchs. In S. B. Ortner and H. Whitehead, eds., *Sexual meanings: The cultural construction of gender and sexuality,* 331–357. Cambridge: Cambridge University Press.

Llinas, Rodolfo, R. 2001. *I of the vortex.* Cambridge, MA: MIT Press.

Lloyd, G. E. R. 1975. *Greek science after Aristotle.* London: Chatto and Windus.

Locke, J. 1690 / 1980. *Second treatise of government.* C. B. Macpherson, ed. Indianapolis: Hackett Publishing.

Lohse, K. 2017. Come on feel the noise—from metaphors to null models. *Journal of Evolutionary Biology* 30 (8): 1506–1508.

Loomis, Jack M., and Andy B. Beall. 1998. Visually controlled locomotion: Its dependence on optic flow, three-dimensional space perception, and cognition. *Ecological Psychology* 10 (3 / 4): 271–285.

Lorblanchet, Michel. 1995. *Les grottes ornées de la préhistoire: Nouveaux regards.* Paris: Editions Errance.

Lowenstein, George. 1994. The psychology of curiosity: A review and reinterpretation. *Psychological Bulletin* 116 (1): 75–98.

Luria, A. R. 1972. *The man with a shattered world: The history of a brain wound.* Cambridge, MA: Harvard University Press.

Lurz, Robert W. 2006. Conscious beliefs and desires: A same-order approach. In Uriah Kriegel and Kenneth Williford, eds., *Self-representational approaches to consciousness.* Cambridge, MA: MIT Press.

Lutz, Catherine A. 1986. Emotion, thought, and estrangement: Emotion as a cultural category. *Cultural Anthropology* 1 (3): 287–309.

———. 1988. *Unnatural emotions: Everyday sentiments on a Micronesian atoll and their challenge to Western theory.* Chicago: University of Chicago Press.

Macdonald, Cynthia and Graham Macdonald, eds. 2010. *Emergence in mind.* Oxford: Oxford University Press.

Malinowski, Bronislaw. 1935. *Coral gardens and their magic.* London. Routledge.

Malraux, André. 1964 / 1978. *The voices of silence: Man and his art.* Stuart Gilbert, trans. Princeton, NJ: Princeton University Press.

Mandler, George. 1980. Recognizing: The judgment of previous occurrence. *Psychological Review* 87: 252–271.

———. 1984. *Mind and body: Psychology of emotion and stress.* New York: W. W. Norton.

Marcel, Anthony J. 1983. Conscious and unconscious perception: An approach to the relations between phenomenal experience and perceptual processes. *Cognitive Psychology* 15: 238–300.

Marcus, Gary. 2008. *Kluge.* Boston: Houghton Mifflin.

Maren, S. 2001. Neurobiology of Pavlovian fear conditioning. *Annual Review of Neuroscience* 24: 897–931.

Marlowe, Frank W., and Julia C. Berbesque. 2009. Tubers as fallback foods and their impact on Hadza hunter gatherers. *American Journal of Physical Anthropology* 140 (4): 751–758.

Marquez, C., G. L. Poirier, M. I. Cordero, M. H. Larsen, A. Groner, J. Marquis, P. J. Magistretti, D. Trono, and C. Sandi. 2013. Peripuberty stress leads to abnormal aggression, altered amygdala and orbitofrontal reactivity and increased prefrontal MAOA gene expression. *Translational Psychiatry* 3 (1): e216.

Marr, David. 1982. *Vision: A computational investigation into the human representation and processing of visual information.* Cambridge, MA: MIT Press.

Marsh, Kerry L., Lucy Johnston, Michael J. Richardson, and R. C. Schmidt. 2009. Toward a radically embodied, embedded social psychology. *European Journal of Social Psychology* 39 (7): 1217–1225.

Marsh, Kerry L., Michael J. Richardson, and Reuben M. Baron. 2006. Contrasting approaches to perceiving and acting with others. *Ecological Psychology* 18 (1): 1–38.

Marshall, L. 1976. *The !Kung of Nyae Nyae.* Cambridge, MA: Harvard University Press.

Martin, R. D. 1983. *Human brain evolution in an ecological context: Fifty-second James Arthur lecture on the evolution of the human brain.* American Museum of Natural History, New York.

Martinsson, Peter, Kristian Ove R. Myrseth, and Conny Wollbrant. 2012. Cooperation in social dilemmas: The necessity of seeing self-control conflict. ESMT Working Paper, 10–004 (R1).

Mason, W. A. 1978. Ontogeny of social systems. In D. J. Chivers and J. Herbert, eds., *Recent advances in primatology,* 1:5–14. London: Academic.

Mauss, Marcel. 1930. Les civilisations, elements et formes. In *Civilisation, le mot et l'idee,* 81–106. Centre international de synthese. Premiere semaine, 2e fascicule. La Renaissance du Livre, Paris.

———. 1954. *The gift: Forms and functions of exchange in archaic societies.* Glencoe, IL: Free Press

Mcbrearty, Sally, and Alison Brooks. 2000. The revolution that wasn't: A new interpretation of the origin of modern human behavior. *Journal of Human Evolution* 39: 453–563. doi: 10.1006 / jhev.2000.0435.

McCall, Cade, and Tania Singer. 2012. The animal and human neuroendocrinology of social cognition, motivation and behavior. *Nature Neuroscience* 15 (5): 681–688.

McCracken, Grant D. 1988. *Culture and consumption: New approaches to the symbolic character of consumer goods and activities.*Bloomington: Indiana University Press.

McDowell, John. 2011. *Perception as a capacity for knowledge.* Aquinas Lecture 75.

McElreath, R., T. H. Clutton-Brock, Ernst Fehr, Daniel Fessler, E. H. Hagen, and P. Hammerstein. 2003. The role of cognition and emotion in cooperation. In *Genetic and Cultural Evolution of Cooperation,* 125–152. Cambridge, MA: MIT Press.

McGinn, Colin. 2011. *The meaning of disgust.* New York: Oxford University Press.

McNamara, J. M., and A. Houston. 2009. Integrating function and mechanism. *Trends in Ecology & Evolution* 24 (12): 670–675.

Mead, George H. 1913. The social self. *Journal of Philosophy, Psychology, and Scientific Methods* 10: 374–380.

Meggitt, M. 1965. *Desert people.* Sydney: Angus and Roberston.

Merleau-Ponty, Maurice. 1942. *La structure du comportement.* Paris: Presses Universitaires de France.

———. 1945 / 1962. *Phenomenology of perception.* Colin Smith, trans. London: Routledge and Kegan Paul.

Metcalfe, Janet. 2008. Evolution of metacognition. In John Dunlosky and Robert A. Bjork, eds., *Handbook of metamemory and memory,* 29–46. New York: Psychology Press.

Metcalfe, Janet, and Hedy Kober. 2005. Self-reflective consciousness and the projectable self. In H. Terrace and J. Metcalfe, eds., *The missing link in cognition: Origins of self-reflective consciousness,* 57–83. New York: Oxford University Press.

Michaels, Claire, Rob Withagen, David M. Jacobs, Frank Zaal, and Raoul Bongers. 2001. Information, perception, and action: A reply to commentators. *Ecological Psychology* 13: 227–244. doi: 10.1207/S15326969ECO1303_3.

Mignault, Alain, and Avi Chaudhuri. 2003. The many faces of a neutral face: Head tilt and perception of dominance and emotion. *Journal of Nonverbal Behavior* 27 (2): 111–132.

Mill, James. 1829a / 1869. *Analysis of the phenomena of the human mind,* 2 vols., 2nd ed. London: Longman, Green, Reader and Dyer.

Mill, John Stuart. 1829 / 2002. *A system of logic.* Honolulu: University Press of the Pacific.

Miller, George A. 1956. The magical number seven, plus or minus two: Some limits on our capacity for processing information. *Psychological Review* 63 (2): 81–97.

Miller, L., and C. Hou. 2004. Portraits of artists: Emergence of visual creativity in dementia. *Archives of Neurology* 61: 842–844. doi: 10.1001 / archneur.61.6.842.

Millikan, Ruth G. 1984. Language, thought, and other biological categories. *Behaviorism* 14 (1): 51–56.

———. 1989. Biosemantics. *Journal of Philosophy* 86 (6): 281–297.

———. 1996. Pushmi-pullyu representations. In James Tomberlin, ed., *Philosophical Perspectives,* 9:185–200. Atascadero, CA: Ridgeview.

———. 2004. On reading signs: Some differences between us and the others. In D. Kimbrough Oller, Ulrike Griebel, Gerd B. Muller, Gunter P. Wagner, and Werner Callebaut, eds., *Evolution of communication systems: A comparative approach,* 15–29. Cambridge, MA: MIT Press.

Mills, John A. 1998. *Control: A history of behavioral psychology.* New York: New York University Press.

Milner, A. David, and Melvyn A. Goodale. 1993. Visual pathways to perception and action. In T. P. Hicks, S. Molotchnikoff, and T. Ono, eds., *Progress in brain research,* 95:317–337. Amsterdam: Elsevier.

———. 1995. *The visual brain in action.* Oxford: Oxford University Press.

Milner, Murray. 1987. Theories of inequality: An overview and a strategy for synthesis. *Social Forces* 65 (4): 1053–1089.

Mitchell, W. J. T., ed. 1981. *On narrative.* Chicago: University of Chicago Press.

Mithen, S. J. 2006. *The singing Neanderthals: The origins of music, language, mind and body.* Cambridge, MA: Harvard University Press.

Moffett, M. W. 2013. Human identity and the evolution of societies. *Human Nature.* doi: 10.1007/s12110-013-9170-3.

Montagu, Ashley. 1989. *Growing young,* 2nd ed. Granby, MA: Bergin and Garvey.

Morgan, C. Lloyd. 1923. *Emergent evolution. The Gifford lectures, 1922.* London: Williams and Norgate.

Morgan, Dave, and Crickette Sanz. 2003. Naïve encounters with chimpanzees in the Goualougo Triangle, Republic of Congo. *International Journal of Primatology* 24 (2): 369–381.

Morris, Desmond, ed. 1969. *Primate ethology: Essays on the socio-sexual behavior of apes and monkeys.* Garden City, NY: Anchor Books.

Mulder, Monique Borgerhoff. 1988. Behavioural ecology in traditional societies. *Trends in Ecology and Evolution* 3 (10): 260–264.

Muller, Martin N., and Richard W. Wrangham. 2009. *Sexual coercion in primates and humans: An evolutionary perspective on male aggression against females.* Cambridge, MA: Harvard University Press.

Mumford, Lewis. 1961. *The city in history: Its origins, its transformations, and its prospects.* New York: Harcourt, Brace and World.

Munro, N., E. Baker, K. K. McGregor, J. Arciuli, and K. Docking. 2011. Iconic gestures support toddlers' retention of word-referent pairings. *12th International Congress for the Study of Child Language.* Montreal.

Nader, L., and O. Hardt. 2009. Update on memory systems and processes. *Neuropsychopharmacology* 36 (1): 251–273.

Nagel, Thomas. 2012. *Mind and cosmos: Why the materialist neo-Darwinian conception of nature is almost certainly false.* Oxford: Oxford University Press.

Nakamura M., P. G. Nestor, J. J. Levitt, A. S. Cohen, T. Kawashima, M. E. Shenton, and R. W. McCarley. 2008. Orbitofrontal volume deficit in schizophrenia and thought disorder. *Brain* 131: 180–195. doi: 10.1093/brain/awm265.

Narvaez D., J. Panksepp, A. N. Schore, and T. R. Gleason. 2013. The value of using an evolutionary framework for gauging children's wellbeing. In D. Narvaez, J. Panksepp, A. N. Schore, T. R. Gleason, eds., *Evolution, early experience and human development: From research to practice and policy,* 3–30. New York: Oxford University Press.

Neisser, Ulrich. 1976. *Cognition and reality: Principles and implications of cognitive psychology.* New York: W. H. Freeman.

Neitzel, Jill, and Timothy Earle. 2014. Dual-tier approach to societal evolution and types. *Journal of Anthropological Archaeology* 36: 181–195.

Newton, Stephen J. 2008. *Art and Ritual: A painter's journey.* London: Ziggurat Books International.

Nielsen, J., T. H. C. Kruger, U. Hartmann, T. Passie, T. Fehr, and M. Zedler. 2013. Synaesthesia and sexuality: The influence of synaesthetic perceptions on sexual experience. *Frontiers in Psychology* 4: 751. doi: 10.3389/fpsyg.2013.00751.

Noble, W., and I. Davidson. 1994. *Human evolution, language and the mind: A psychological and archaeological inquiry.* Cambridge: Cambridge University Press.

Noë, Alva. 2005. *Action in perception.* Cambridge, MA: MIT Press.

———. 2009. Extending our view of mind. *Trends in Cognitive Sciences* 13 (6): 237–238.

———. 2015. *Strange tools: Art and human nature.* New York: Hill and Wang.

Norenzayan, A. 2013. *Big gods: How religion transformed cooperation and conflict.* Princeton, NJ: Princeton University Press.

Norenzayan, A., A. Sharrif, W. Gervais, A. Willard, R. A. McNamara, E. Slingerland, and J. Henrich. 2016. The cultural evolution of prosocial religions. *Behavioral and Brain Sciences* 39: 10–65.

Norman, Joel. 2002. Two visual systems and two theories of perception: An attempt to reconcile the constructivist and ecological approaches. *Behavioral and Brain Sciences* 25: 73–144.

Nowak, Andrzej, and Robin R. Vallacher, eds. 1998. *Dynamical social psychology,* vol. 647. New York: Guilford Press.

Nowak, Martin A. 2006. Five rules for the evolution of cooperation. *Science* 314 (5805): 1560–1563.

Nowak, Martin A., and K. Sigmund. 1998. Evolution of indirect reciprocity by image scoring. *Nature* 393 (6685): 573–577.

Nussbaum, Martha. 2001. *Upheavals of thought: The intelligence of emotions.* Cambridge: Cambridge University Press.

———. 2003. *Upheavals of thought.* Cambridge: Cambridge University Press.

Okasha, S., and K. Binmore. 2012. *Evolution and rationality: Decisions, co-operation and strategic behaviour.* Cambridge: Cambridge University Press.

Olff, Miranda, et al. 2013. The role of oxytocin in social bonding, stress regulation and mental health: An update on the moderating effects of context and interindividual differences. *Psychoneuroendocrinology* 38.9 (2013): 1883–1894.

Olsen, M. A., and R. H. Fazio. 2001. Implicit attitude formation through classical conditioning. *Psychological Science* 12 (5): 413–417.

Olson, Mansur. 1965. *The logic of collective action: Public goods and the theory of groups,* revised ed. Cambridge, MA: Harvard University Press.

Olsson, Erik J. 2011. The value of knowledge. *Philosophy Compass* 6 (12): 874–883.

Orchard, Andy. 1995. *Pride and Prodigies: Studies in the Monsters of the Beowulf-Manuscript,* chap. 4. New York: D. S. Brewer.

O'Regan, J. Kevin. 2001. What it is like to see: A sensorimotor theory of perceptual experience. *Synthese* 192 (1): 79–103.

O'Regan, J. Kevin, and Alva Noë. 2001. A sensorimotor account of vision and visual consciousness. *Behavioral and Brain Sciences* 24 (5): 939–973.

Orpen, J. M. 1874. A glimpse into the mythology of the Maluti Bushmen. *Cape Monthly Magazine* 9: 1–11.

O'Shaughnessy, Brian. 1980. *The will: A dual aspect theory.* Cambridge: Cambridge University Press.

———. 2003. *Consciousness and the world.* Oxford: Clarendon Press.

Ospovat, D. 1995. *The development of Darwin's theory.* Cambridge: Cambridge University Press.

Ostrom, Elinor. 2000. Collective action and the evolution of social norms. *Journal of Economic Perspectives* 14 (3): 137–158.

———. 2015. *Governing the commons.* Cambridge: Cambridge University Press.

O'Wolff, Jerry, and Paul Sherman. 2007. *Rodent societies: An ecological and evolutionary perspective.* Chicago: University of Chicago Press.

Oyegbile, O. Temitayo, and Catherine Marler. 2005. Winning fights elevates testosterone levels in California mice and enhances future ability to win fights. *Hormones and Behavior* 48: 259–267. doi: 10.1016/j.yhbeh.2005.04.007.

Paabo, S. 2014. The human condition: A molecular approach. *Cell* 157 (1): 216–226.

Pacherie, Elisabeth. 2000. The content of intentions. *Mind and Language* 15 (4): 400–432.

———. 2008. The phenomenology of action. *Cognition* 107 (1): 179–217.

———. 2011. Nonconceptual representations for action and the limits of intentional control. *Social Psychology* 42 (1): 67–73.

———. 2015. Conscious intentions: The social creation myth. In T. Metzinger and J. M. Windt, eds., *Open Mind.* Frankfurt am Main: Mind Group. doi: 10.15502/9783958570122.

Pagani, Luca, Stephan Schiffels, Deepti Gurdasani, Petr Danecek, . . . et al. 2015. Tracing the route of modern humans out of Africa by using 225 human genome sequences from Ethiopians and Egyptians. *American Journal of Human Genetics* 96 (6): 986–991.

Pagel, Mark. 2012. *Wired for culture: Origins of the human social mind.* New York: W. W. Norton.

Paley, William. 2008. *Natural theology; or evidences of the existence and attributes of the deity,* reprint ed. Oxford: Oxford University Press.

Palmer, S. E. 1999. *Vision science: Photons to phenomenology.* Cambridge, MA: MIT Press.

Palombit, R. 1999. Infanticide and the evolution of pair bonds in nonhuman primates. *Evolutionary Anthropology* 7 (4).

Panchanathan, K., and R. Boyd. 2003. A tale of two defectors: The importance of standing for evolution of indirect reciprocity. *Journal of Theoretical Biology* 224 (1): 115–126.

Panksepp, Jaak. 1971. Aggression elicited by electrical stimulation of the hypothalamus in albino rats. *Physiology and Behavior* 6: 311–316.

———. 1996. Affective neuroscience: A paradigm to study the animate circuits for human emotions. In Robert D. Kavanaugh, Betty Zimmerberg, and Steven Fein, eds., *Emotion: Interdisciplinary Perspectives,* 29–60. Mahwah, NJ: Lawrence Erlbaum.

———. 1998. *Affective neuroscience: The foundations of human and animal emotions.* New York: Oxford University Press.

———. 2011. Cross-species affective neuroscience decoding of the primal affective experiences of humans and related animals. *PLoS ONE* 6 (8): e21236.

———. 2012. Naturalizing the mammalian mind. *Journal of Consciousness Studies* 19 (3 / 4): 1–19.

———. 2013. How primary-process emotional systems guide child development: Ancestral regulators of human happiness, thriving, and suffering. In D. Narvaez, J. Panksepp, A. N. Schore, and T. R. Gleason, eds., *Evolution, early experience and human development: From research to practice and policy,* 74–94. New York: Oxford University Press.

Panksepp, Jaak, and Lucy Biven. 2012. *The archaeology of mind: Neuroevolutionary origins of human emotions.* New York: W. W. Norton.

Parker, S. W., and C. A. Nelson. 2005. The impact of early institutional rearing on the ability to discriminate facial expressions of emotion: An event related potential study. *Childhood Development* 76: 54–72. doi: 10.1111/j.1467-8624.2005.00829.x.

Peacocke, Christopher. 1992. *A study of concepts.* Cambrige, MA: MIT Press.

———. 1998. Nonconceptual content defended. *Philosophy and Phenomenological Research* 58 (2): 381–388.

Pepper, G. V., and D. Nettle. 2017. The behavioural constellation of deprivation: Causes and consequences. *Behavioral and Brain Sciences* 11: 1–72.

Perugini, M., M. Gallucci, F. Presaghi, and A. P. Ercolani. 2003. The personal norm of reciprocity. *European Journal of Personality* 17 (4): 251–283. doi: 10.1002/per.474

Pessoa, Luiz. 2013. *The cognitive-emotional brain.* Cambridge, MA: MIT Press.

Pettitt, P., and N. Bader. 2000. Direct AMS radiocarbon dates for the Sungir mid upper palaeolithic burials. *Antiquity* 74 (284): 269–270. doi: 10.1017/S0003598X00059196.

Pham, Michale T., Joel B. Cohen, John W. Pracejus, and G. David Hughes. 2001. Affect monitoring and the primacy of feelings in judgment. *Journal of Consumer Research* 28 (2001): 167–188.

Phelps, Elizabeth. 2006. Emotion and cognition: Insights from studies of the human amygdala. *Annual Review of Psychology* 57: 27–53.

Phelps, Elizabeth, and Michael Gazzaniga. 1992. Hemispheric differences in mnemonic processing: the effects of left hemisphere interpretation. *Neuropsychologia* 30 (3): 293–297.

Piaget, Jean. 1923. *The language and thought of the child.* New York: Psychology Press.

Piaget, Jean, and B. Inhelder. 1969. *The psychology of the child.* New York: Basic Books.

Pinker, Steven. 2010. The cognitive niche: Coevolution of intelligence, sociality, and language. *Proceedings of the National Academy of Sciences* 107, supp. 2: 8993–8999.

———. 2011. *The better angels of our nature.* New York: Viking.

Pinker, Steven, Werner Kalow, and Harold Kalant. 1997. Evolutionary psychology: An exchange. Reply by Stephen Jay Gould in response to *Darwinian Fundamentalism* (June 12, 1997). *New York Review of Books,* October 9.

Pinkham, A. E., D. L. Penn, D. O. Perkins, and J. Lieberman. 2003. Implications for the neural basis of social cognition for the study of schizophrenia. *American Journal of Psychiatry* 160: 815–824. doi: 10.1176/appi.ajp.160.5.815.

Pizarro, D., and P. Bloom. 2003. The intelligence of the moral intuitions: Comments on Haidt. *Psychological Review* 110 (1): 193–196.

Plantinga, Alvin. 2010. *Science and religion* (debate with Daniel Dennett). Oxford: Oxford University Press.

Plato. 1992. *The Republic.* G. M. A. Grube, trans. Indianapolis: Hackett Publishing.

Podrouzek, W., and D. Furrow. 1988. Preschoolers' use of eye contact while speaking: The influence of sex, age, and conversational partner. *Journal of Psycholinguistic Research* 17: 89–98.

Pollan, Michael. 2018. *How to change your mind: What the new science of psychedelics teaches us about consciousness, dying, addiction, depression, and transcendence.* New York: Penguin Press.

Portmann, A. 1990. *A zoologist looks at human kind.* New York: Columbia University Press.

Potts, R. 1996. Evolution and climate variability. *Science* 273: 922–923.

Price, J. L., S. T. Carmichael, and W. C. Drevets. 1996. Networks related to the orbital and medial prefrontal cortex: A substrate for emotional behavior? *Progress in Brain Research* 107: 523–536. doi: 10.1016/S0079-6123(08)61885-3.

Prinz, Jesse. 2004. *Gut reactions: A perceptual theory of emotion.* Oxford: Oxford University Press.

———. 2011. Against empathy. *Southern Journal of Philosophy* 49 (Spindel supp.): 214–233.

Proctor, Rebecca Darby, A. Williamsona, Frans B. M. de Waal, and Sarah F. Brosnana. 2013. Chimpanzees play the ultimatum game. *Proceedings of the National Academy of Sciences of the United States of America* 110 (6).

Proust, Joelle. 2007. Metacognition and metarepresentation: Is a self-directed theory of mind a precondition for metacognition? *Synthese* 2: 271–295.

———. 2009. The representational basis of brute metacognition: A proposal. In Robert Lurz, ed., *Philosophy of animal minds,* 165–183. Cambridge: Cambridge University Press:

———. 2015. The representational structure of feelings. In T. Metzinger and J. M. Windt, eds., *Open Mind,* 31:1–25. Frankfurt am Main: Mind Group.

————. 2016. The evolution of communication and metacommunication in primates. *Mind and Language* 31 (2): 177–203.

Pusey, A. 2001. Of genes and apes. In F. de Waal, ed., *Tree of origin*. Cambridge, MA: Harvard University Press.

Putnam, H. 1975 / 2003. *Mind, language and reality. Philosophical papers,* vol. 2. Cambridge: Cambridge University Press.

Quine, W. V. 1960. *Word and object*. Cambridge, MA: MIT Press.

Raftopoulos, Athanassios, and Vincent C. Müller. 2006. The phenomenal content of experience. *Mind and Language* 21 (2): 187–219.

Raghanti, Mary Ann, Melissa K. Edler, Alexa R. Stephenson, Emily L. Munger, Bob Jacobs, Patrick R. Hof, Chet C. Sherwood, Ralph L. Holloway, and C. Owen Lovejoy. 2018. Neurochemical origin of hominids. *Proceedings of the National Academy of Sciences Feb 2018* 115 (6): E1108–E1116. doi:10.1073/pnas.1719666115.

Raine, A., M. Buchsbaum, and L. LaCasse. 1997. Brain abnormalities in murderers indicated by positron emission tomography. *Biological Psychiatry* 42 (6): 495–508.

Raine, A., and Y. Yang. 2006. Neural foundations to moral reasoning and antisocial behavior. *Social Cognitive Affective Neuroscience* 1 (3): 203–213. doi: 10.1093/scan/nsl033.

Ramnani, N. 2006. The primate cortico-cerebellar system. *Nature Reviews Neuroscience* 7 (7): 511–522.

————. 2012. Frontal lobe and posterior parietal contributions to the cortico-cerebellar system. *Cerebellum* 11 (2): 366–383.

Rand, David, Joshua Greene, and Martin Nowak. 2012. Spontaneous giving and calculated greed. *Nature* 489: 427–430.

Rappaport, R. A. 1979. *Ecology, meaning and religion*. Richmond, VA: North Atlantic Books.

Ravosa, W., and M. Dagosta. 2007. *Primate origins: Adaptations and evolution*. New York: Springer.

Ray, E., and C. Heyes. 2011. Imitation in infancy: The wealth of the stimulus. *Developmental Science* 14 (1): 92–105.

Read, Dwight. 2016. *How culture makes us human: Primate social evolution and the formation of human societies*. London: Routledge.

Reddy, V. 2008. *How infants know minds*. Cambridge, MA: Harvard University Press.

Reed, Edward S. 1986. James Gibson's ecological revolution in perceptual psychology: A case study in the transformation of scientific ideas. *Studies in the History and Philosophy of Science* 17: 65–99.

————. 1987. Why do things look the way they do? The implications of J. J. Gibson's "The ecological approach to visual perception." In Alan Costall, ed., *Cognitive Psychology in Question,* 90–114. New York: St. Martin's Press.

————. 1996. *Encountering the world: Toward an ecological psychology*. Oxford: Oxford University Press.

Reed, Edward S., and Rebecca K. Jones. 1982. Perception and cognition: A final reply to Heil. *Journal for the Theory of Social Behaviour* 12: 223–224.

Reno, P., R. Meindl, M. McCollum, and C. Lovejoy. 2003. Sexual dimorphism in *Australopithecus afarensis* was similar to that of modern humans. *Proceedings of the National Academy of Science USA* 100 (16): 9404–9409.

Rescorla, Michael. 2009. Cognitive maps and the language of thought. *British Journal for the Philosophy of Science* 60 (2): 377–407.

Rhodes, R. 1994. Aural images. In J. Ohala, L. Hinton, and J. Nichols, eds., *Sound symbolism*. Cambridge: Cambridge University Press.

Richards, R. 1989. *Darwin and the emergence of evolutionary theories of mind and behavior.* Chicago: University of Chicago Press.

Richards, Ruth, Dennis K. Kinney, Inge Lunde, Maria Benet, and Ann P. C. Merzel. 1988. Creativity in manic-depressives, cyclothymes, their normal relatives, and control subjects. *Journal of Abnormal Psychology* 97 (3): 281–288.

Richardson, Michael J., Kerry L. Marsh, and Reuben M. Baron. 2007. Judging and actualizing intrapersonal and interpersonal affordances. *Journal of Experimental Psychology* 33 (4): 845–859.

Richerson, Peter J., and Robert Boyd. 2001. The evolution of subjective commitment to groups: A tribal instincts hypothesis. In Ralph Nesse, ed., *Evolution and the capacity for commitment*. New York: Russell Sage Foundation.

———. 2008. *Not by genes alone: How culture transformed human evolution.* Chicago: University of Chicago Press.

Ridley, Matt. 1996. *The origins of virtue: Human instincts and the evolution of cooperation.* New York: Penguin.

Rietveld, Erik, Sanneke de Haan, and Damiaan Denys. 2013. Social affordances in context: What is it that we are bodily responsive to? *Behavioral and Brain Sciences* 36: 436. doi: 10.1017/S0140525X12002038.

Rietveld, Erik, and Julian Kiverstein. 2014. A rich landscape of affordances. *Ecological Psychology* 26 (4): 325–352. doi: 10.1080/10407413.2014.958035.

Rilling, J., D. Gutman, T. Zeh, G. Pagnoni, G. Berns, and C. Kilts. 2002. A neural basis for social cooperation. *Neuron* 35 (2): 395–405.

Robbins, Philip, and Murat Aydede. 2009. A short primer on situated cognition. In Murat Aydede and P. Robbins, eds., *The Cambridge handbook of situated cognition*, 3–10. Cambridge: Cambridge University Press.

Rogoff, Barbara. 2003. *The cultural nature of human development*. Oxford: Oxford University Press.

Rohrer, Tim. 2006. Image schemata in the brain. In Beate Hampe and Joe Grady, eds., *From perception to meaning: Image schemas in cognitive linguistics*, 165–196. Berlin: Mouton de Gruyter.

Rolls, E. T., J. Hornak, D. Wade, and J. McGrath. 1994. Emotion-related learning in patients with social and emotional changes associated with frontal lobe damage. *Journal of Neurology, Neurosurgery, and Psychiatry* 57: 1518–1524. doi: 10.1136/jnnp.57.12.1518.

Rooney, David, and Bernard McKenna. 2007. Wisdom in organizations: Whence and whither. *Social Epistemology* 21 (2): 113–138. doi: 10.1080/02691720701393434

Rorty, Richard. 1979. *Philosophy and the mirror of nature*. Eugene: University of Oregon Press.

Rosenberg, Alexander. 1994. *Instrumental biology, or the disunity of science*. Chicago: University of Chicago Press.

Rosenberg, Gregg. 2004. *A place for consciousness: Probing the deep structure of the natural world.* Oxford: Oxford University Press.

Rosenberg, K., and W. Trevathan. 2002. Birth, obstetrics and human evolution. *International Journal of Obstetrics and Gynecology* 109: 1199–1206.

Rosenthal, D. 2008. Consciousness and its functions. *Neuropsychologia* 46: 829–840.

Rostovtzeff, M. 1927 / 1960. *A history of the ancient world,* vol. 2, *Rome.* Oxford: Clarendon Press.

Rowe, J. 1946. Inca culture at the time of the Spanish conquest. In Julian H. Steward, ed., Bulletin 143, *Handbook of South American indians,* 2:183–330. Smithsonian Institution: Bureau of American Ethnology.

Rowlands, Mark. 2006. *Body language: Representation in action.* Cambridge, MA: MIT Press.

Rumelhart, D. E. 1980. Schemata: The building blocks of cognition. In Rand J. Spiro, Bertram C. Bruce, and William F. Brewer, eds., *Theoretical issues in reading comprehension.* Hillsdale, NJ: Lawrence Erlbaum.

Russell, Bertrand. 1905. On denoting. *Mind, New Series* 14 (56): 479–493.

Russell, James A. 2003. Core affect and the psychological construction of emotion. *Psychological Review* 110 (1): 145–172.

Ryan, R. M., J. Kuhl, and E. Deci. 1997. Nature and autonomy: An organizational view of social and neurobiological aspects of self-regulation in behavior and development. *Development and Psychopathology* 9: 701–728.

Ryle, G. 1949. *The concept of mind.* London: Hutchinson House.

Saarinen, J. A. 2015. A conceptual analysis of the oceanic feeling with a special note on painterly aesthetics. *Jyväskylä Studies in Education, Psychology and Social Research* 518.

Sachs, J. L., U. G. Mueller, T. P. Wilcox, and J. J. Bull. 2004. The evolution of cooperation. *The Quarterly Review of Biology* 79 (2): 135–160.

Sachs, Matthew, Antonio Damasio, and Assal Habibi. 2015. The pleasures of sad music. *Frontiers in Human Neuroscience* 9: 404.

Sahlins, Marshall. 1972. *Stone Age economics.* Chicago: Aldine Atherton.

Sakai T., M. Matsui, A. Mikami, L. Malkova, Y. Hamada, M. Tomanaga, J. Suzuki, M. Tanaka, M. Takako, T. Miyabi-Nishawaki, M. Nakatsukasa, and T. Matsuzawa. 2012. Developmental patterns of chimpanzee cerebral tissues provide important clues for understanding the remarkable enlargement of the human brain. *Current Biology* 22: R791–R792. doi: 10.1098/rspb.2012.2398.

Sanfey, Alan G., James K. Rilling, Jessica A. Aronson, Leigh E. Nystrom, and Jonathan D. Cohen. 2003. The neural basis of economic decision-making in the ultimatum game. *Science* 300: 1755–1758. doi: 10.1126/science.1082976.

Sapolsky, R. 2004. *Why zebras don't get ulcers,* 3rd ed. New York: Times Books.

———. 2006. Social cultures among nonhuman primates. *Current Anthropology* 47 (4): 645–654.

Satpute, Ajay B., and Matthew D. Leiberman. 2006. Integrating automatic and controlled processes into neurocognitive models of social cognition. *Brain Research* 1079: 86–97.

Saver, J. L., and A. R. Damasio. 1991. Preserved access and processing of social knowledge in a patient with acquired sociopathy due to ventromedial frontal damage. *Neuropsychologia* 29: 1241–1249. doi: 10.1016/0028-3932(91)90037-9.

Savin-Williams, Ritch C. 1979. Dominance hierarchies in groups of early adolescents. *Child Development* 50: 923–935. doi: 10.2307/1129316.

Sayer, A. 1992. *Method in social science: A realist approach.* London: Routledge.

Scarantino, Andrea. 2003. Affordances explained. *Philosophy of Science* 70 (5): 949–961.

Schachtman, Todd R., and Steve S. Reilly, eds. 2011. *Associative learning and conditioning theory: Human and non-human applications.* New York: Oxford University Press.

Scherer, Klaus R. 2009. The dynamic architecture of emotion: Evidence for the component process model. *Cognition and Emotion* 23: 1307–1351.

Schiller, Herbert I. 1971. *Mass communications and American empire.* New York: Augustus Kelley.

Schore, A. N. 1994. *Affect regulation and the origin of the self.* Mahweh, NJ: Lawrence Erlbaum.

———. 2001. Effects of a secure attachment relationship on right brain development, affect regulation, and infant mental health. *Infant Mental Health Journal* 22 (1 / 2): 7–66.

———. 2013. Bowlby's "environment of evolutionary adaptedness": Recent studies on the interpersonal neurobiology of attachment and emotional development. In D. Narvaez, J. Panksepp, A. N. Schore, and T. R. Gleason, eds., *Evolution, early experience and human development: From research to practice and policy,* 31–67. New York: Oxford University Press.

Schwartz, B., A. Ward, J. Monterosso, S. Lyubomirsky, K. White, and D. R. Lehman. 2002. Maximizing versus satisficing: Happiness is a matter of choice. *Journal of Personality and Social Psychology* 83 (5): 1178–1197.

Seabright, Paul. 2013. The birth of hierarchy. In Kim Sterelny, Richard Joyce, Brett Calcott, and Ben Fraser, eds., *Cooperation and its evolution,* 109–116. Cambridge, MA: MIT Press.

Searle, John R. 1983. *Intentionality: An essay in philosophy of mind.* Cambridge: Cambridge University Press.

———. 1992. *The rediscovery of the mind.* Cambridge, MA: MIT Press.

Seed, Amanda, and Michael Tomasello. 2010. Primate cognition. *Topics in Cognitive Science* 2 (3): 407–419.

Semrud-Clikeman M. and G. W. Hynd. 1990. Right hemisphere dysfunction in nonverbal learning disabilities: Social, academic, and adaptive functioning in adults and children. *Psychological Bulletin* 107: 196–209. doi: 10.1037/0033-2909 .107.2.196.

Seneca. 1962. On benefits (Seneca's epistle LXXXI). In Seneca, *Ad lucilium epistulae morales,* vol. 2 (Loeb Classical Library). Cambridge, MA: Harvard University Press.

Sergent, J., and J. L. Signoret. 1992. Varieties of functional deficits in prosopagnosia. *Cerebral Cortex* 2: 375–388.

Seyfarth, R. M., D. L. Cheney, and R. A. Hinde. 1978. Some principles relating social interactions and social structure among primates. In D. J. Chivers and J. Herbert, eds., *Recent advances in primatology,* 1:39–51. London: Academic Press.

Shapiro, D., L. D. Jamner, and S. Spence. 1997. Cerebral laterality, repressive coping, autonomic arousal, and human bonding. *Acta Physiologica Scandinavica Supplementum* 640: 60–64.

Sheffield, F. S. 1965. Relation between classical conditioning and instrumental learning. In W. F. Prokasy, ed., *Classical conditioning: A symposium*. New York: Appleton-Century-Crofts.

Shermer, Michael. 2015. *The moral arc: How science and reason lead humanity toward truth, justice, and freedom*. New York: Henry Holt.

Siegel, D. J. 1999. *The developing mind: Toward a neurobiology of interpersonal experience*. New York: Guilford Press.

Silberbauer, G. 1981. *Hunter and habitat in the Central Kalahari Desert*. Cambridge: Cambridge University Press.

Simmora, D., Kiki Chang, Lawrence Strong, and Terence Ketter. 2005. Creativity in familial bipolar disorder. *Journal of Psychiatric Research* 39: 623–631.

Skinner, Burrhus F. 1953. *Science and human behavior*. New York: Simon and Schuster.
———. 1974. *About behaviorism*. New York: Alfred A. Knopf.

Slingerland, Edward, and Brenton Sullivan 2016. Durkheim with data: The database of religious history. *Journal of the American Academy of Religion* 85 (2): 1–38.

Sloan, P. 2006. Kant on the history of nature: The ambiguous heritage of the critical philosophy for natural history. *Cambridge Studies in the History of the Biological and Medical Sciences* 37 (4): 627–648.

Sloman, Aaron. 1978. *The computer revolution in philosophy: Philosophy, science, and models of the mind*. Hassocks, UK: Harvester Press.

Smaldino, Paul E., and Joshua M. Epstein. 2015. Social conformity despite individual preferences for distinctiveness. *Royal Society Open Science* 2: 140437.

Smith, D. D. 1973. Teaching introductory sociology by film. *Teaching Sociology* 1: 48–61.

Smith, J. D., W. E. Shields, and D. A. Washburn. 2003. The comparative psychology of uncertainty monitoring and metacognition. *Behavioral and Brain Sciences* 26: 317–339.

Snowdon, C. T. 2001. From primate communication to human language. In F. de Waal, ed., *Tree of origin: What primate behavior can tell us about human social evolution*. Cambridge, MA: Harvard University Press.

Snyder, Bob. 2000. *Music and memory*. Cambridge, MA: MIT Press.

Sober, Eliot. 2000. *Philosophy of biology*, 2nd ed. Boulder, CO: Westview Press.

Sokoloff, Kenneth L., and Stanley Engerman. 2000. Institutions, factor endowments, and paths of development in the new world. *Journal of Economic Perspectives* 14: 217–232.

Solms, Mark. 2000. Dreaming and REM sleep are controlled by different brain mechanisms. *Behavioral and Brain Sciences* 23 (6): 843–850 (discussion 904).

Solms, Mark, and Karl Friston. 2018. How and why consciousness arises: Some considerations from physics and physiology. *Journal of Consciousness Studies* 25 (5 / 6): 202–238.

Solomon, Robert. 2003. *What is an emotion? Classic and contemporary readings*. Oxford: Oxford University Press.

————. 2006. *True to our feelings: What our emotions are really telling us.* New York: Oxford University Press.

Soltis, J., R. Boyd, and P. J. Richerson. 1995. Can group-functional behaviours evolve by cultural group selection? An empirical test. *Current Anthropology* 36 (3): 473–494.

Sosis, Richard. 2000. Religion and intragroup cooperation: Preliminary results of a comparative analysis of utopian communities. *Cross-Cultural Research* 34 (1): 70–87.

Sosis, Richard, and Candace Alcorta. 2003. Signaling, solidarity, and the sacred: The evolution of religious behavior. *Evolutionary Anthropology* 12 (6): 264–274.

Spelke, E., and K. Kinzler. 2007. Core knowledge. *Developmental Science* 10: 89–96.

Spelke, Elizabeth. 2010. Innateness, choice, and language. In J. Bricmont and J. Franck, eds., *Chomsky Notebook,* 203–210. New York: Columbia University Press.

Sperber, Dan 2002. In defense of massive modularity. In I. Dupoux, ed., *Language, brain, and cognitive development,* 47–57. Cambridge, MA: MIT Press.

Sperber, Dan, and Nicolas Baumard. 2012. Moral reputation: An evolutionary and cognitive perspective. *Mind and Language* 27 (5): 495–518. doi: 10.1111/mila.12000.

Sperber, Dan, and Deidre Wilson. 2002. Pragmatics, modularity and mind-reading. *Mind and Language* 17: 3–23.

Spikins, Penny, Gail Hitchens, Andy Needham, and Holly Rutherford. 2014. The cradle of thought: Growth, learning, play and attachment in Neanderthal children. *Oxford Journal of Archaeology* 33 (2): 111–134.

Stammbach, E. 1978. On social differentiation in groups of captive female hamadryas baboons. *Behaviour* 67: 322–338.

Stanford, C. 2001. The ape's gift: Meat eating, meat sharing, and human evolution. In F. de Waal, ed. *Tree of origin.* Cambridge, MA: Harvard University Press.

Stanley, D. A., and R. Adolphs. 2013. Toward a neural basis for social behavior. *Neuron* 80 (3): 816–826.

Sterelny, Kim. 2003. *Thought in a hostile world.* Oxford: Blackwell.

————. 2012a. *The evolved apprentice: How evolution made humans unique.* Cambridge, MA: MIT Press.

————. 2012b. From fitness to utility. In Samir Okasha and Ken Binmore, eds., *Evolution and rationality: Decisions, co-operation and strategic behavior,* 246–273. Cambridge: Cambridge University Press.

Sterelny, Kim, ed., with Richard Joyce, Brett Calcott, and Ben Fraser. 2013. *Cooperation and its evolution.* Cambridge, MA: MIT Press.

Stern, D. 1985. *The interpersonal world of the infant.* New York: Basic Books.

Stoneman, Richard, ed. 1994. *Legends of Alexander the Great.* London: Everyman Dent.

Storr, Anthony. 1989. *Solitude: A return to the self.* New York: Ballantine.

Stout, D. B. 1971. Aesthetics in "primitive societies." In C. F. Joplin, ed., *Art and aesthetics in primitive societies.* New York: E. P. Dutton.

Strawson, Galen. 2006. Realistic monism: Why physicalism entails panpsychism. *Journal of Consciousness Studies* 13 (10 / 11): 3–31.

Strongman, Kenneth T., and Brian G. Champness. 1968. Dominance hierarchies and conflict in eye contact. *Acta Psychologica* 28: 376–386.

Stotz, Karola. 2014. Extended evolutionary psychology: The importance of transgenerational developmental plasticity. *Frontiers in Psychology* 5: 908.

Strum, Shirley C. 2001. *Almost human: A journey into the world of baboons*. Chicago: University of Chicago Press.

Sturge-Apple, Melissa L., and Dante Cicchetti. 2012. Differential susceptibility in spillover between interparental conflict and maternal parenting practices: Evidence for OXTR and 5-HTT genes. *Journal of Family Psychology* 26 (3): 431.

Szamado, Szabolcs, and E. Szathmary. 2006. Selective scenarios for the emergence of natural language. *Trends in Ecology and Evolution* 21 (10): 555–561.

Szpunar, Karol K. 2010. Episodic future thought: An emerging concept. *Perspectives on Psychological Science* 5 (2): 142–162.

Tajfel, Henri. 1982. Social psychology of intergroup relations. *Annual Review of Psychology* 33: 1–39. doi: 10.1146/annurev.ps.33.020182.000245.

Talle, A. 1988. *Women at a loss: Changes in Maasai pastoralism and their effects on gender relations*. Stockholm: Stockholm Studies in Social Anthropology.

Tan, J., and B. Hare. 2013. Bonobos share with strangers. *PLoS ONE* 8 (1): e51922.

Tangney, June P., and Ronda L. Dearing. 2003. *Shame and guilt*. New York: Guilford Press.

Tarski, A. 1956 / 1984. *Logic, semantics, metamathematics: Papers 1923–38*. Woodger, J., and Corcoran, J., eds. and trans. Indianapolis: Hackett Publishing.

Thaler, R. H., and C. R. Sunstein. 2008. *Nudge: Improving decisions about health, wealth, and happiness*. New Haven, CT: Yale University Press.

Theodoridou, Angeliki, Angela C. Rowe, Ian S. Penton-Voak, and Peter J. Rogers. 2009. Oxytocin and social perception: Oxytocin increases perceived facial trustworthiness and attractiveness. *Hormones and Behavior* 56: 128–132.

Thomas, Nigel T. J. 1999. Are theories of imagery theories of imagination? An active perception approach to conscious mental content. *Cognitive Science* 23: 207–245.

Thompson, Evan. 1995. *Colour vision: A study in cognitive science and the philosophy of perception*. New York: Routledge.

———. 2007. *Mind in life: Biology, phenomenology, and the sciences of mind*. Cambridge, MA: Harvard University Press.

Thompson, Evan, and Dan Zahavi. 2007. Philosophical issues: Phenomenology. In Morris Moscovitch, Philip Zelazo, and Evan Thompson, eds., *Cambridge handbook of consciousness*, 67–87. Cambridge: Cambridge University Press.

Thurnwald, Richard. 1932. The psychology of acculturation. *American Anthropologist* 34 (4): 557–569.

Tiegerman, E., and L. H. Primavera. 1984. Imitating the autistic child: Facilitating communicative gaze behavior. *Journal of Autism and Developmental Disorders* 14: 27–38.

Tolman, E. C. 1948. Cognitive maps in rats and men. *Psychological Review* 55 (4): 189–208.

Tomasello, M. 2000. Culture and cognitive development. *Current Directions in Psychological Science* 9 (2): 37–40.

———. 2009. *Why we cooperate*. Cambridge, MA: MIT Press.

———. 2014. *A natural history of human thinking.* Cambridge, MA: Harvard University Press.

Tomkins, Silvan S. 1962. *Affect imagery consciousness,* vol. 1, *The positive affects.* London: Tavistock.

Tooby, John, and Leda Cosmides. 2001. Does beauty build adapted minds? Toward an evolutionary theory of aesthetics, fiction and the arts. *SubStance* 94 / 95, 30(1): 6–27.

Tooby, John, and Irven DeVore. 1987. The reconstruction of hominid behavioral evolution through strategic modeling. In Warren G. Kinzey, ed., *The evolution of human behavior: Primate models,* 183–237. Albany, NY: SUNY Press.

Tops, M., S. L. Koole, H. IJzerman, and F. T. Buisman-Pijlman. 2014. Why social attachment and oxytocin protect against addiction and stress: Insights from the dynamics between ventral and dorsal corticostriatal systems. *Pharmacology Biochemistry and Behavior* 119: 39–48.

Tramo, M. J. 2001. Biology and music: Music of the hemispheres. *Science* 291 (5501): 54–56.

Trivers, Robert L. 1971. The evolution of reciprocal altruism. *Quarterly Review of Biology* 46 (1): 35–57.

———. 2005. Reciprocal altruism: 30 years later. In C. P. van Schaik and P. M. Kappeler, eds., *Cooperation in primates and humans: Mechanisms and evolution,* 67–83. Berlin: Springer-Verlag.

Tronick, Ed, and Jeffrey Cohn. 1989. Infant-mother face-to-face interaction: Age and gender differences in coordination and the occurrence of miscoordination. *Child Development* 60: 85–92. doi: 10.1111/j.1467–8624.1989.tb02698.x.

Tronson, N. C., and J. R. Taylor. 2007. Molecular mechanisms of memory reconsolidation. *Nature Reviews Neuroscience* 8 (4): 262–275.

Truppa, V., E. Piano Mortari, D. Garofoli, S. Privitera, and E. Visalberghi. 2011. Same / different concept learning by capuchin monkeys in matching-to-sample tasks. *PLoS ONE* 6 (8): e23809.

Tulving, Endel. 1983. *Elements of episodic memory.* Oxford: Clarendon Press.

———. 2005. Episodic memory and autonoesis: Uniquely human? In H. S. Terrace and J. Metcalfe, eds., *The missing link in cognition,* 4–56. New York: Oxford University Press.

Turvey, M. T. 1992. Affordances and Prospective Control: An outline of the ontology. *Ecological Psychology* 4 (3): 173–187.

Tversky, Amos, and Daniel Kahneman. 1973. Availability: A heuristic for judging frequency and probability. *Cognitive Psychology* 5 (2): 207–232.

Tye, Michael. 1984. The adverbial approach to visual experience. *Philosophical Review* 93: 195–226.

———. 2006. The truth about true blue. *Analysis* 66 (4): 340–344

Tylor, Edward B. 1899. *The early history of mankind.* New York: Henry Holt.

Ungerleider, L., and M. Mishkin. 1982. Two cortical visual systems. In D. J. Ingle, M. A. Goodale, and R. J. W. Mansfield, eds., *Analysis of visual behaviour.* Cambridge, MA: MIT Press.

Uttal, William R. 2001. *The new phrenology: The limits of localizing cognitive processes in the brain.* Cambridge, MA: MIT Press.

———. 2005. *Mind and brain: A critical appraisal of cognitive neuroscience.* Cambridge, MA: MIT Press.

Vallacher, Robin R., Stephen J. Read, and Andrzej Nowak. 2002. The dynamical perspective in personality and social psychology. *Personality and Social Psychology Review* 6 (4): 264–273.

Van Acker, R., and S. S. Valenti. 1989. Perception of social affordances by children with mild handicapping conditions: Implications for social skills research and training. *Ecological Psychology* 1 (4): 383–405.

van de Waal, Erica and Christele Borgeaud, and Andrew Whiten. 2013. Potent social learning and conformity shape a wild primate's foraging decisions. *Science* 340 (6131): 483–485.

van Fraassen, Bas C. 1980. *The scientific image.* Oxford: Oxford University Press.

Van IJzendoorn, Marinus. 1995. Adult attachment representations, parental responsiveness, and infant attachment: A meta-analysis on the predictive validity of the adult attachment interview. *Psychological Bulletin* 117 (3): 387.

Van Ijzendoorn, Marinus H., and Pieter M. Kroonenberg. 1998. Cross-cultural patterns of attachment: A meta-analysis of the strange situation. *Child Development* 59: 147–156.

Varela, F., with Humberto Maturana. 1973/1980. *Autopoiesis and cognition: The realization of the living.* Boston: Reidel.

Varela, F. J., E. Rosch, and E. Thompson. 1991. *The embodied mind: Cognitive science and human experience.* Cambridge, MA: MIT Press.

Vialou, Denis. 1996. *Au coeur de la prehistoire.* Paris: Gallimard.

Vollenweider, Franz X. 2001. Brain mechanisms of hallucinogens and entactogens. *Dialogues in Clinical Neuroscience* 3 (4): 265–279.

Vollenweider, Franz X., Alex Gamma, and Margreet Fi Vollenweider-Scherpenhuyzen. 1999. Neural correlates of hallucinogen-induced altered states of consciousness. In S. Hameroff, A. Kaszniak, and David Chalmers, eds., *Toward a science of consciousness III: The third tucson discussions and debates.* Cambridge, MA: MIT Press.

Von Neumann, John, and Oskar Morgenstern. 1944. *Theory of games and economic behavior.* Princeton, NJ: Princeton University Press.

Von Overwalle, Frank. 2011. A dissociation between social mentalizing and general reasoning. *Neuroimage* 54 (2): 1589–1599.

Vrtička, Pascal, and Patrik Vuilleumier. 2012. Neuroscience of human social interactions and adult attachment style. *Frontiers in Human Neuroscience* 6 (212): 1–17.

Vygotsky, Lev. 1962. *Thought and language.* Cambridge, MA: MIT Press.

Wagner, A. 2008. Robustness and evolvability: A paradox resolved. *Proceedings of the Royal Society London Series B* 275: 91–100.

Walker, M. P., T. Brakefield, J. A. Hobson, and R. Stickgold. 2003. Dissociable stages of human memory consolidation and reconsolidation. *Nature* 425 (6958): 616–620.

Walum, H., I. D. Waldman, and L. J. Young. 2015. Statistical and methodological considerations for the interpretation of intranasal oxytocin studies. *Biological Psychiatry* 79 (3): 251–257.

Warnecken, F. 2013. Altruistic behaviors from a developmental and comparative perspective. In Kim Sterelny, Richard Joyce, Brett Calcott, and Ben Fraser, eds., *Cooperation and its evolution.* Cambridge, MA: MIT Press.

Warnecken, F., and M. Tomasello. 2006. Altruistic helping in human infants and young chimpanzees. *Science* 311 (5765): 1301–1303.

Warren, W. H. 1984. Perceiving affordances: Visual guidance of stair climbing. *Journal of Experimental Psychology* 10 (5): 683–703.

Watson, John B. 1930. *Behaviorism,* revised ed. Chicago: University of Chicago Press.

Weaver, I. C. G., N. Cervoni, F. A. Champagne, A. C. D'Alessio, S. Sharma, J. R. Seckl, S. Dymov, M. Szyf, and M. J. Meaney. 2004. Epigenetic programming by maternal behavior. *Nature and Neuroscience* 7: 847–854. doi: 10.1038/nn1276.

Webb, Barbara. 2012. Cognition in insects. *Philosophical Translations of the Royal Society of London B* 367 (1603): 2715–2722.

Weber, Max. 1946 / 1921. Politics as a vocation. In H. H. Gerth and C. Wright Mills, trans., *From Max Weber: Essays in sociology.* New York: Oxford University Press.

Weiner, A. 1976. *Women of value, Men of renown.* Austin: University of Texas Press.

Weiner, N. 1948 / 1965. *Cybernetics: Or control and communication in the animal and the machine.* Cambridge, MA: MIT Press.

Weiskrantz, Lawrence. 1986. *Blindsight: A case study and implications.* New York: Oxford University Press.

West-Eberhard, M. 2003. *Developmental plasticity and evolution.* New York: Oxford University Press.

Westermarck, Edward. 1894. *The history of human marriage.* New York: MacMillan.

Wheatley, Thalia, and Jonathan Haidt. 2005. Hypnotic disgust makes moral judgments more severe. *Psychological Science* 16: 780–784. doi: 10.1111/j.1467-9280.2005.01614.x.

Wheeler, Michael. 2005. *Reconstructing the cognitive world: The next step.* Cambridge, MA: MIT Press.

———. 2008. Cognition in context: Phenomenology, situated robotics and the frame problem. *International Journal of Philosophical Studies* 16 (3): 323–349.

White, R. 2003. The epistemic advantage of prediction over accommodation. *Mind* 112 (448): 653–683.

Whitely, David. 2009. *Cave paintings and the human spirit: The origin of creativity and belief.* Amherst, NY: Prometheus books.

Whiten, Andrew, and David Erdal. 2012. The human socio-cognitive niche and its evolutionary origins. *Philosophical Transactions of the Royal Society of London B* 367 (1599): 2119–2129.

Wiessner, Polly. 1983. Style and social information in Kalahari San projectile points. *American Antiquity* 48: 253–276. doi: 10.2307/280450.

———. 1984. Reconsidering the behavioral basis for style: A case study among the Kalahari San. *Journal of Anthropological Archaeology* 3: 190–234. doi: 10.1016/0278-4165(84)90002-3.

————. 2009. Parent-offspring conflict in marriage: Implications for social evolution and material culture among the Ju/'hoansi Bushmen. In Stephen Shennan, ed., *Patterns and process in cultural evolution (origins of human behavior and culture).* Berkeley: University of California Press.

————. 2014 The embers of society: Firelight talk among the Ju/'hoansi Bushmen. *Proceedings of the National Academy of Sciences* 111 (39): 14013–14014.

Williams, D. R., and H. Williams. 1969. Auto-maintenance in the pigeon: Sustained pecking despite contingent non-reinforcement. *Journal of Experimental Analysis and Behavior* 12: 511–520.

Willner, Paul, Jack Bergman, Louk Vanderschuren, and Bart Ellenbroek, eds. 2015. "Pharmacological Approaches to the Study of Social Behavior, Part I: Reviews." *Behavioral Pharmacology* 26, no. 6.

Wills, Julian, Oriel Feldman Hall, Michael R. Meager, and Jay J. Van Bavel. 2018. Dissociable contributions of the prefrontal cortex in group-based cooperation. *Social Cognitive and Affective Neuroscience* 13 (4): 349–356.

Wilson, David S. 2002. *Darwin's cathedral.* Chicago: University of Chicago Press.

Wilson, David S., and Eliot Sober. 1994. Reintroducing group selection to the human behavioral sciences. *Behavioral Brain Sciences* 17: 585–654.

Wilson, Edmund O. 1975. *Sociobiology: The new synthesis.* Cambridge, MA: Harvard University Press.

Wilson, M. 2002. Six views of embodied cognition. *Psychonomic Bulletin and Review* 9 (4): 625–636.

Wimsatt, William C., and Jeffery C. Schank. 2004. Generative entrenchment, modularity and evolvability: When genic selection meets the whole organism. In Gerhard Schlosser and Günter P. Wagner, eds., *Modularity in development and evolution.* Chicago: University of Chicago Press.

Winkielman, P., and K. C. Berridge. 2004. Unconscious emotion. *Current Directions in Psychological Science* 13 (3): 120–123.

Wismer-Fries, A. B., T. E. Ziegler, J. R. Kurian, S. Jacoris, and S. Pollak. 2005. Early experience in humans is associated with changes in neuropeptides critical for regulating social behavior. *Proceedings of the National Academy of Sciences* 102: 17237–17240. doi:10.1073/pnas.0504767102.

Withagen, Rob, Harjo J. de Poel, Duarte Araujo, and Gert-Jan Pepping. 2012. Affordances can invite behavior: Reconsidering the relationship between affordances and agency. *New Ideas in Psychology* 30: 250–258. doi: 10.1016/j.newideapsych.2011.12.003.

Withagen, Rob, and Claire Michaels. 2005. On ecological conceptualizations of perceptual systems and action systems. *Theory and Psychology* 15 (5): 603–620.

Wong, K. 2012. Why humans give birth to helpless babies. http://blogs.scientific american.com/observations/2012/08/28/why-humans-give-birth-to-helpless -babies/, accessed October 27, 2013.

Wout, M. V. T., R. S. Kahn, A. G. Sanfey, and A. Aleman. 2005. Repetitive transcranial magnetic stimulation over the right dorsolateral prefrontal cortex affects strategic decision-making. *Neuroreport* 16: 1849–1852.

Wright, Robert. 2018. *Why Buddhism is true: The science and philosophy of meditation and enlightenment.* New York: Simon and Schuster.

Wynn, Karen, and Paul Bloom. 2014. The moral baby. In Melanie Killen and Judith Smetana, eds., *Handbook of moral development,* 2nd ed. New York: Psychology Press.

Yoffee, Norman. 2005. *Myths of the archaic state: Evolution of the earliest cities, states, and civilizations.* Cambridge: Cambridge University Press.

Young, Larry J. The neural basis of pair bonding in a monogamous species: A model for understanding the biological basis of human behavior. In *National Research Council (US) Panel for the Workshop on the Biodemography of Fertility and Family Behavior.* Washington, DC: National Academies Press.

Young, M. E., and E. A. Wasserman. 2003 / 1997. Entropy detection by pigeons: Response to mixed visual displays after same-different discrimination training. *Journal of Experimental Psychology* 23: 157–170.

Zaidel, Dahlia W. 2005. *Neuropsychology of art: Neurological, cognitive, and evolutionary perspectives.* New York: Psychology Press.

Zajonc, R. B. 1968. Attitudinal effects of mere exposure. *Journal of Personality and Social Psychology Monograph Supplement* 9 (2 / 2): 1–27.

———. 1980. Feeling and thinking: Preferences need no inferences. *American Psychologist* 35 (2): 151–175.

Zelazo, P. D., and W. A. Cunningham. 2007. Executive function: Mechanisms underlying emotion regulation. In J. J. Gross, ed., *Handbook of emotion regulation,* 135–158. New York: Guilford.

Zimmermann, Peter, Cornelia Mohr, and Gottfried Spangler. 2009. Genetic and attachment influences on adolescents' regulation of autonomy and aggressiveness. *Journal of Child Psychology and Psychiatry* 50 (11): 1339–1347.

Zanna, Mark P., Charles A. Kiesler, and Paul A. Pilkonis. 1970. Positive and negative attitudinal affect established by classical conditioning. *Journal of Personality and Social Psychology* 14 (4): 321–328.

Acknowledgments

Co-writing a book of this scope has been a remarkable odyssey, and sometimes we felt stranded, sometimes lashed to the mast as the Sirens sang deafeningly, and sometimes as if we had navigated Scylla and Charybdis and hit the high water, heading home.

Stephen's acknowledgments: I would like to express gratitude to my parents, Edward and Carol, for their love and inspiration. Deepest thanks to my generous brothers Dave and Dan. It is wonderful to have best friends and brothers, and rare when they are the same. Thanks to my whole extended family for so much support: Keaton, Jackson, Maddy, Garrison, Nicole, Elaine, Waigong, and Waipo. Special thanks to Wen Jin, mother of my pride and joy. I am very thankful to my co-author, Rami Gabriel, for his clarity, integrity, insight, and friendship. I am also especially grateful to my son, Julien. *Fortis est non pertubaris in rebus asperis. Filius meus vita mea sunt esque semper protegam.*

Rami's acknowledgments: I owe everything and more to my parents, Olfat and Hani. To my sister Dina, Jeremy, and nephews Roger and Francis, my love and deep appreciation. Thank you to my family for kindness and support: Mihad, Layal, Nada, Widad, Khaled, Kais, Sandra, Kamel, Dahlia, Karim, Christiane, Magdi, and all my kin in Toronto. Respect and gratitude to my friends and fellow musicians:

Gary M. F., Alfonso, Steve, Beau, Scott, Alex, Casey, Tim, Jake, Jon, Ronnie, Miles, Michael F. B., and Nick. I am very thankful to my co-author and friend, Steve Asma, for his wisdom, patience, and curiosity. I am also especially grateful to my wife, Noor, who gracefully tuned my heart to the emotions in and around love. *Contra mundum!*

. . .

We are fortunate to have many warm friends and brilliant colleagues. Some have helped directly with this project, while others have supported our work respectively and together. We would like to thank Kim McCarthy, Andrew Causey, Brigid Hains, Anesa Miller, Peter Catapano, Greg Brandenburgh, Alex Kafka, Robert Wright, Stanley B. Klein, Michael Gazzaniga, Jack Loomis, Peter Khooshabeh, Aaron Zimmerman, Cathy Lynn Grossman, Qiuxuan Lu, Steve Corey, Rich King, Steve Mirsky, Teresa Prados-Torreira, Kate Hamerton, Rojhat Avsar, Robert Watkins, Joan Erdman, Carmelo Esterrich, Michelle Yates, Bob Vallier, Scott Wolfman, Elif Allenfort, Dave Edington, Michael Sims, Peter Altenberg, Ted Di Maria, Joanna Ebenstein, Peter Olson, Doug Johnson, Abbas Raza, Jim Christopulos, Howard McCullum, Oscar Valdez, Joseph Carroll, Krista Rogers, Brett King, Gerald Rizzer, Mathias Clasen, Lester Friedman, John Kaag, Kim Sterelny, Mark Johnson, William Irwin, Dario Maestripieri, Bob Zellman, Coltan Scrivner, Cheryl Johnson-Odim, Thomas Kristjansen, Owen Flanagan, Elizabeth Branch-Dyson, Josh Summers, Ted Hardin, Gregory F. Tague, Steven Barrie-Anthony, Nevi Cline, Elmo Painter, Steve Daoud, Robyn Whatley, Kevin Henry, Robert Hanserd, Anthony Madrid, Marya Schechtman, K.K. Szpunar, Daniel Cervone, Phillip O'Sullivan, Mary Hegarty, Sebastian Huydts, Todd Cambio, Eric Ederer, MJH, Mark Solms, Agneta Fischer, Deborah Holdstein, Erin McCarthy, Colin Klein, Frans De Waal, and Baheej Khleif.

Special thanks for indexical support provided by the Department of Humanities, History, and Social Sciences at the School of Liberal Arts and Sciences at Columbia College Chicago, and to Noor Shawaf for her indelible clinch work editing and tracking down references. We would also like to thank students over the last five years in our courses "Evolution of the Mind" and "Emotions" for their insights and forbearance as we waxed philosophical. Elements of this work have been presented at the following conferences: Cognitive Science Society in Portland, 2010; Southern Society for Philosophy and Psychology in Savannah, 2012; Social and Affective Neurosciences in Boston, 2015; Animal and Human Emotions in Erice, 2016; University of Illinois, Chicago, 2017; and Society for Philosophy and Psychology in Ann Arbor, 2018. Chapter 1 builds on ideas first discussed in "Modularity in Affective Neuroscience and Cognitive Psychology," published in *Journal of Consciousness Studies* 19, no. 3–4 (2012): 19–25. We would also like to acknowledge the Chicago Cultural Center, where we initially presented our ideas to the public in a series of panels.

Many thanks to our excellent editor at Harvard, Andrew Kinney. Thanks also to Olivia Woods, Stephanie Vyce, Tim Jones, Lynn Everett, Debbie Masi, and everyone at Harvard University Press. We would also like to non-anonymously thank several anonymous referees for their valuable suggestions.

. . .

We are indebted to Glennon Curran, who started this journey with us, and whose perspective, dedication, and enthusiastic thirst for knowledge enlivened the early Research Group. The late great Tom Greif, with whom we were absorbed in vigorous weekly debates was a huge inspiration as an engaged and empathetic individual who believed there was an intrinsic connection between understanding the mind and leading an ethical life. He is missed. Our greatest intellectual debt is to the father of affective neuroscience and all-around mensch, Jaak Panksepp, who spent his life indefatigably doing research and writing an infectious, exhilarating oeuvre. Sadly, Jaak was writing the foreword to this book when he passed away, and while we do not have his foreword, we are grateful to have his influence baked into every chapter.

Index

Abrutyn, Seth, 207

action-oriented representations (AORs), 80–81, 85, 90, 288, 330n52; cognitive maps and, 162, 175; evaluative appraisal of, 251; reflection-oriented and, 193; social affordance model of, 87. *See also* pushmi-pullyu representations

adaptationism, 41, 62–63, 264, 313

adornment, 231, 234–236, 271, 358n42

Aeschylus, 281–282

Aesop, 278, 282

aesthesis, 168–169

affect, 8, 153, 197; Barrett on, 10–13, 145, 197–198; Berridge on, 44, 61, 332n17; cognitive maps and, 161–163; conceptualization of, 27–28; Damasio on, 2, 20, 27, 32, 166; decoupling of, 159–161, 164; domestication of, 228–231; dream state and, 164–167; evolution of, 205–209, 243–244, 251–257; functions of, 159–160; Panksepp on, 245, 336n28; Proust on, 340n12; sensory, 8

affective neuroscience paradigm, 27–29, 41, 232; of Panksepp, 7–8, 28, 37–38

affective-sensorimotor complex, 84

affordances, 15, 76, 157–158, 320n62; as communication, 81–83; definition of, 26; emotional learning and, 119; enculturation and, 136–138; perception and, 80, 88, 156–159; Proust on, 340n10, 341n28; proxemics and, 86, 221; social grooming and, 97–98; social intelligence and, 83–85; stair-climbing and, 158

Agassiz, Louis, 56

agent / patient sentence structure, 150–152

Ainsworth, Mary, 129–130

Alexander the Great, 310

Al-Ghazzali, 266

allocentric maps, 161–162, 341n34. *See also* cognitive maps

altruism, 261; cooperation and, 223, 235, 290–291; cost-benefit analysis of, 355n161; Darwin on, 238; egalitarian, 247; empathy and, 117, 235, 243; Gould on, 354n155; within kinship groups, 249–250, 354n155, 354n158; reciprocal, 216, 226, 228–230, 243, 249, 290; Wilson on, 354n158

grooming, social, 67, 81, 86, 186; in
baboons, 79, 86–87; in chimpanzees,
239; in humans, 97, 242
guilt, 230, 307; altruism and, 228;
morality and, 222, 230; shame and,
240, 251–252, 258, 301

Hadza people, 333n37
Haidt, Jonathan, 278
haptic perception, 128–129, 132, 159,
186, 337n55, 338n103
Harris, Marvin, 247, 269
Hauser, Marc, 222
Hebb, Donald, 23
Heft, Harry, 77, 331n63, 358n19
Hegel, G. W. F., 51, 63
Heidegger, Martin, 185
Henrich, Joe, 116–118
Hercules, 309, 310
Hermann, Esther, 126
heroism, 270, 278, 309–310
Heyes, Cecilia, 114
hierarchies. See dominance hierarchies
Hinduism, 236, 247, 278–279, 282, 305
hippocampus, 262, 323n75; cognitive
maps and, 165; dream state and, 164,
165
Hobbes, Thomas, 214, 311, 355n159
Hofstadter, Douglas, 201, 202
Hohman, Gottfried, 108
holism, 43, 47, 51–55, 59, 68, 193;
definition of, 53; embodiment as, 78,
186; emergent, 5; philosophy of
language and, 194; Plato and, 311;
Putnam and, 348n36
Homo economicus, 213–218, 224, 227
Homo erectus, 111, 226, 263
Hox gene, 57
Hrdy, Sarah B., 15, 111, 226
Hume, David, 46, 177, 248
hunter-gatherers, 17–18, 211, 216, 221,
224–225, 257–258
hunting, 110, 258, 332n28; agriculture
and, 216; by primates, 97–101,
113–114, 247
Husserl, Edmund, 157
Hutto, Daniel D., 329n29, 330n52,
339n3, 341n17, 342n36

hybrids, 17, 52, 103, 273, 343n77
hypothalamic-pituitary-adrenal axis, 130
hypothalamus, 104, 109–110, 262,
346n10

image grammar, 175–179
imagination, 16–20, 159, 168–172,
178–183, 244; in animals, 171, 244;
Aristotle on, 169–170; cognitive
architecture of, 171–172; complexity
of, 155; Darwin on, 171, 181–182;
dreams and, 153, 161, 164–167, 171;
evolution of, 268–269; judgment and,
168; Kant on, 170, 343n66; linguistic,
190–195; music and, 283; night talk
and, 256, 257, 259, 263, 357n195;
Plato on, 168–169; representation and,
153–154; social identity and, 246;
theory of mind and, 260
imitation. See mimicry
Implicit Attitudes Task (IAT), 322n66
impulse control, 10, 127, 227, 244, 251,
258–260; gratitude and, 303; kinship
bonding and, 206; Sandi on, 134;
Seneca on, 196
Incas, 208
incentive salience, 61, 332n17
incest, 222, 256, 313
infant-caregiver relationship, 15,
100–103, 109–113, 127–133, 223;
emotional-response systems in,
129–130, 249; neural pathways of,
346n10; social intelligence and,
122–124, 133–138. See also attachment
infanticide, as reproductive strategy,
102, 105, 114, 334n61
"informavores," 100, 101, 333n31
Inhelder, B., 305
intelligence quotient (IQ), 127
Intelligent Design, 46–48, 324n5
"intensity calibration," 258–259
"intentional arc," 31
intentionality, 62–67, 250; Arnold on,
68; consciousness and, 67–71;
non-representational forms of, 45;
Searle on, 30; shared, 223, 244, 245;
teleology and, 46
Islam, 237, 266, 280–281, 302

Mahabharata, 278–279, 282
Marquez, C., 133–134
Marr, David, 37
marriage. See pair-bonding
Marsh, Kerry L., 84
Marx, Karl, 296
Maturana, Humberto, 58
Mauss, Marcel, 240
McGinn, Colin, 145
Mead, G. H., 231
melanin, 54–55
memory, 159, 203; art of, 163; cognitive maps of, 161–163; decoupling of, 160, 166; as dynamic process, 167; forms of, 166; image symbols of, 175–178; Plato on, 168–169; reconsolidation of, 165–166; selfhood and, 232, 233; somatic markers of, 166; stages of, 115
mereology, 51
Merleau-Ponty, Maurice, 29, 31, 78, 157, 178
Metcalfe, Janet, 77
Mill, John Stuart, 22–23, 260
Millikan, Ruth, 15–16, 66, 68, 98; pushmi-pullyu representations of, 79–80, 83–85, 148, 330n52
Milner, A. David, 30, 81
mimicry, 31, 114–118, 123, 178; tool creation and, 94, 100, 101
mirror (F5) neurons, 31–32, 114–115, 119, 137, 248, 357n197
Mishima, Yukio, 97, 281
Mishkin, M., 30
Mithen, Steven, 19, 188
Modell, Arnold, 19
morality, 222–228, 255–256, 260–261, 286; adaptive benefits of, 242–250; Darwin on, 247, 251–252, 355n160; Freud on, 355n160; neural pathways of, 161–162. See also virtue
mother-child bonding. See infant-caregiver relationship
motherese, 346nn9–10
music, 187–188, 267, 282–285
Myin, Erik, 329n29, 330n52, 339n3, 341n17
myth, 20, 264–265, 270, 278, 282, 309–310

Nagel, Thomas, 45, 67–71, 325n9
narcissism, 238
Narvaez, Darcia, 15, 127
nationalism, 208–209, 211–212, 242
natural selection, 110, 217, 218, 221
natural theology, 46–47, 59, 67, 324n5
Neanderthals, 113, 189
Nelson, C. A., 135–136
neocortex, 266, 323n75; cerebellum and, 172–174, 187; rational, 2–3, 250
neoteny, 125, 226
nested brainmind hierarchies, 8–9
Nettle, Daniel, 143
neuroplasticity, 5, 139–144
Newton, Isaac, 45, 223–224
Nietzsche, Friedrich, 308, 356n176
Noë, Alva, 30, 313–314
nonconceptual content, 155–156, 159–161; definition of, 156
nonverbal communication, 34, 85–89, 174, 234; autism and, 136
Norenzayan, A., 286
norepinephrine, 104, 145
nucleus accumbens, 61–62, 262, 332n17
Nussbaum, Martha, 145

objective-reflective-normative thinking, 244
obsessive-compulsive disorder (OCD), 289, 342n47
Okasha, Samir, 204, 217
Olson, Mancur, 214
opioids, 130, 219, 250, 296
optic flow, 158
orangutans, 108, 126
orbitofrontal cortex (OFC), 127–128, 133–136, 260, 340n9
Orchard, Andy, 310
origin stories, 213, 282, 355n159
ornaments. See adornment
Ostrom, Elinor, 214–215
Ovid, 272, 278
Owen, Richard, 56
oxytocin, 105, 109–113, 219, 250, 334n57; neural pathways of, 346n10; social bonding and, 112, 130–132; touch and, 337n55